Amarjit S. Basra
Editor

Crop Sciences:
Recent Advances

Crop Sciences: Recent Advances has been co-published simultaneously as *Journal of Crop Production,* Volume 1, Number 1 (#1) 1998.

Pre-publication
REVIEWS,
COMMENTARIES,
EVALUATIONS . . .

"Provides an authoritative coverage of a wide range of topics that relate to modern crop production–from conservation biology and genetics to crop improvement, physiology, agronomy and the production environment. The papers, which present both original research findings as well as scientific reviews, aim to put the chosen subjects in a broad and practical crop production perspective. The volume is international in scope and includes articles by authors from many parts of the world. It covers research of relevance to tropical, subtropical and temperate agriculture. I would recommend it to any scientist or student interested in the improvement of crop productivity."

Geoffrey Hawtin, PhD
Director General
International Plant Genetic Resources Institute
Rome, Italy

Crop Sciences:
Recent Advances

Crop Sciences:
Recent Advances

Amarjit S. Basra
Editor

Crop Sciences: Recent Advances has been co-published simultaneously as *Journal of Crop Production,* Volume 1, Number 1 (#1) 1998.

The Food Products Press
An Imprint of
The Haworth Press, Inc.
New York • London

Published by

The Food Products Press, 10 Alice Street, Binghamton, NY 13904-1580

The Food Products Press is an imprint of The Haworth Press, Inc., 10 Alice Street, Binghamton, NY 13904-1580 USA.

Crop Sciences: Recent Advances has been co-published simultaneously as *Journal of Crop Production,* Volume 1, Number 1 (#1) 1998.

The development, preparation, and publication of this work has been undertaken with great care. However, the publisher, employees, editors, and agents of The Haworth Press and all imprints of The Haworth Press, Inc., including The Haworth Medical Press and The Pharmaceutical Products Press, are not responsible for any errors contained herein or for consequences that may ensue from use of materials or information contained in this work. Opinions expressed by the author(s) are not necessarily those of The Haworth Press, Inc.

Cover design by Thomas J. Mayshock Jr.

Library of Congress Cataloging-in-Publication Data

Crop sciences : recent advances / Amarjit S. Basra, editor.
    p. cm.
Also published as Journal of crop production, vol. 1, no. 1, 1998.
Includes bibliographical references and index.
ISBN 1-56022-060-0 (alk. paper).–ISBN 1-56022-059-7 (alk. paper)
    1. Crop science. I. Basra, Amarjit S.
SB91.C783 1997
631.5–dc21
                          97-18156
                          CIP

INDEXING & ABSTRACTING

Contributions to this publication are selectively indexed or abstracted in print, electronic, online, or CD-ROM version(s) of the reference tools and information services listed below. This list is current as of the copyright date of this publication. See the end of this section for additional notes.

- *AGRICOLA Database,* National Agricultural Library, 10301 Baltimore Boulevard, Room 002, Beltsville, MD 20705

- *CNPIEC Reference Guide: Chinese National Directory of Foreign Periodicals*, P.O. Box 88, Beijing, People's Republic of China

- *Crop Physiology Abstracts,* c/o CAB International/CAB ACCESS . . . available in print, diskettes updated weekly, and on INTERNET. Providing full bibliographic listings, author affiliation, augmented keyword searching, CAB International, P.O. Box 100, Wallingford, Oxon OX10 8DE, UK

- *Derwent Crop Production File,* Derwent Information Limited, Derwent House, 14 Great Queen Street, London WC2B 5DF, England

- *Environmental Abstracts,* Congressional Information Service, Inc., 4520 East-West Highway, Suite 800, Bethesda, MD 20814-3389

- *Field Crop Abstracts,* c/o CAB International/CAB ACCESS . . . available in print, diskettes updated weekly, and on INTERNET. Providing full bibliographic listings, author affiliation, augmented keyword searching, CAB International, P.O. Box 100, Wallingford, Oxon OX10 8DE, UK

(continued)

- *Foods Adlibra,* Foods Adlibra Publications, 9000 Plymouth Avenue North, Minneapolis, MN 55427

- *Grasslands & Forage Abstracts,* c/o CAB International/CAB ACCESS . . . available in print, diskettes updated weekly, and on INTERNET. Providing full bibliographic listings, author afilliation, augmented keyword searching, CAB International, P.O. Box 100, Wallingford Oxon 0X10 8DE, UK

- *Plant Breeding Abstracts,* c/o CAB International/CAB ACCESS . . . available in print, diskettes updated weekly, and on INTERNET. Providing full bibliographic listings, author afilliation, augmented keyword searching, CAB International, P.O. Box 100, Wallingford Oxon 0X10 8DE, UK

- *Seed Abstracts,* c/o CAB International/CAB ACCESS . . . available in print, diskettes updated weekly, and on INTERNET. Providing full bibliographic listings, author afilliation, augmented keyword searching, CAB International, P.O. Box 100, Wallingford Oxon 0X10 8DE, UK

- *Soils & Fertilizers Abstracts,* c/o CAB International/CAB ACCESS . . . available in print, diskettes updated weekly, and on INTERNET. Providing full bibliographic listings, author afilliation, augmented keyword searching, CAB International, P.O. Box 100, Wallingford Oxon 0X10 8DE, UK

- *Weed Abstracts,* c/o CAB International/CAB ACCESS . . . available in print, diskettes updated weekly, and on INTERNET. Providing full bibliographic listings, author afilliation, augmented keyword searching, CAB International, P.O. Box 100, Wallingford Oxon 0X10 8DE, UK

(continued)

SPECIAL BIBLIOGRAPHIC NOTES

related to special journal issues (separates)
and indexing/abstracting

☐ indexing/abstracting services in this list will also cover material in any "separate" that is co-published simultaneously with Haworth's special thematic journal issue or DocuSerial. Indexing/abstracting usually covers material at the article/chapter level.

☐ monographic co-editions are intended for either non-subscribers or libraries which intend to purchase a second copy for their circulating collections.

☐ monographic co-editions are reported to all jobbers/wholesalers/approval plans. The source journal is listed as the "series" to assist the prevention of duplicate purchasing in the same manner utilized for books-in-series.

☐ to facilitate user/access services all indexing/abstracting services are encouraged to utilize the co-indexing entry note indicated at the bottom of the first page of each article/chapter/contribution.

☐ this is intended to assist a library user of any reference tool (whether print, electronic, online, or CD-ROM) to locate the monographic version if the library has purchased this version but not a subscription to the source journal.

☐ individual articles/chapters in any Haworth publication are also available through the Haworth Document Delivery Service (HDDS).

ABOUT THE EDITOR

Amarjit Singh Basra, PhD, is an eminent Plant Physiologist (Associate Professor) at Punjab Agricultural University in Ludhiana, India. The author of numerous research and professional publications, Dr. Basra has received coveted scientific awards and honors in recognition of his original and outstanding contributions to crop science. He is a member of several professional societies including the American Society of Agronomy, Crop Science Society of America, Australian Society of Plant Physiologists, and International Society for Plant Molecular Biology. He has made scientific trips to several countries and provides leadership in organizing and fostering cooperation in crop research at the international level.

Crop Sciences:
Recent Advances

CONTENTS

Preface

Crops provide food, feed, fiber and other important products for sustaining human life. Advances in crop sciences have led to phenomenal increases in food production during the past few decades. However, the accelerated use of natural resources, high external inputs, intensive cropping patterns and monoculture practices have caused numerous problems in almost every continent. In the face of burgeoning populations, diminishing per capita availability of land and water resources, ecological imbalances and environmental degradations, we need a new paradigm for harnessing crop production research to provide sustained food security throughout the world.

The purpose of this book is to present an authoritative review of some of the latest advances in crop sciences covering a wide range of topics related to crop production and having far-reaching implications for sustainable crop productivity and environmental sustainability. Given the fact that multidisciplinary approaches are gaining momentum into the future, this volume emphasizes the integration of basic and applied research, encompassing diverse disciplines of crop sciences.

Individual articles written by an international group of experts identify the important advances that have been made in the field of crop sciences, addressing key gaps in the knowledge base, and offering technical solutions to critical global problems of crop production. The first three chapters discuss how the wheat gene pool is being enriched with genes from other species–with an emphasis on amphiploidization and direct backcrossing; the agronomic and physiological consequences of incorporating dwarf height genes into wheat; and the relationship between vertical leaf nitrogen distribution and canopy photosynthesis to maximize radiation use efficiency, biomass production, and yield potential in winter cereals. Subsequent chapters address the genetic diversity and phylogenetic relationships among 16 diploid species of cotton based on isozyme markers; the mechanisms of heterosis in crop plants focusing on molecular studies; the use of microsatellite markers for molecular breeding; the role of weed seed and seed bank dynamics in developing improved weed management systems; the occurrence and implications of allelopathic interactions in

xi

agroecosystems; the hormonal control of dormancy induction in developing seeds; the sensitivity of crop plants to water stress during meiosis and flowering and their consequences for crop productivity; and the effect of low temperature on germination and early seedling growth, particularly canola, in relation to biochemical and molecular processes. The last two articles critically evaluate the impact of changing global environment on world crop production, and the current situation of crop production in China (with a focus on grain production), along with strategies to meet future food needs of its population.

The book is wide ranging in scope and should prove extremely useful to students, teachers, researchers and practitioners in the fields of agronomy, botany, genetics, plant breeding, soil science, horticulture, forestry, agroclimatology, plant physiology, biochemistry, biotechnology, plant pathology, seed science and agribiodiversity.

I wish to express my special thanks and appreciation to all the contributing authors for their splendid collaboration. I am grateful to Bill Cohen, Publisher of The Haworth Press, for his inspiring support and to Richard Rockman for his meticulous work as the Production Editor. This volume has been published simultaneously as the *Journal of Crop Production*, Vol. 1, No. 1 [#1] which aims to develop a series of thematic monographs in the future to serve the international crop science community in a unique way.

Amarjit S. Basra

Deepening the Wheat Gene Pool

T. S. Cox

SUMMARY. For many crop species, genetic uniformity is a result of modern breeding and farming practices. Common wheat (*Triticum aestivum* L.) is an exception, having been a genetically narrow species throughout its entire existence. This paper discusses the evolutionary bottlenecks through which today's wheat germplasm has descended, and the ways in which the wheat gene pool is being enriched with genes from other species. The worldwide gene pool of common wheat is descended from a very small number of spontaneous interspecific hybrids, which originated as a result of two natural amphiploidization events. In the more recent event, plant(s) of emmer wheat, which were being cultivated at the time by early Neolithic farmers, were fertilized by a weedy diploid goatgrass, *Aegilops tauschii,* producing primitive common wheat. Because of the rarity of this event, today's common wheat has extremely low levels of polymorphism at enzyme, storage protein, and DNA marker loci, compared with its parent species, especially *Ae. tauschii.* In fact, the bulk of evolutionary evidence suggests that common wheat began its existence as a highly monomorphic species and that its genetic variation was reduced further by domestication. Despite common wheat's narrow genetic base, human-guided evolution has produced a profusion of distinct land races over a period of five or more millenia, and modern breeding has maintained steady genetic improvement throughout the current century. To protect these gains, humans have resorted to interspecific crossing to improve wheat's pest resistance.

T. S. Cox, formerly Research Geneticist, USDA-ARS, Manhattan, KS 66506, USA.

Address correspondence to (current address): T. S. Cox, Institute for Rural Health Studies, Road 4, Banjara Hills, Hyderabad, 500034, India.

[Haworth co-indexing entry note]: "Deepening the Wheat Gene Pool." Cox, T. S. Co-published simultaneously in *Journal of Crop Production* (The Food Products Press, an imprint of The Haworth Press, Inc.) Vol. 1, No. 1 (#1), 1998, pp. 1-25; and: *Crop Sciences: Recent Advances* (ed: Amarjit S. Basra) The Food Products Press, an imprint of The Haworth Press, Inc., 1998, pp. 1-25. Single or multiple copies of this article are available for a fee from The Haworth Document Delivery Service [1-800-342-9678, 9:00 a.m. - 5:00 p.m. (EST). E-mail address: getinfo@haworth.com].

1

But why should wheat's progenitors not be regarded as sources of useful genetic variation for all economic traits? Humans would have been fortunate indeed if the rare amphiploid(s) that gave rise to common wheat carried the ideal allele at every locus. This paper provides a critical analysis of methodologies for bypassing wheat's genetic bottleneck and deepening its primary gene pool. It emphasizes the two general approaches to expanding the wheat gene pool: amphiploidization and direct backcrossing. Advantages and disadvantages of both methodologies are provided, drawing on examples of gene transfer from the various subspecies of *T. monococcum, T. urartu, T. turgidum, T. timopheevii, ssp. monococcum, Ae. speltoides,* and *Ae. tauschii. [Article copies available for a fee from The Haworth Document Delivery Service: 1-800-342-9678. E-mail address: getinfo@haworth.com]*

KEYWORDS. Bread wheat, common wheat, evolution, diversity, biodiversity, interspecific, introgression, breeding, *Aegilops, Triticum*

Every year, on every continent, breeders of common wheat (*Triticum aestivum*) produce thousands–perhaps hundreds of thousands–of hybrids among cultivars and breeding lines. This massive recombination of genetic material produces the genetically diverse populations that are needed for continued improvement of productivity, pest resistance, and other characters. Yet the worldwide gene pool of common wheat is descended from a very small number of spontaneous interspecific hybrids. Herein, I will discuss the evolutionary bottlenecks through which today's wheat germplasm has descended, and the ways in which the wheat gene pool is being enriched with genes from other species.

THE GENETIC BOTTLENECKS

Ladizinsky (1984) summarized evidence for the "founder effect" in crop plant evolution–the restriction of genetic variability that often results from domestication–and discussed its consequences. He pointed out that the founder effect is especially important in allopolyploid crops such as wheat. Allopolyploids arise from processes of interspecific hybridization and chromosome doubling and contain the entire genomes of two or more species in homozygous condition. Hexaploid, or common, wheat (*T. aestivum,* genomic formula BBAADD) originated as a result of two natural amphiploidization events (Figure 1). The caption describes wheat evolution in greater detail, and Table 1 lists synonyms for botanical names).

FIGURE 1. Diagrammatic representation of the evolution of wheat. Solid arrows indicate hybridization followed by chromosome doubling. Dashed arrows indicate domestication or direct selection within a species. Boxes indicate cultivated taxa.

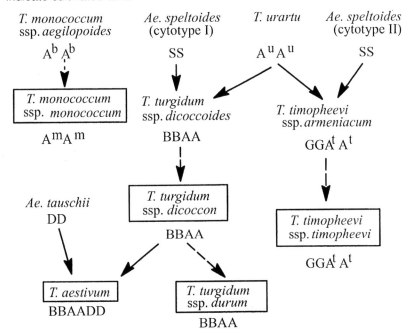

The wild tetraploids *T. turgidum* ssp. *dicoccoides* and *T. timopheevi* ssp. *armeniacum* arose millions of years ago as chromosome-doubled hybrids between the wild diploid *T. urartu* (Dvorak et al., 1993; Takumi et al., 1993) and two different cytotypes of another wild diploid, *Ae. speltoides* (Jiang and Gill, 1994). These tetraploids were domesticated as ssp. *dicoccon* and ssp. *timopheevi*, respectively. The cultivated diploid wheat, *T. monococcum* ssp. *monococcum*, arose as a domesticated form of *T. aegilopoides*, a relative of *T. urartu*. Common wheat evolved from a spontaneous amphiploid between a tetraploid wheat and the diploid goatgrass *Ae. tauschii*. The tetraploid parent probably was cultivated emmer wheat, ssp. *dicoccon* (Kihara, 1944; McFadden and Sears, 1946), because the range of the wild progenitor, ssp. *dicoccoides*, does not overlap with that of *Ae. tauschii* (Zohary and Hopf, 1988). In support of this thesis, Jaaska (1978) found that common wheats tended to contain alcohol dehydrogenase isozyme bands from ssp. *dicoccon* rather than ssp. *dicoccoides*. The cultivated and wild forms of *T. timopheevi* can be eliminated from common wheat's family tree based on cytological evidence (Jiang and Gill, 1994). The direct diploid ancestor of wheat was *Ae. tauschii* ssp. *strangulata*, based on morphological traits of newly synthesized hexaploids (Watanabe, 1983) and occurrence in common wheat of α-amylase (Nishikawa, Furuta, and Wada, 1980) and aspartate amino transferase (Jaaska, 1980) isozymes typical of ssp. *strangulata* and rare in ssp. *typica*. Subspecies *strangulata* has a limited geographical range near the southern shores of the Caspian Sea and shows less genetic polymorphism than does ssp. *typica* (Lubbers et al., 1991). According to Jaaska (1980), ssp. *strangulata* is probably the ancestral form of *Ae. tauschii*, and ssp. *typica* evolved traits that allowed it to occupy a much wider geographical and ecological range, from eastern Turkey to western China and Pakistan.

TABLE 1. Specific and subspecific designations of common wheat and related species, according to van Slageren (1994), along with common names and previously used botanical nomenclature.

Species	Subspecies	Common name	Previous designations[a]
Aegilops speltoides Tausch	both	–	*T. speltoides*
Ae. tauschii Coss.	both	goatgrass	*T. tauschii, Ae. squarrosa*
Triticum urartu Tumanian ex Gandylian	–	–	–
T. monococcum L.	*monococcum* *aegilopoides* (Link) Thell.	einkorn wild einkorn	– *T. boeoticum* *T. baeoticum* *T. thaoudar*
T. turgidum L.	*dicoccoides* (Körn. ex Asch. & Graebn.)Thell.	wild emmer	*T. dicoccoides*
	dicoccon (Schrank) *durum* (Desf.)Husn. Husn. *carthlicum* (Nevski) A. Löve & D. Löve	emmer durum durum	*T. dicoccum* *T. durum* *T. persicum*
T. timopheevi L.	*armeniacum* (Jakubz.) van Slageren *timopheevi*	timopheevi	*T. araraticum*
T. aestivum L.	*aestivum*	common wheat	

[a] Many other botanical names have been used in the past for some of these taxa or portions thereof; however, only the more commonly used ones are listed here.

Common wheat's tetraploid ancestor, emmer wheat (*T. turgidum* ssp. *dicoccon,* genomic formula BBAA) was domesticated about 9000 years ago, at the dawn of agriculture. Its direct progenitor, wild emmer (ssp. *dicoccoides*), despite having emerged through the bottleneck of amphiploidy (Figure 1) is highly variable genetically. Having existed for millions of years, wild emmer has had time to accumulate variation through mutation as well as possible introgression from diploid relatives (Vardi and Zohary, 1967). Although domestication resulted in low polymorphism for endosperm proteins in cultivated emmer relative to its wild ancestor (Levy, Galili, and Feldman, 1988), the A and B genomes of common wheat–which were derived from cultivated emmer–are more genetically variable than is the D genome (Metakovsky et al., 1984).

The most recent step in wheat evolution, which occurred in southwest Asia about 7000 years ago (Zohary and Hopf, 1988), led directly to reduced polymorphism in the D genome. Plant(s) of emmer wheat, which were being cultivated at the time by early Neolithic farmers, were fertilized by windborne pollen from plant(s) of a weedy diploid goatgrass, *Aegilops tauschii* (genomic formula DD). The resulting hybrid seeds produced partially fertile plants the following season, and progeny of those plants were incorporated by farmers into the domestication process.

Since Kihara (1944) and McFadden and Sears (1946) first determined that *T. turgidum* and *Ae. tauschii* are the parents of common wheat, wheat geneticists have reproduced such amphiploids many times. To do so, they usually have found it necessary to rescue the F_1 embryo on an artificial culture medium. Production of viable *T. turgidum* × *Ae. tauschii* F_1 seed must have been extremely rare on Neolithic farms; however, once it survived to reproductive stage, a triploid hybrid easily could have produced hexaploid seed through spontaneous formation of unreduced male and female gametes (Xu and Joppa, 1995). The many wild-type traits exhibited by its raw allopolyploid progeny would have been eliminated by farmers whenever mutants with alternative traits arose.

For the early farmers who developed common wheat as a crop, the process of breeding via selection of spontaneous mutants was long and slow. Once more desirable wheats were developed, any newly arising allopolyploids would have been rejected (Ladizinsky, 1984). Such selection would have severely limited the scope of the new species' gene pool. Furthermore, no wild hexaploid wheat species existed; therefore, enrichment of the gene pool via natural introgression as has occurred in other crops (Harlan, 1965) would have been insignificant.

Even if repeated natural hybridization between *Ae. tauschii* and emmer had occurred, as suggested by Zohary, Harlan, and Vardi (1969), genetic drift apparently has eliminated much of the introgressed variation. Common wheat has extremely low levels of polymorphism at enzyme, storage protein, and DNA marker loci, compared with its parent species, especially *Ae. tauschii*. Nishikawa et al. (1980) found seven α-amylase isozyme alleles at a locus on chromosome 6D and two alleles at another locus on 7D in a collection of 60 *Ae. tauschii* accessions. Only one allele occurred at each locus in common wheat. Among 79 *Ae. tauschii* accessions, Lagudah and Halloran (1988) identified no fewer than 72 alleles at the complex gliadin storage-protein locus on chromosome 1D and 57 alleles at a second locus on 6D. At a third locus coding for high molecular-weight glutenin proteins, 14 alleles occurred. In contrast, among 29 wheat genotypes of diverse origin, only one allele was identified at each D-genome gliadin locus and two at the glutenin locus.

Lubbers et al. (1991) found high levels of polymorphism among 102 *Ae. tauschii* accessions at all 20 RFLP loci that they identified with genomic probes, but when A.K. Fritz, B.S. Gill, and I (unpublished) surveyed 21 common-wheat cultivars from around the world using 33 genomic probes (each locus potentially identifying 3 loci across wheat's 3 genomes), only 10 probes showed any polymorphism. Over all loci, *Ae. tauschii* accessions had a mean of 3.4 alleles per locus (compared with 1.1 for wheat, in all 3 genomes) and a mean adjusted polymorphic index of 0.41 (compared with 0.04 for wheat).

Whereas natural introgession of genes from *T. tauschii* into wheat was rare, gene flow from emmer and even timpoheevi (GGAA) wheats into common wheat may have occurred at a higher rate. Throughout its existence, *T. aestivum* has been grown in close proximity to *T. turgidum* in southwest Asia and the Mediterranean, and alongside the wild and cultivated forms of *T. timopheevi* in Iraq, Turkey, and the Caucasus. Hexaploid × tetraploid hybrid seeds often are viable. The resulting pentaploid plants are female-fertile and may set backcross seed when exposed to pollen from hexaploid plants. As a result, the genetic bottleneck through which the A and B genomes emerged has been widened during the several millenia that have passed since the origin of common wheat. Cultivated emmer would have been the primary source of such introgression, because of the very limited geographic range of wild emmer and timopheevi wheats. On the other hand, direct introgression of genes from the diploid species *Ae. tauschii, T. monococcum,* or *T. urartu* into common wheat would not have occurred naturally. The endosperm in hybrid seeds from such hexaploid × diploid crosses never develops, necessitating embryo rescue when artificial crosses are made (Gill and Raupp, 1987; Cox et al., 1991).

In summary, the bulk of evolutionary evidence suggests that common wheat began its existence as a highly monomorphic species and that its genetic variation was reduced further by domestication. Augmentation of its genetic base via introgression was limited to the A and B genomes, and the only significant source of naturally introgressed alleles would have been tetraploid wheats. Despite common wheat's narrow genetic base, human-guided evolution has produced a profusion of distinct land races over a period of five or more millenia, and modern breeding has maintained steady genetic improvement throughout the current century (Cox et al., 1988). Over the past few decades, humans have used interspecific crossing to widen wheat's genetic base at an ever-increasing rate. For example, only 2 of the first 20 genes to receive symbols (*Lr1* to *Lr20*) for resistance to leaf rust (caused by *Puccinia recondita* Rob. ex Desm.) were introduced from outside *T. aestivum,* whereas 18 more recently named genes in the range *Lr21* to *Lr44* were introgressed from other species (McIntosh, Wellings, and Park, 1995). In wheat as in most species, transfer of resistance genes always has been the most prominent objective of interspecific crossing. But why should wheat's progenitors not be regarded as sources of useful genetic variation for all economic traits? Humans would have been fortunate indeed if the rare amphiploid(s) that gave rise to common wheat carried the ideal allele at every locus. In fact, the existence of genotype × environment interaction rules out the very existence of an "ideal allele," leading to the inevitable conclusion that the ancestors of wheat can be sources of genes for improvement of many traits.

DEEPENING THE GENE POOL

Zohary et al. (1969) urged wheat breeders and geneticists to make greater use of wheat's diploid and tetraploid progenitors as sources of economically valuable genes. At that time, very little screening of ancestral species for resistances or other traits had been done. Intraspecific variation in particular was largely neglected, most studies concentrating instead on morphological or genetic differences among species. But in the past two decades, evaluation and utilization of wild or primitive wheat collections has become widespread (Tables 2 and 3).

Gene transfer into *T. aestivum* from progenitor species–including *T. monococcum, Ae. speltoides, T. turgidum,* and its cousin *T. timopheevi*–is often described as "wide crossing" (Sharma and Gill, 1983) because genes are being moved between ploidy levels; however, Jiang, Friebe, and Gill (1994) placed the progenitor species in common wheat's primary gene pool, based on chromosome homology. Wheat also has a secondary

TABLE 2. Examples of gene transfer into common wheat via synthetic amphiploids.

Primary hybrid			
Pedigree	Genomic constitution	Trait(s)	Reference
Ae. speltoides X *T. monococcum*	SSA^mA^m	Leaf rust, stem rust	Kerber and Dyck (1990)
T. turgidum ssp. *durum* X *T. monococcum*	$BBAAA^mA^m$	Leaf rust	Valkoun, Kucerova, and Bartos (1986)
T. aestivum ('Tetra Canthatch') X *Ae. tauschii*	BBAADD	Leaf rust	Innes and Kerber (1994); Kerber and Dyck (1969); Kerber (1987)
T. turgidum ssp. *dicoccoides* X *Ae. tauschii*	BBAADD	General	Lange and Jochemsen (1992)
T. turgidum ssps. *carthlicum*, *dicoccon*, and *durum* X *Ae. tauschii*	BBAADD	Septoria leaf blotch	May and Lagudah (1992)
T. turgidum ssp. *dicoccum* X *Ae. tauschii*	BBAADD	Stagonospora glume blotch	Nicholson, Rezanoor, and Worland (1993)
T. turgidum ssp. *durum* X *Ae. tauschii*	BBAADD	Karnal bunt / Wheat curl mite	Villareal et al. (1994) / Thomas and Connor (1986)

gene pool, which comprises annual and perennial taxa (so-called "alien species") with which it has much less chromosomal homology. Gene transfer from alien species has been reviewed by many workers, most recently Jiang, Friebe, and Gill (1994), and will not be discussed herein. I also do not intend to catalog all interspecific gene transfers in wheat to date; rather, I offer herein a critical analysis of methodologies for bypassing wheat's genetic bottleneck and deepening its primary gene pool.

Simmonds (1993) argued that introgression, which "focuses on the use

TABLE 3. Examples of gene transfer into common wheat via direct backcrossing.

Primary hybrid Pedigree	Genomic constitution	Traits	Reference
T. turgidum ssp. *durum* X *T. monococcum* ssp. *boeoticum*	BAA^m	Stem rust	Gerechter-Amitai et al. (1971)
T. turgidum ssp. *durum* X *Ae. speltoides*	BAS	General	Vardi and Zohary (1967)
T. aestivum X *Ae. speltoides*	BADS	Leaf rust	Dvorak (1977)
T. aestivum X *T. monococcum* ssp. *monococcum*	BAA^{m}D	Leaf rust	Cox et al. (1994a)
T. aestivum X *Ae. tauschii*	BADD	Leaf rust	Cox, Raupp, and Gill (1994); Innes and Kerber (1994)
		Hessian fly	Cox and Hatchett (1994); Gill et al. (1987)
		Soilborne viruses	Cox et al. (1994b)
T. aestivum X *T. turgidum* ssp. *durum*	BBAAD	Stem rust Hessian fly	Stebbins, Patterson, and Gallun (1983); Patterson, Foster, and Ohm (1988)
T. turgidum ssp. *durum* X *T. timopheevi* ssp. *timopheevi*	BGAA^t	Stagonospora glume blotch	Ma and Hughes (1995)
T. aestivum X *T. timopheevi* ssp. *timopheevi*	BGAA^t	Stem rust and powdery mildew	Allard and Shands (1954)
T. aestivum X *T. timopheevi* ssp. *armeniacum*	BGAA^{t}D	Stem rust	Dyck (1992)
		Leaf rust, powdery mildew, and tan spot	Brown-Guedira (1995)

of small numbers of entries explicitly tested for desired characters/traits introduced by crossing and backcrossing into adapted stocks," has received the lion's share of attention from crop breeders and geneticists, who have largely neglected what he termed "incorporation," or selection within exotic populations themselves to produce elite, locally adapted genotypes or populations. Simmonds (1993) regarded introgression as effective only for transfer of genes with large effects and/or "highish" heritability, whereas "incorporation," in his view, has the potential to accelerate genetic gain for all traits.

All hexaploid cultivars and landraces are products of the same genetic bottleneck and represent unlikely sources of truly new alleles to fuel "incorporation" in Simmonds' sense. Conceivably, incorporation could be done instead with related cultivated species. Durum, and to a lesser extent, emmer, has a long breeding history. Einkorn wheat (*T. monococcum* ssp. *monococcum*) also may have considerable breeding potential as a crop in its own right (Rafi, Ehdaie, and Waines, 1992; Vallega, 1992). In a comparison of modern and landrace cultivars of common wheat under low-fertility conditions in Kansas, USA, Shroyer and Cox (1993) evaluated one randomly chosen accession of einkorn wheat, which was not included in the published study. The einkorn entry yielded as well as standard hard winter wheat cultivars that were grown widely in Kansas during the 1940s and produced better yields than did common-wheat land races from Asia, even when the weight of einkorn's adhering hulls was subtracted. In Italy, one cycle of crossing and selection was sufficient to produce einkorn lines with hull-corrected grain yields as high as those of elite common-wheat cultivars and higher than those of elite durums under low fertility conditions (Vallega, 1992).

Any "incorporation" work done in einkorn, emmer, durum, or wild wheat species still must be followed by introgression into common wheat. Despite attempts at direct evaluation (e.g., Rafi, Ehdaie, and Waines, 1992), the agronomic value of genes carried by the wild diploids and tetraploids, excepting genes for resistance to pests or specific soil conditions, can be determined only after transfer to the hexaploid level. Useful genes may be masked by undesirable traits in the wild species themselves, and may interact unpredictably once transferred. For example, suppressors of rust-resistance genes are widespread and well-documented (Bai and Knott, 1992). In the presence of such interactions, one must take a "dialectical" approach to breeding (Bramel-Cox and Cox, 1989), bringing together adapted and exotic gene pools in order to evaluate the value of their genes in combination.

The typical wheat introgression program is intended to transfer one or more resistance genes, but a considerable amount of other genetic material is introgressed at the same time, much of it entirely unique within *T. aestivum*. In several notorious cases, genes with negative effects on grain yield or end-use quality were transferred along with resistance genes from alien species (Martin and Stewart, 1986; The et al., 1988; Knott, 1989; Marais, 1992). Because the chromosomes of alien species normally do not recombine with those of wheat, vast stretches of DNA, some of it deleterious, may remain perpetually linked to an alien-derived resistance gene. On the other hand, genes transferred from most progenitor species may be separated from unwanted genes via normal recombination, and some of the new allelic variation may be useful in breeding, for reasons outlined above.

There are two general approaches to expanding the wheat gene pool: amphiploidization and direct backcrossing. Both methods are analogous to events that occurred during the evolution of wheat. Dvorak (1977) noted that in nature, there is an inverse relationship between occurrence of the two processes. In hybrids between two species with at least one set of homologous chromosomes (e.g., BBAA × AA), direct introgression via natural backcrossing to the higher-ploidy parent occurs most often (Vardi and Zohary, 1967). On the other hand, in a hybrid between two parental species that have no chromosome homology with each other (e.g., BBAA × DD), lack of pairing at meiosis results in formation of unreduced gametes that can unite to form natural amphiploids (Xu and Joppa, 1995).

The relationship between chromosome homology and production of unreduced gametes is somewhat less well-defined in artificial breeding. Hybrids between parents with nonhomologous genomes vary widely in their tendency to produce hexaploid seed from unreduced gametes, depending upon the genotype of the tetraploid parent (Xu and Joppa, 1995). Conversely, BBAADD × DD and BBAADD × AA hybrids, despite their parents' sharing one homologous genome, can produce unreduced female gametes at a relatively high frequency (Alonso and Kimber, 1984; Gill and Raupp, 1987; Cox et al., 1991); however, such hybrids are male-sterile and must be backcrossed in order to transfer genes. Both amphiploidy and direct backcrossing have been used to transfer genes from progenitor species to common wheat. The two methods are complementary, as the following examples illustrate.

Introgression via Amphiploids

Because amphiploids usually are highly fertile and true-breeding, they have obvious appeal as germplasm sources. Wheat's evolution may be extended and redirected by artificial formation of new amphiploids, a process dubbed "radical breeding" by McFadden and Sears (1947). They crossed each of two tetraploid wheats (*T. turgidum* ssp. *dicoccoides* and ssp. *timopheevi*) with eight diploid species and artificially doubled the chromosome numbers of the resulting hybrids. McFadden and Sears (1947) advocated the use of these synthetic amphiploids–especially those incorporating the A, B, and D (then called "C") genomes–as "basic material for the transfer of characters from other species and genera."

Synthetic amphiploids derived from various parental species often have been used to transfer genes for disease and insect resistance from diploid to tetraploid or hexaploid wheats (Table 2). Some synthetics, such as ones derived from crosses between *T. turgidum* (BBAA) and *T. monococcum* (AA), do not occur naturally, because pairing of A-genome chromosomes precludes formation of unreduced gametes. These amphiploids can be produced only with the use of colchicine. In contrast, the amphiploid *Ae. speltoides* (SS) × *T. monococcum* (AA) (Kerber and Dyck, 1990) is comparable to one of the natural amphiploids that occurred in wheat's evolutionary tree (Figure 1). But *Ae. speltoides* carries genes that permit pairing of homoeologous (i.e., similar but not homologous) chromosomes (Riley, Kimber, and Chapman, 1961), as well as gametocidal genes (Tsujimoto and Tsunewaki, 1985). Homoeologous pairing is useful for allowing recombination and transfer of target genes to wheat chromosomes (Chen, Tsujimoto, and Gill, 1994), but both the pairing and the gametocidal genes can prevent recovery of agronomically suitable, fully fertile progeny (Marais and Pretorius, 1996). Some genes transferred from *Ae. speltoides* are associated with negative effects on yield, even after repeated back-crossing (The et al., 1988; Knott, 1989).

Among many possible amphiploids, synthetic hexaploids derived from *T. turgidum* and *Ae. tauschii* (Table 2; Figure 2) recently have received the closest attention from breeders, not only as sources of new resistance genes but also as building blocks of a new hexaploid germplasm base. They produce completely fertile hybrids with most wheat cultivars and, potentially, carry alleles in all three genomes that have never before been introduced into common wheat. In choosing a tetraploid parent to hybridize with *Ae. tauschii*, breeders or geneticists can select from among four subspecies–*durum, dicoccon, carthlicum,* or *dicoccoides*–as well as tetraploid stocks derived by eliminating the D genome from common wheat. For all of these reasons, synthetic hexaploids may be useful for improving agronomic and quality traits as well as resistances in wheat.

Testing 50 synthetic hexaploids derived from crosses between elite durum wheat genotypes and *Ae. tauschii,* Villareal et al. (1994) identified one line that outyielded an elite Mexican spring wheat, Seri 82, and two

FIGURE 2. Alternative methods for producing synthetic hexaploids from hybrids between *T. turgidum* and *Ae. tauschii*. When the triploid (BAD) hybrid produces unreduced male and female gemetes, it may be selfed or crossed to *T. aestivum* (BBAADD) to produce hexaploid progeny. Alternatively, triploid seedlings may be treated with colchicine to produce fertile hexaploid plants. Synthetic amphiploids may be constructed in a similar way with other combinations of species, generally ones that do not share a homologous genome (Table 1).

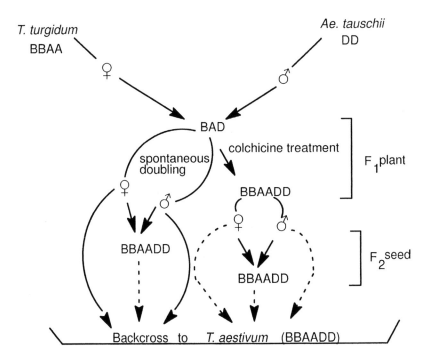

that had higher biomass. Several entries had extremely large kernels, with 1000-kernel weights exceeding 60 g, compared with 41 g for Seri 82; however, most had very low numbers of kernels per spike or per square meter of plot area, possibly indicating partial sterility or small numbers of functional florets per spike. In the process of transferring resistance to cereal cyst nematode from *Ae. tauschii* to common wheat via similar durum-derived synthetic hexaploids, Eastwood, Lagudah, and Young (1994) selected BC_1-derived progenies that outyielded their recurrent parent by 10 to 24% in trials conducted in Australia.

It has not yet been demonstrated whether agronomic improvements obtained in backcross populations derived from synthetic hexaploids are attributable to genes introgressed from *Ae. tauschii, T. turgidum,* or both. The tetraploid parents of the synthetics used by Eastwood, Lagudah, and Young (1994), Villareal et al. (1994), and Eastwood (1995) were elite durum wheat cultivars or breeding lines, and likely contributed many favorable alleles to the A and B genomes of their backcross progenies. As L.R. Joppa (USDA-ARS, Fargo, ND, USA, pers. commun.) has pointed out, genetic buffering may have permitted genetic atrophy within the A and B genomes of common wheat; as a consequence, wheat might benefit from an infusion of new genes. Joppa's prediction is based on the fact that the BBAA components of several common wheats have been re-extracted, and the resulting plants often are weak to the point of near-inviability (Kerber, 1964; Joppa, pers. commun.). [Nevertheless, it should be recognized that the extracted tetraploid 'TetraCanthatch' has been used extensively as a parent of synthetic hexaploids (Innes and Kerber, 1994.)]

Among the other cultivated subspecies of *T. turgidum,* there is little current interest in producing new synthetics with cultivated emmer (ssp. *dicoccon*), but the rare ssp. *carthlicum* has been used with some success. A free-threshing wheat grown in the Caucasus region, ssp. *carthlicum* has been considered by some workers to be a candidate for status as the evolutionary parent of bread wheat (MacKey, 1966; Kerber and Bendelow, 1977; Bushuk and Kerber, 1978); however, it more likely evolved through introgression of the Q gene (which conditions common wheat-type spike morphology and threshability) from *T. aestivum* into *T. turgidum* (Morrison, 1993). Whatever its evolutionary status, ssp. *carthlicum* may be artificially hybridized with *Ae. tauschii* to produce triploid hybrids that have a very high rate of unreduced gamete formation. For example, we have used a selection from a ssp. *dicoccoides* × ssp. *carthlicum* cross (obtained from A.J. Lukaszewski, Univ. California Riverside) to produce BAD triploids that exhibit up to 60% hexaploid seed-set when pollinated by *T. tauschii* in our greenhouse. The tendency of this tetraploid to produce unreduced gametes is also observed in crosses with rye (*Secale cereale* L.) and is under genetic control (Lukaszewski, pers. commun.). Futhermore, synthetic hexaploids produced from ssp. *carthlicum* have grain that threshes free of the lemma and palea, unlike the tough-threshing, hulled grain of synthetics that are produced by ssp. *durum* (Villareal, Mujeeb-Kazi, and Rajaram, 1996). An ideal tetraploid parent for production of synthetic hexaploids might be an elite durum wheat cultivar into which the Q gene and gene(s) for unreduced gamete formation had been backcrossed from ssp. *carthlicum.*

McFadden and Sears (1947) concluded that their *T. turgidum* ssp. *dicoccoides* × *Ae. tauschii* synthetic hexaploid "probably has no value from a breeding standpoint," based on its lack of resistance to rusts. But more recent research has demonstrated extensive genetic variability within germplasm collections of ssp. *dicoccoides* for end-use quality (Ciaffi et al., 1991), cold tolerance (Tahir and Ketata, 1993), and disease resistance (Moseman et al., 1984, 1985) and within *Ae. tauschii* for disease and insect resistance (Gill et al., 1986; Cox et al., 1992). Synthetics combining the genomes of both species might prove to be a valuable genetic resource, but would be agronomically inferior to those derived from ssp. *carthlicum* or ssp. *durum.*

Introgression via Direct Hybrids

Direct interploidy hybridization followed by backcrossing has been used often in wheat improvement (Table 3; Figure 3). Hybrids between common or durum wheat and related species generally are male-sterile but usually will set seed when backcrossed. Gerechter-Amitai et al. (1971) demonstrated that a stem rust resistance gene could be transferred from *T. monococcum* ssp. *aegilopoides* into durum wheat by simple backcrossing. At the next-higher ploidy level, several important resistance genes have been transferred directly into common wheat from cultivated durum wheat (Table 3). When individual chromosomes from durum wheat's wild progenitor, ssp. *dicoccoides,* were substituted into the durum cultivar Langdon, chromosome 6B increased grain protein concentration, and 1B improved cooking quality (Joppa, Hareland, and Cantrell, 1991); furthermore, the genes conditioning protein and quality were expressed in crosses to an elite durum (Steiger et al., 1996). In the same experiments, chromosome 4B from ssp. *dicoccoides* had a positive effect on grain yield (Elias, Steiger, and Cantrell, 1996). Genes from ssp. *dicoccoides* for high protein and quality also were transferred directly into hard red spring wheat (Khan et al., 1989).

The more distantly related tetraploid *T. timopheevi* ssp. *timopheevii* contributed important genes for powdery mildew and stem rust resistance (Allard and Shands, 1954), and its wild progenitor ssp. *armeniacum* has contributed genes for resistance to leaf and stem rust, powdery mildew, and tan spot via direct backcrossing to wheat (Dyck, 1992; Brown-Guedira, 1995). Dvorak (1977) backcrossed genes for leaf rust resistance into hard red spring wheat cultivars from the diploid *Ae. speltoides.*

Direct backcrossing, like amphiploid breeding, has been used heavily in

FIGURE 3. Direct backcrossing, with *T. aestivum* as recurrent and *Ae. tauschii* as donor parent. It is also possible to use *T. aestivum* as the male in making the first cross and/or the second backcross; however, the F_1 plant is male sterile, *T. aestivum* <u>must</u> be used as the male in making the first backcross. Direct backrossing may be done in a similar way with other combinations species that share a homologous genome (Table 2.)

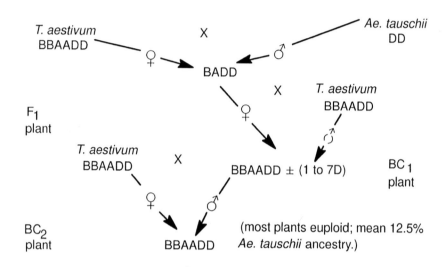

recent years to exploit genes from *Ae. tauschii*. Gill and Raupp (1987) described and demonstrated the methodology by which *T. aestivum* × *Ae. tauschii* F_1 embryos (genomic constitution ABDD) can be rescued on an artificial medium and backcrossed directly to the hexaploid parent. When pollinated by wheat, F_1 plants set approximately one seed per spike, of which about 50% are germinable. Chromosome numbers of BC_1 plants range from 35 to over 50 (Gill and Raupp, 1987; Cox et al., 1991); of these BC_1 plants, 25% (Cox et al., 1991) to 50% (Gill and Raupp, 1987) arise from fertilization of an unreduced female gamete. Resulting 49-chromosome (BBAADDD) plants carry at least one copy of every parental gene. They are more female-fertile than the F_1 and sometimes set selfed seed as well. Cox et al. (1991) provided a formula for computing approximate numbers of BC_1 plants needed to recover a target gene, given average chromosome-number distribution. Gill and Raupp (1987) and Innes and Kerber (1994) predicted that backcrossing BC_1 plants as males would screen out aneuploid pollen and produce primarily 42-chromosome BC_2 progeny; however, among all BC_2 plants produced in this way by Cox, Sears, and Bequette (1995a), only 60% had 42 chromosomes, and not all of those were necessarily euploid. Rates of transmission of alleles from the diploid parent to BC_2 progenies and recombination among loci are generally at the rates expected under the assumptions of nonpreferential segregation and complete chromosome homology (Fritz et al., 1995a).

Whereas synthetic hexaploids have been used successfully in spring-wheat improvement (Eastwood, Lagudah, and Young, 1994; Villareal et al., 1994 and pers. commun.; Eastwood, 1995) and may be the more efficient method for gene transfer, we have used the direct backcrossing approach to improve winter wheats. Several considerations have influenced this decision. Only one gene conditioning nonthreshability, *Tg* from *Ae. tauschii,* segregates in direct-backcross populations, whereas the additional gene *q* from the durum parent segregates in crosses between common wheat and synthetic hexaploids (Villareal et al., 1996). The difference between the two methodologies in frequency of threshable progeny is considerable. Synthetic hexaploids have been useful as donor parents in northern Mexico and Australia, where durum and common spring wheats both are well-adapted. In contrast, durum wheat is not productive in the central plains of the USA, where hard winter wheat is produced. Introgression of novel genes from the nonadapted durum parent of a synthetic into the A and B genomes of hard winter wheats, at the same time genes are being introduced from the *Ae. tauschii* parent into the D genome, disrupts adaptation to a much greater extent than do direct crosses between hard winter wheats and *Ae. tauschii.* And despite the fact that synthetic hexaploids are true-breeding, they and their progeny can be chromosomally unstable (Gill, Hatchett, and Raupp, 1987). Finally, genes may be backcrossed directly from *Ae. tauschii* into almost any elite winter wheat cultivar or breeding line, and progeny similar to the winter wheat parent can be selected from BC_2 populations (Cox et al., 1995b).

Cox, Raupp, and Gill (1994) transferred three leaf rust-resistance genes from *Ae. tauschii* into several hard winter wheat cultivars by direct back-crossing. Resistant BC_2 lines had up to double the grain yield of their respective recurrent parents under heavy leaf-rust infection, and similar yields with no leaf rust (Cox, Sears, and Bequette, 1995a; Cox et al., 1996). Overall, susceptible sister lines had mean yields equal to those of their recurrent parents. The results, which suggested that genes from *Ae. tauschii* have small effects on quantitative traits, were supported by a more detailed molecular-marker study of one parental combination (Fritz et al., 1995b). The relative proportion of *Ae. tauschii* chromatin incorporated into an individual line–estimated as the proportion of marker loci carrying the *Ae. tauschii* allele–had no detectable effect on grain yield, kernel weight, kernel protein concentration, or kernel hardness. On the other hand, there does appear to be a gene linked to *Lr41* from *Ae. tauschii* that increases green leaf-area duration even when disease is absent (Cox et al., 1996). Genes for kernel softness (Cox et al., 1995a, b); increased kernel weight with no reduction in kernel number (Fritz et al., 1995b); and decreased dough-mixing time with no reduction in mixing tolerance (Cox et al., 1996; Knackstedt, 1995) also have been transferred.

Triticum urartu, the donor of wheat's A genome, has been little utilized in germplasm breeding. Direct introgression from the wild A-genome nonprogenitor, *T. monococcum* ssp. *aegilopoides,* may be accomplished as with *Ae. tauschii,* and we have released leaf rust-resistant germplasms KS94WGRC32 and KS96WGRC34 derived from ssp. *aegilopoides* (unpublished). But direct backcrossing with the cultivated einkorn, ssp. *monococcum,* often is impossible, because the F_1 hybrid is both male- and female-sterile (Cox et al., 1991). We identified one accession of ssp. *monococcum,* PI355520, that produced female-fertile progeny, and test-cross ratios originally led us to believe that PI355520 carried two duplicate genes that conditioned the crossability trait. Further studies (Cox, Jellen, and Gill, 1993) showed that a random amplified polymorphic DNA (RAPD) marker was associated with a single gene governing production of female-fertile F_1s. The hybrid-fertility gene from PI355520 has been used in a bridge cross to transfer a gene for leaf-rust resistance directly into the A genome of winter wheat from a noncrossable accession of ssp. *monococcum* (Cox et al., 1994a).

FUTURE OF THE GENE POOL

For many crop species, genetic uniformity is a result of modern breeding and farming practices (Raeburn, 1995). Common wheat is an exception, having been a genetically narrow species throughout its entire existence. It is no wonder that wheat breeders in most regions of the world expend much of their effort on attempts to protect the crop against diseases and insects, or that many of the genes they deploy have originated from outside *T. aestivum*. Although introduction of resistance genes is still the primary motive for interspecific crossing, every new cross inevitably introduces allelic variation that never existed previously in common wheat. The average effects of these new alleles is likely to be negative for most agronomic traits (e.g., Cox, Sears, and Bequette, 1995a), but continued introgression and selection will retain alleles with positive or neutral effects.

Availability of molecular techniques and their more widespread use in wheat research will in no way eliminate the need for introduction of new variation from other species. In fact, molecular mapping in standard common wheats was hindered by lack of polymorphism, necessitating the use of diploid populations (Gill et al., 1991) and crosses with synthetic hexaploids (Nelson et al., 1995a, b). As techniques are refined, asexual gene transfer will make possible the introduction of individual genes from far outside wheat's natural gene pools. But pleiotropic effects of transferred genes, the rigors of the transformation process itself, or the necessity of using transformation-friendly recipient genotypes can result in poorly adapted or low-yielding end-products. Moving introduced genes into elite backgrounds will continue to depend on the availability of a diverse gene pool. And the gene pool itself will become stagnant without input from active, field-oriented germplasm and breeding programs.

In the past, wheat researchers often concentrated on alien species, passing over the more accessible ancestral taxa. McIntosh (1991) neatly summed up the reasons for this, noting that ancestral wheat species have been neglected as germplasm "because cytogeneticists consider the work to be routine, and breeders tend to leave it for others." Today, the spotlight is trained on molecular techniques, but wheat's progenitors already have moved well out of the shadows and may have found a permanent place in the gene pool of common wheat.

REFERENCES

Allard, R.W. and R.G. Shands. (1954). Inheritance of resistance to stem rust and powdery mildew in cytologically stable spring wheats derived from *Triticum timopheevi*. *Phytopathology* 44:266-274.

Alonso, L.C. and G. Kimber. (1984). Use of restitution nuclei to introduce alien genetic variation into hexaploid wheat. *Zeitschrift für Pflanzenzuchtung* 92:185-189.

Bai, D. and D.K. Knott. (1992). Suppression of rust resistance in bread wheat (*Triticum aestivum* L.) by D-genome chromosomes. *Genome* 35:276-282.

Bramel-Cox, P.J. and T.S. Cox. (1989). Use of wild germplasm in sorghum improvement. In *Proceedings of the 43rd Annual Corn and Sorghum Industry Research Conference,* D. Wilkinson, ed. Washington, D.C.: American Seed Trade Association. pp. 13-26.

Brown-Guedira, G.L. (1995). Breeding value and cytogenetic structure of *Triticum timopheevii* var. *araraticum*. Manhattan, Kansas, USA: Kansas State University. Unpublished PhD Dissertation.

Bushuk, W. and E.R. Kerber. (1978). The role of *Triticum carthlicum* in the origin of bread wheat based on gliadin electrophoregrams. *Canandian Journal of Plant Science* 58:1019-1024.

Chen, P.D., H. Tsujimoto, and B.S. Gill. (1994). Transfer of *Ph1* genes promoting homoeologous pairing from *Triticum speltoides* to common wheat. *Theoretical and Applied Genetics* 88:97-101.

Ciaffi, M., S. Benedetelli, B. Giorgi, E. Porceddu, and D. Lafiandra. (1991). Seed storage proteins of Triticum *turgidum* ssp. *dicoccoides* and their effect on the technological quality in durum wheat. *Plant Breeding* 107:309-319.

Cox, T.S., R.K. Bequette, R.L. Bowden, R.G. Sears. (1996). Grain yield and breadmaking quality of wheat lines with the leaf rust resistance gene *Lr41*. *Crop Science* 36:(in press).

Cox, T.S., L.G. Harrell, P. Chen, and B.S. Gill. (1991). Reproductive behavior of hexaploid/diploid wheat hybrids. *Plant Breeding* 107:105-118.

Cox, T.S. and J.H. Hatchett (1994). Hessian fly-resistance gene *H26* transferred from *Triticum tauschii* to common wheat. *Crop Science* 34:958-960.

Cox, T.S., E.N. Jellen, and B.S. Gill. (1993). Development of a genetic map for A-genome wheat. *Agronomy Abstracts* 1993:174.

Cox, T.S., W.J. Raupp, and B.S. Gill. (1994). Leaf rust-resistance genes *Lr41*, *Lr42*, and *Lr43* transferred from *Triticum tauschii* to common wheat. *Crop Science* 34:339-349.

Cox, T.S., W.J. Raupp, D.L. Wilson, B.S. Gill, S. Leath, W.W. Bockus, and L.E. Browder. (1992). Resistance to foliar diseases in a collection of *Triticum tauschii* germ plasm. *Plant Disease* 76:1061-1064.

Cox, T.S., R.G. Sears, and R.K. Bequette. (1995a). Use of winter wheat × *Triticum tauschii* backcross populations for germplasm evaluation. *Theoretical and Applied Genetics* 90:571-577.

Cox, T.S., R.G. Sears, R.K. Bequette, and T.J. Martin. (1995b). Germplasm enhancement in winter wheat × *Triticum tauschii* backcross populations. *Crop Science* 35:913-919.

Cox, T.S., R.G. Sears, B.S. Gill, and E.N. Jellen (1994a). Registration of KS91WGRC11, KS92WGRC15 and KS92WGRC23 leaf rust-resistant hard red winter wheat germplasms. *Crop Science* 34:546-547.

Cox, T.S., J.P. Shroyer, Liu Ben-Hui, R.G. Sears, and T.J. Martin. (1988). Genetic improvement in agronomic traits of hard red winter wheat cultivars from 1919 to 1987. *Crop Science* 28:756-760.

Cox, T.S., M.E. Sorrells, G.C. Bergstrom, R.G. Sears, B.S. Gill, E.J. Walsh, S. Leath, and J.P. Murphy (1994b). Registration of KS92WGRC21 and KS92WGRC22 hard red winter wheat germplasms resistant to wheat spindle-streak mosaic virus, wheat soilborne mosaic citrus, and powder mildew. *Crop Science* 34:546.

Dvorak, J. (1977). Transfer of leaf rust resistance from *Aegilops speltoides* to *Triticum aestivum*. *Canadian Journal of Genetics and Cytology* 19:133-141.

Dvorak, J. and P. DiTerlizzi, H.B. Zhang, and P. Resta. (1993). The evolution of polyploid wheat: Identification of the A genome donor species. *Genome* 36:21-31.

Dyck, P.L. (1992). Transfer of a gene for stem rust resistance from *Triticum araraticum* to hexaploid wheat. *Genome* 35:788-792.

Eastwood, R.F. (1995). Genetics of resistance to *Heterodera avanae* in *Triticum tauschii* and its transfer to bread wheat (*Triticum aestivum*). Melbourne: University of Melbourne. Unpublished PhD Dissertation.

Eastwood, R.F., E.S. Lagudah, and R.M. Young. (1994). Utilization of *Triticum tauschii* as a source of CCN resistance. In *Proceedings of the Seventy Assembly, Wheat Breeding Society of Australia,* ed. J. Paull, I.S. Dundas, K.J. Sheperd, and G.J. Hollamby. Adelaide: University of Adelaide, pp.133-136.

Elias, E.M., D.K. Steiger, and R.G. Cantrell. (1996). Evaluation of lines derived from wild emmer chromosome substitutions: II.Agronomic traits. *Crop Science* 36:228-233.

Fritz, A.K., T.S. Cox, B.S. Gill and R.G. Sears. (1995a). Molecular marker-facilitated analysis of introgression in winter wheat × *Triticum tauschii* populations. *Crop Science* 35:1691-1695.

Fritz, A.K., T.S. Cox, B.S. Gill, and R.G. Sears. (1995b). Marker-based analysis of quantitative traits in winter wheat × *Triticum tauschii* populations. *Crop Science* 35:1695-1699.

Gerechter-Amitai, A.K., I. Wahl, A. Vardi, and D. Zohary. (1971). Transfer of stem rust seedling resistance from wild diploid einkorn to tetraploid durum wheat by means of a triploid hybrid bridge. *Euphytica* 20:281-285.

Gill, B.S., J.H. Hatchett, and W.J. Raupp. (1987). Chromosome mapping of Hessian fly-resistance gene *H13* in the D genome of wheat. *Journal of Heredity* 78:97-100.

Gill, B.S., and W.J. Raupp. (1987). Direct genetic transfers from *Aegilops squarrosa* L. to hexaploid wheat. *Crop Science* 27:445-450.

Gill, B.S., W.J. Raupp, H.C. Sharma, L.E. Browder, J.H. Hatchett, T.L. Harvey, J.G. Moseman, and J.G. Waines. (1986). Resistance in *Aegilops squarrosa* to wheat leaf rust, wheat powdery mildew, greenbug, and Hessian fly. *Plant Disease* 70:553-556.

Gill, K. S., E. L. Lubbers, B. S. Gill, W. J. Raupp, and T. S. Cox. 1991. A genetic linkage map of *Triticum tauschii* (DD) and its relationship to the D genome of bread wheat (AABBDD). *Genome* 34:362-374.

Harlan, J.R. (1965). The possible role of weed races in the evolution of cultivated plants. *Euphytica* 14:173-176.

Innes, R.L. and E.R. Kerber. (1994). Resistance to wheat leaf rust and stem rust in *Triticum tauschii* and inheritance in hexaploid wheat of resistance transferred from *T. tauschii*. *Genome* 37:813-822.

Jaaska, V. (1978). NADP-dependent aromatic alcohol dehydrogenase in polyploid wheats and their diploid relatives. *Theoretical and Applied Genetics* 53:209-217.

Jaaska, V. (1980). Aspartate aminotransferase and alcohol dehydrogenase isoenzymes: intraspecific differentiation in *Aegilops tauschii* and the origin of the D genome polyploids in the wheat group. *Plant Syst. Evol.* 137:259-273.

Jiang, J., B. Friebe, and B.S. Gill. (1994). Recent advances in alien gene transfer in wheat. *Euphytica* 73:199-212.

Jiang, J. and B.S. Gill. (1994). Different species-specific chromosome transloca-tions in *Triticum timopheevii* and *T. turgidum* support the diphyletic origin of polyploid wheats. *Chromosome Research* 2:59-64.

Joppa, L.R., G.A. Hareland, and R.G. Cantrell. (1991). Quality characteristics of Langdon durum-dicoccoides chromosome substitution lines. *Crop Science* 31:1513-1517.

Kerber, E.R. (1964). Wheat: reconstitution of the tetraploid component (AABB) of hexaploids. *Science* 143:253-255.

Kerber, E.R. (1987). Resistance to leaf rust in hexaploid wheat: Lr32, a third gene derived from *Triticum tauschii. Crop Science* 27:204-206.

Kerber, E.R. and V.M. Bendelow. (1977). The role of *Triticum carthlicum* in the origin of bread wheat based on comparative milling and baking properties. *Canadian Journal of Plant Science* 57:367-373.

Kerber, E.R. and P.L. Dyck. (1969). Inheritance in hexaploid wheat of leaf rust resistance and other characters derived from *Aegilops squarrosa. Canadian Journal of Genetics and Cytology* 11:639-640.

Kerber, E.R. and P.L. Dyck. (1990). Transfer to hexaploid wheat of linked genes from adult-plant leaf rust and seedling stem rust resistance from an amphiploid of *Aegilops speltoides* × *Triticum monococcum. Genome* 33:530-537.

Khan, K., R. Frohberg, T. Olson, and L. Huckle. (1989). Inheritance of gluten protein components of high-protein hard red spring wheat lines derived from *Triticum turgidum* var. *dicoccoides. Cereal Chemistry* 66:397-401.

Kihara, H. (1944). Discovery of the DD-analyser, one of the ancestors of *vulgare* wheats. *Agriculture and Horticulture (Tokyo)* 19:889-890.

Knackstedt, M.A. (1995). Bread making quality associated with novel gliadin and glutenin proteins and overall bread making quality in *Triticum tauschii* × *Triticum aestivum* derived lines. Manhattan, Kansas, USA: Kansas State Uni-versity. Unpublished Ph.D. Dissertation.

Knott, D.R. (1989). The effect of alien transfers of genes for leaf rust resistance on the agronomic and quality characteristics of wheat. *Euphytica* 44:65-72.

Ladizinsky, G. (1984). Founder effect in crop plant evolution. *Economic Botany* 39:191-199.

Lagudah, E.S. and G.M. Halloran. (1988). Phylogenetic relationships of *Triticum tauschii,* the D genome donor to hexaploid wheat. *Theoretical and Applied Genetics* 75:592-598.

Lange, W. and G. Jochemsen. (1992). Use of the gene pools of *Triticum turgidum* ssp. *dicoccoides* and *Aegilops squarrosa* for the breeding of common wheat (*T. aestivum*), through chromosome-doubled hybrids. *Euphytica* 59:197-212.

Levy, A.A., G. Galili, and M. Feldman. (1988). Polymorphism and genetic control of high molecular weight glutenin subunits in wild tetraploid wheat *Triticum turgidum* var. *dicoccoides. Heredity* 61:63-72.

Lubbers, E.L., K.S. Gill, T.S. Cox, and B.S. Gill. (1991). Variation of molecular markers among geographically diverse accessions of *Triticum tauschii. Genome* 34:354-361.

Ma, H. and G.R. Hughes. (1995). Genetic control and chromosomal location of *Triticum timopheevii*-derived resistance to *Septoria nodorum* blotch in durum wheat. *Genome* 38:332-338.

MacKey, J. (1966). Species relationships in *Triticum. Hereditas (Suppl.)* 2:237-275.

Marais, G.F. (1992). The modification of a common wheat-*Thinopyrum distichum* translocated chromosome with a locus homoeoallelic to *Lr19. Theoretical and Applied Genetics* 85:73-78.

Marais, G.F. and Z.A. Pretorius. (1996). Gametocidal effects and resistance to wheat leaf rust and stem rust in derivatives of a *Triticum turgidum* ssp. *durum / Aegilops-speltoides* hybrid. *Euphytica* 88:117-124.

Martin, D.J., and B.G. Stewart. (1986). Dough-mixing properties of a wheat-rye derived cultivar. *Euphytica* 35:225-232.

May, C.E. and E.S. Lagudah. (1992). Inheritance in hexaploid wheat of *Septoria tritici* blotch resistance and other characteristics derived from *Triticum tauschii. Australian Journal of Agricultural Research* 43:433-442.

McFadden, E.S. and E.R. Sears. (1946). The origin of *Triticum speltoides* and its free-threshing hexaploid relatives. *Journal of Heredity* 37:81-89, 107-116.

McFadden, E.S. and E.R. Sears. (1947). The genome approach in radical wheat breeding. *Journal of the American Society of Agronomy* 39:1011-1026.

McIntosh, R.A. (1991). Alien sources of disease resistance in bread wheats. In *Proceedings of Dr. Kihara Memorial International Symposium on Cytoplasmic Engineering in Wheat: Nuclear and Organellar Genomes of Wheat Species,* ed. T. Saskuma and T. Kinoshita. Yokohama, Japan: Kyoto Univ., pp. 320-332.

McIntosh, R.A., C.R. Wellings, and R.F. Park. (1995). *Wheat Rusts: An Atlas of Resistance Genes.* East Melbourne, Australia: CSIRO Publications.

Metakovsky, E.V., A. Yu. Novoselskaya, M.M. Kopus, T.A. Sobko, and A.A. Sozinov. (1984). Blocks of gliadin components in winter wheat detected by one-dimensional polyacrylamide gel electrophoresis. *Theoretical and Applied Genetics* 67:559-568.

Morrison, L.A. (1993). *Triticum-Aegilops* systematics: taking an integrative approach. In *Biodiversity in Wheat Improvement,* ed. A.B. Damania. New York: John Wiley & Sons, pp. 59-66.

Moseman, J.G., E. Nevo, Z.K. Gerechter-Amatai, M.A. El-Morshidy, and D. Zohary. (1985). Resistance of *Triticum dicoccoides* collected in Israel to infection with *Puccinia recondita. Crop Science* 25:262-265.

Moseman, J.G., E. Nevo, M.A. El-Morshidy, and D. Zohary. (1984). Resistance of *Triticum dicoccoides* to infection with *Erysiphe graminis tritici. Euphytica* 33:41-47.

Nelson, J.C., A.E. Van-deynze, E. Autrique, M.E. Sorrells, Y.H. Lu, M. Merlino, M. Atkinson, and P. Leroy. (1995a). Molecular mapping of wheat: Homoeologous group 2. *Genome* 38:516-524.

Nelson, J.C., A.E. Van-Deynze, E. Autrique, M.E. Sorrells, Y.H. Lu, S. Negre, M. Bernard, and P. Leroy. (1995b). Molecular mapping of wheat: Homoeologous group 3. *Genome* 38:525-533.

Nicholson, P., H.N. Rezanoor, and A.J. Worland. (1993). Chromosomal location of resistance to *Septoria nodorum* in a synthetic hexaploid wheat determined by the study of chromosomal substitution lines in 'Chinese Spring' wheat. *Plant Breeding* 110:177-184.

Nishikawa, K., Y. Furuta, and T. Wada. (1980). Genetic studies on ∝-amylase isozymes in wheat. III. Intraspecific variation in *Aegilops squarrosa* and birthplace of hexaploid wheat. *Japanese Journal of Genetics,* 55:325-336.

Patterson, F.L., J.E. Foster, and H.W. Ohm. (1988). Gene H16 in wheat for resistance to Hessian fly. *Crop Science* 28:652-654.

Raeburn, P. (1995). *The Last Harvest.* New York: Simon and Schuster.

Rafi, M.M, B. Ehdaie, and J.G. Waines. (1992). Quality traits, carbon isotope discrimination and yield components in wild wheats. *Annals of Botany* 69:467-474.

Riley, R., G. Kimber, and V. Chapman. (1961). Origin of genetic control of diploid-like behavior of polyploid wheat. *Journal of Heredity* 52:22-25.

Sharma, H.C. and B.S. Gill. (1983). Current status of wide hybridization in wheat. *Euphytica* 32:17-31.

Shroyer, J.P. and T.S. Cox. (1993). Productivity and adaptive capacity of winter wheat landraces and modern cultivars grown under low-fertility conditions. *Euphytica* 70:27-33.

Simmonds, N.W. (1993). Introgression and incorporation strategies for the use of crop genetic resources. *Biological Reviews* 68:539-562.

Stebbins, N.B., F.L. Patterson, and R.L. Gallun. (1983). Inheritance of resistance of PI 94587 wheat to biotypes B and D of Hessian fly. *Crop Science* 23:251-253.

Steiger, D.K., E.M. Elias, and R.G. Cantrell. (1996). Evaluation of lines derived from wild emmer chromosome substitutions: I. Quality traits. *Crop Science* 36:223-227.

Tahir, M. and H. Ketata. (1993). Use of landraces, primitive forms, and wild species for the development of winter and facultative durum wheat germplasm. In *Biodiversity in Wheat Improvement,* ed. A.B. Damania. New York: John Wiley & Sons, pp. 341-351.

Takumi, S., S. Nasuda, Y.G. Liu, and K. Tsunewaki. (1993). Wheat phylogeny determined by RFLP analysis of nuclear DNA. 1. Einkorn wheat. *Japanese Journal of Genetics* 68:73-79.

The, T.T., B.D.H. Latter, R.A. McIntosh, F.W. Ellison, P.S. Brennan, J. Fisher, G.J. Hollamby, A.J. Rathjen, and R.E. Wilson. (1988). Grain yields of near-isogenic lines with added genes for stem rust resistance. In *Proceedings of the Seventh International Wheat Genetics Symposium,* ed. T.M. Miller and RMD Koebner. Cambridge, U.K.: Institute of Plant Science Research.

Thomas, J.B., and R.L. Connor. (1986). Resistance to colonization by the wheat curl mite in *Aegilops squarrosa* and its inheritance after transfer to common wheat. *Crop Science* 26:527-530.

Tsujimoto, H. and K. Tsunewaki. (1985). Hybrid dysgenesis in common wheat caused by gametocidal genes. *Japanese Journal of Genetics* 60:565-578.

Valkoun, J., D. Kucerova and P. Bartos. (1986). Transfer of leaf rust resistance from *Triticum monococcum* L. to hexaploid wheat. *Zeitschrift für Planzenzuchtung* 96:271-278.

Vallega, V. (1992). Agronomical performance and breeding value of selected strains of diploid wheat, *Triticum monococcum*. *Euphytica* 61:13-23.

van Slageren, M.W. (1994). Wild Wheats: A Monograph of *Aegilops* L. *and Amblyopyrum* (Jaub. & Spach) Eig. Wageningen, The Netherlands: Wageningen Agricultural University/ICARDA.

Vardi, A. and D. Zohary. (1967). Introgression in wheat via triploid hybrids. *Heredity* 22:541-560.

Villareal, R.L., A. Mugeeb-Kazi, G. Fuentes-Davila, S. Rajaram, and E. del Toro. (1994). Resistance to Karnal Bunt (*Tilletia indica* Mitra) in synthetic hexaploid wheats derived from *Triticum turgidum* × *T. tauschii*. *Plant Breeding* 112:63-69.

Villareal, R.L., A. Mujeeb-Kazi and S. Rajaram. (1996). Inheritance of threshability in synthetic hexaploid (*Triticum turgidum* × *T. tauschii*) by *T. aestivum* crosses. *Plant Breeding* (in press).

Watanabe, N. (1983). Variation of D genomes affecting the morphological characters of common wheat. *Japanese Journal of Breeding* 33:296-301.

Xu, S.J. and L.R. Joppa. (1995). Mechanisms and inheritance of first division restitution in hybrids of wheat, rye, and *Aegilops squarrosa*. *Genome* 38:607-615.

Zohary, D., J.R. Harlan, and A. Vardi. (1969). The wild diploid progenitors of wheat and their breeding value. *Euphytica* 18:58-65.

Zohary, D. and M. Hopf. (1988). *Domestication of Plants in the Old World*. Oxford, U.K.: Clarendon Press.

SUBMITTED: 09/13/96
ACCEPTED: 12/08/96

Physiological and Agronomic Consequences of Rht Genes in Wheat

Martin P. N. Gent
Richard K. Kiyomoto

SUMMARY. Yield of wheat has risen dramatically and world-wide in the last two decades, in part because of the widespread introduction of Rht genes that reduce the length of tillers of wheat. We review the physiological consequences of incorporating reduced height genes into wheat. The Rht_1 or Rht_2 genes modulate the morphology and physiology of wheat in a manner that involves compensation among several physiological processes. For instance, Rht genes decrease leaf area, but photosynthesis per unit area increases, so biomass accumulation is rarely altered. Although Rht genes increase leaf permeability to water vapor, plant water status changes in compensation to minimize differences in water use efficiency. Perhaps due to less competition for carbohydrate during stem elongation, semidwarf wheat has greater harvest index than tall wheat at maturity. Although tall wheat has a greater weight per kernel at maturity, this does not completely compensate for greater grain number per spike in semidwarf wheat. The compensation that leads to greater yield in semidwarf wheat appears to fail in Rht_1Rht_2 or Rht_3 dwarf wheat. Either specific photosynthesis does not com-

Martin P. N. Gent, Associate Scientist, Richard K. Kiyomoto, Associate Scientist, Department of Forestry and Horticulture, The Connecticut Agricultural Experiment Station, POB 1106, New Haven, CT 06504-1106, USA.

Address correspondence to: Martin P. N. Gent, Department of Forestry and Horticulture, The Connecticut Agricultural Experiment Station, POB 1106, New Haven, CT 06504-1106, USA (E-mail: mgent@caes.state.ct.us).

[Haworth co-indexing entry note]: "Physiological and Agronomic Consequences of Rht Genes in Wheat." Gent, Martin P. N., and Richard K. Kiyomoto. Co-published simultaneously in *Journal of Crop Production* (The Food Products Press, an imprint of The Haworth Press, Inc.) Vol. 1, No. 1 (#1), 1998, pp. 27-46; and: *Crop Sciences: Recent Advances* (ed: Amarjit S. Basra) The Food Products Press, an imprint of The Haworth Press, Inc., 1998, pp. 27-46. Single or multiple copies of this article are available for a fee from The Haworth Document Delivery Service [1-800-342-9678, 9:00 a.m. - 5:00 p.m. (EST). E-mail address: getinfo@haworth.com].

pletely compensate for decreased area per leaf, or the reduction in plant height retards canopy closure and efficient interception of solar radiation, resulting in lower biomass and yield of dwarf wheat. Drought stress reduces yield and harvest index of dwarf wheat more than tall wheat. Rht genes have insignificant effects on rate of development or winter hardiness. The effects of Rht genes on physiology appear to be similar in winter, spring and durum wheat. *[Article copies available for a fee from The Haworth Document Delivery Service: 1-800-342-9678. E-mail address: getinfo@haworth.com]*

KEYWORDS. *Triticum,* dwarfing, harvest index, height, kernel weight, yield, carbohydrate, leaf area, leaf permeability, light interception, nitrogen, partitioning, photosynthesis, respiration, water use

INTRODUCTION

Yield of wheat has risen dramatically and world-wide in the last two decades, in part because of the widespread introduction of dwarfing genes. This has reduced the height or stem length of the tillers of the wheat plant. The most important or practical benefit of reduced height is more erect straw, which is not prone to collapse or 'lodging' before harvest. Resistance to lodging may be simply due to shorter straw, as the mechanical strength of dwarf wheat is not inherently greater than of tall wheat (Paollilo and Niklas, 1996). Lodging reduces the efficiency of mechanical harvesting of grain, and grain quality is reduced due to humid conditions when the spike lays on the ground or is covered by other plants. In part, the increase in yield has been due to greater use of fertilizer and irrigation in wheat production. These practices increase plant height and the likelihood of lodging. Thus, the benefit of increased fertilizer and irrigation necessitated the introduction of varieties with reduced height.

In this review, we discuss the physiological consequences of incorporating the reduced height genes, Rht_1, Rht_2 and Rht_3 into wheat. We comment on developmental and morphological traits related to yield of wheat as they are affected by plant height. These genes confer a stem length of 1.0 to 1.5 meters for tall wheat, 0.7 to 0.9 meters for semidwarf wheat with Rht_1 or Rht_2 genes, and 0.4 to 0.6 meters for dwarf wheat with Rht_1Rht_2 or Rht_3 genes. Environment and genetic background affect plant height within these ranges. The genes Rht_1 and Rht_2, usually incorporated singly and not together, are responsible for most of the commercially successful wheat varieties in use today. Rht_3 is another dwarfing gene that confers gibberellin insensitivity, but reduces stem length more that Rht_1 or

Rht_2. Rht_3 is often used in physiological comparisons, but has not been incorporated into commercial varieties.

This review will summarize the finding of comparisons of wheat varieties that differ in these genes, or Rht substitutions in near isogenic backgrounds, referred to as reduced height isolines. About 10 other genes that confer short stature have been identified in wheat, however, they have not been used in breeding programs, nor have many physiological consequences of these genes been examined. This review does not cover the effects of these other reduced height genes. The identification and characterization of the genes that reduce stem length in wheat have been reviewed elsewhere (Gale and Youssefian, 1985; Yamada, 1990), and a new terminology was proposed recently (Borner et al., 1996). Several recent reviews have discussed physiological characteristics related to yield of wheat with emphasis on response to dry climates (Ludlow and Muchow, 1990; Loss and Siddique, 1994) and on growth before anthesis (Slafer, Andrade, and Sattore, 1990).

GIBBERELLIN METABOLISM

The dwarfing genes Rht_1, Rht_2, and Rht_3 confer insensitivity of stem elongation to the hormone gibberellin (Gale and Youssefian, 1985). The Rht_3 gene confers a lack of response to many other physiological processes normally sensitive to gibberellin (Ho, Nolan, and Shute, 1981). The rate of metabolism of gibberellin conjugate GA_1 is not altered by Rht_3, but GA_1 accumulates 10 fold more in leaves of Rht_3 wheat than that of tall wheat (Stoddart, 1984). A lack of regulation of an early step in gibberellin metabolism, from conjugate GA_{19} to GA_{20}, leads to an accumulation of GA_1 (Appleford and Lenton, 1991).

HISTORICAL TRENDS

Agronomic characteristics have been compared for varieties that have been developed and released over the last 50 to 100 years, when they were grown side by side in yield trials in productive environments, or under stress. In various wheat production areas, breeding has resulted in shorter stems and earlier spike emergence or anthesis: in England (Austin et al., 1980), central USA (Cox et al., 1988), Japan (Hoshino and Seko, 1996), and Australia (Perry and D'Antuono, 1989). Breeding did not alter plant biomass; the substantial increases in yield were entirely due to an increase in harvest index. When the distribution of weight among plant parts was measured at the time of maximum biomass accumulation, modern semi-dwarf varieties of winter wheat in England had lower stem weights but equal leaf area index and leaf weights as older tall varieties (Austin et al., 1980). For both winter and spring wheat grown in central USA, in a favorable environment, the increase in yield due to breeding was twice that in a drought-stressed environment (Feyerherm, Paulsen, and Sebaugh, 1984).

These studies raise the question, what plant height is optimal for yield? Apparently, some reduction of plant height improves yield substantially. In studies of reduced height isolines in a variety of backgrounds, the best yields were produced by isolines of intermediate height. When crosses that resulted in Rht_1 or Rht_2 semidwarfs were segregated by height, yield did not vary with height (Busch and Rauch, 1993). A second question concerns the physiological basis for the increase in harvest index, or partitioning of biomass to grain, in varieties of wheat with reduced height. This phenomenon is a third reason why Rht genes have increased wheat yields. We evaluate the various physiological processes that may affect harvest index, and the possibility of compensation among these processes. Finally, stressful environments alter the yield benefit of reduced height genes. We discuss the physiological basis of this environment × genotype interaction.

LEAF AREA INDEX

If leaf area were reduced as much as plant height, then reduced height genes could result in less light interception, and biomass would decrease for this reason. In a comparison of historical varieties of winter wheat in England, there was no trend with stem height in leaf area index or leaf weight per plot area (Austin et al., 1980). In a comparison of varieties of spring wheat in Australia, older taller varieties tended to have greater leaf area index than modern shorter varieties (Yunusa et al., 1993). In Yugoslavia, a 10-year study of varieties differing in height found a negative relation overall between yield and leaf area index and yield, but a positive relation to leaf area duration (Borojevic and Williams, 1982).

Specific effects of Rht genes on leaf area were examined in isogenic backgrounds. Both Rht_1 and/or Rht_2 genes decreased cell length of flag leaves of field plants and sheath and blade of leaf 1 of plants in controlled conditions, and their effects were additive (Keyes, Paolillo, and Sorrells, 1989). Leaves were smaller due to a decreased cell length, and not due to fewer cells. In a controlled environment, Rht_3 reduced leaf length by 0.5, compared to 0.15 for Rht_1 or Rht_2, and Rht_3 reduced cell number as well as cell dimensions (Hoogendoorn, Rickson, and Gale, 1990). In another study, the flag leaf of dwarf wheat, with Rht_1 and Rht_2, had smaller cells, but more of them, than an isogenic tall wheat (Lecain, Morgan, and Zerbi, 1989). The effects of Rht genes on leaf area were less consistent in other studies. Plants reduced to a single tiller had less total leaf area with Rht_1 and Rht_2 combined, and especially with Rht_3, but there was no effect of Rht_1 or Rht_2 alone on flag leaf area (King, Gale, and Quarrie, 1983). In northeastern USA, leaf area index and leaf area per unit weight were less for dwarf than for tall reduced height isolines of winter wheat early in stem elongation, but not at anthesis or thereafter (Gent, 1995).

Temperature and light affect leaf elongation. When leaf length was compared for plants grown at 25°C and 10°C, tall genotypes had a greater increase in length than semidwarf, and there was no increase in length for genotypes with Rht_1 and Rht_2, or Rht_3 (Pinthus et al., 1989). Leaves elongated faster in dark than in light, and the limiting effect of dwarfing genes on elongation at 25°C was greater in dark (Pinthus and Abraham, 1996). When grown at different temperatures, area and length of flag leaves were greater for tall than for dwarf isolines at 24/19°C but not at 15/10°C, in two isogenic spring wheat backgrounds (Bush and Evans, 1988). These studies suggest that dwarfing genes set an upper limit on leaf elongation rate, a limit which may not be reached at cool growth temperatures. In the field, a reduction in leaf area index due to Rht genes is more likely to occur at warm than at cool growth temperatures.

When stems elongate, they could contribute to light interception, in a manner other than an increase in leaf and sheath area. Typically, a crop canopy is assumed to intercept light by an array of randomly distributed horizontal leaves, and interception is not a function of angle of solar radiation (Montieth, 1969). Interception of light by vertical stems is ignored because their area is less than that of leaves, and their orientation is nearly parallel to solar radiation at noon. However, when there are gaps in the canopy or between rows, light interception will depend on the angle of incidence and plant height. Around noon early in crop development, tall wheat will intercept more light in gaps or between rows than dwarf wheat (Gent, unpublished calculations). This effect is likely to disappear after spike emergence, in part because the plants are big enough to present a closed canopy, and in part because the spikes intercept a large fraction of solar radiation in a manner that depends more on the density of spikes per unit area than on the angle of incident radiation (Thorne et al., 1988; Yunusa et al., 1993). In northeastern USA, biomass and light interception of tall isolines in the Itana winter wheat background were about 0.2 times greater than in dwarf isolines early in stem elongation, but semidwarf did not differ consistently from tall isolines (Gent, 1995). After anthesis, light interception did not differ between height classes. However among reduced height isolines of spring wheat in Australia, tall isolines had more biomass than dwarf isolines at maturity (Nizam Uddin and Marshall, 1989). Although there is a limited period of benefit of increased height for canopy light interception and biomass accumulation in wheat, yet it may be one reason for low yield in dwarf wheat.

PHOTOSYNTHESIS

The accumulation of biomass depends on photosynthesis and respiration per unit leaf area, as well as the fraction of light intercepted by the canopy. While photosynthesis of leaves has been examined extensively, the relationship between leaf photosynthesis and canopy photosynthesis is not well understood. Among winter wheat varieties differing in height, specific photosynthesis (rate per unit area) was inversely related to area of flag leaves (Gale, Edrich, and Lupton, 1974). Applied gibberellin had only a transient effect on photosynthesis, even in tall, gibberellin-responsive varieties. In a field trial of spring wheat varieties in Mexico, specific photosynthesis was positively correlated with plant height before anthesis, but there was no correlation after anthesis (Fischer et al., 1981). The dwarf varieties, with Rht_1 and Rht_2, or Rht_3 differed significantly from both the semidwarf and tall varieties, but when present singly, the Rht_1 or Rht_2 genes had little effect on photosynthesis. However, when crosses of some of these varieties were grown in a controlled environment, the total carbon assimilated in the third leaf was not related to that of the flag leaf or to yield (Rawson et al., 1983). In England, specific photosynthesis of flag leaves of semidwarf winter wheat varieties was only occasionally greater than the tall varieties (Pearman, Thomas, and Thorne, 1979). When measured by O_2 evolution in a controlled environment, flag leaf photosynthesis was in the order Rht_1 and Rht_2 = Rht_2 > Rht_1 > tall (Kulshrestha and Tsunoda, 1981b). In a single comparison, a Rht_1 and Rht_2 dwarf had faster specific photosynthesis, leaf conductance and stomatal density than a tall isoline (Lecain, Morgan, and Zerbi, 1989). Since Morgan, LeCain, and Wells (1990) found more chlorophyll, protein, and Ribulosebisphosphate carboxylase per unit leaf area in the dwarf than the tall isoline, they ascribed the effect of dwarfing genes on photosynthesis to a greater density of cells capable of photosynthesis, rather than a change in photosynthetic metabolism.

There have been few studies of the effect of height on canopy photosynthesis. In northeastern USA, an old tall winter wheat variety had greater canopy photosynthesis than a modern semidwarf until anthesis, but not thereafter (Gent and Kiyomoto, 1985). Photosynthesis per unit area of flag leaves did not differ in the field or under controlled conditions. In a comparison of reduced height isolines in Burt and Omar winter wheat backgrounds, dwarf wheat, with the combination of Rht_1 and Rht_2 genes, had less canopy and flag leaf photosynthesis than either semidwarf or tall isolines (Kiyomoto and Gent, 1989a). There was no difference due to Rht_1 or Rht_2 present singly in semidwarf, compared to tall isolines. When isolines of Itana winter wheat were grown in northeastern USA, light interception and canopy photosynthesis was about 0.2 times greater in tall than in dwarf wheat only early in stem elongation, but not after anthesis (Gent, 1995). Differences between semidwarf and tall isolines were not consistent from year to year. After anthesis, neither light interception nor canopy photosynthesis differed between height classes. In these studies of Rht_1 and/or Rht_2 in isogenic backgrounds, canopy photosynthesis was never increased but where Rht genes affected canopy photosynthesis, it was reduced. The agronomic effects may be governed by compensation between leaf area and photosynthesis per unit area. Although Rht genes increase photosynthesis per unit leaf area, this increase may never completely compensate for the reduction in area per leaf. The lack of compensation at the canopy level may be due, in part, to the contribution of stem height *per se* on light interception of the canopy, as discussed above.

RESPIRATION

The reduced height genes increase specific respiration (per unit leaf area), as well as photosynthesis. In a comparison among varieties of spring wheat, flag leaf respiration rates were in the order $Rht_2 > Rht_1Rht_2 > Rht_1 >$ tall (Kulshrestha and Tsunoda, 1981b). The relative differences in respiration due to dwarfing genes were greater than those in photosynthesis. Respiration per unit weight of stem was also faster in dwarf compared to tall varieties of spring wheat grown in controlled conditions (Rawson and Evans, 1971). These observations are consistent with the greater density of cells in dwarf than in tall wheat (Keyes, Paolillo, and Sorrells, 1989; Lecain, Morgan, and Zerbi, 1989), which would concentrate the mitochondria responsible for respiration. In addition, endogenous gibberellin should slow metabolism and respiration in tall genotypes more than in dwarf, gibberellin-insensitive genotypes (Ho, Nolan, and Shute, 1981). However, in Canada, varietal differences among winter wheat in respiration per unit of biomass were not expressed per unit of land area (McCullough and Hunt, 1989). In northeastern USA, dark respiration of the canopy of a tall winter wheat variety was less than that of a semidwarf per unit dry weight, but respiration did not differ per unit land area (Gent and

Kiyomoto, 1985). Consequently, there may be little agronomic effect of increased respiration.

TRANSPIRATION AND LEAF PERMEABILITY

Because drought stress has a great effect on biomass accumulation and yield, leaf permeability to water vapor and water relations have been studied extensively in wheat. In spring wheat varieties grown under irrigation in Mexico, leaf permeability to water vapor was in the order $Rht_3 = Rht_1Rht_2 > Rht_1 = Rht_2 >$ tall (Fischer et al., 1981). Yield is correlated to leaf permeability. In central USA, only lower leaves had greater permeability in semidwarf than in tall winter wheat varieties; upper leaves did not differ in this respect (Kirkham and Smith, 1978). In an isogenic spring wheat background, transpiration efficiency and water use efficiency declined with plant height under well-watered but not under drought conditions, but carbon isotope discrimination declined with plant height under both conditions (Ehdaie and Waines, 1994). Dwarf wheat with Rht_3 differed most significantly with respect to its tall counterpart. Under two light intensities and two N fertility regimes, an $Rht_1 Rht_2$ dwarf winter wheat had less water use efficiency than a tall isoline only under high N and low light (Morgan, Lecain, and Wells, 1990). Leaf permeability was always greater for the dwarf than the tall isoline, and the differences were most pronounced under high N.

Water relations within the wheat plant also vary with stem height. In winter wheat varieties grown under irrigation in central USA, the upper leaves of tall wheat had more negative water potential than those of semidwarf wheat (Kirkham and Smith, 1979). Lower leaves did not differ in water potential. The water potential of grain was less negative in tall than in semidwarf varieties, so differences between grain and leaf water potential were less for semidwarf than the tall varieties (Kirkham and Smith, 1979). In comparisons of spring wheat grown in Mexico under drought or well-watered conditions, differences in plant water relations were not related to drought susceptibility, as defined by the ratio of yield under dry and wet conditions (Fischer and Wood, 1979). Weekly measurements of soil water in winter wheat plots in central USA showed no difference in water use between semidwarf and tall varieties (Holbrook and Welsh, 1980). Thus, the generally greater leaf permeability of dwarf wheat was not translated into poor water use efficiency or lower yield, except for $Rht_1 Rht_2$ or Rht_3 dwarfs grown under dry conditions.

HARVEST INDEX

Harvest index is consistently greater in semidwarf compared with tall wheat. This has been noted in yield trials of diverse varieties differing in height. When such factors as stand establishment and lodging were controlled, the historical trend among varieties toward shorter stature and greater yield was correlated with increased harvest index, in England (Austin et al., 1980), central USA (Cox et al., 1988), Japan (Hoshino and Seko, 1996), and Australia (Perry and D'Antuono, 1989). The effect on harvest index of Rht genes is clearly demonstrated in comparisons of isolines in different genetic backgrounds (Allan, 1983, 1986, 1989; Youssefian, Kirby, and Gale, 1992). In comparisons of 7 isogenic winter wheat backgrounds in northwestern USA, the increase in harvest index due to dwarfing genes was greatest in inherently tall backgrounds (Allan, 1983). Dwarf Rht_1Rht_2 wheat had higher harvest index than Rht_1 or Rht_2 semidwarf in 9 of 28 comparisons, and semidwarf had higher harvest index than tall in 20 of 28 comparisons within background, with means of 0.41, 0.38 and 0.32, for dwarf, semidwarf and tall, respectively (Allan, 1983). A single Rht gene changed harvest index more than a second dose, although the effect on plant height was additive (Allan, 1983). In a similar study of Rht_3, it resulted in similar plant height as Rht_1Rht_2, but it resulted in lower yield, and Rht_1Rht_2 isolines yielded less than semidwarf Rht_1 or Rht_2 isolines (Allan, 1986). An extrapolation of the historical trends in varieties suggested that a further decrease in plant height could increase harvest index to 0.6 and increase yield by 0.25 (Austin et al., 1980). However, in isogenic backgrounds, a reduction in plant height to dwarf stature apparently decreased biomass more than it increased harvest index, and thus it did not increase grain yield.

The effect of dwarfing genes on harvest index depends on environment. A comparison of winter wheat varieties released over 50 years in central USA showed a smaller increase in yield with time in drought-stressed compared to highly productive environments (Feyerherm, Paulsen, and Sebaugh, 1984). Among spring wheat varieties grown in Mexico, harvest index was 0.38, 0.45 and 0.49 under wet conditions, and 0.32, 0.34 and 0.38 under dry conditions, for dwarf, semidwarf and tall varieties, respectively (Fischer and Wood, 1979). Among spring wheat backgrounds in Australia, the relative reduction in yield in rain-fed compared to irrigated conditions was 0.22, 0.23 and 0.30, for tall, semidwarf and dwarf isolines, respectively (Nizam Udddin and Marshall, 1989). In tests of Rht_1 and Rht_2 isolines in Centana spring wheat background in locations with diverse yield potential in north central USA, Rht_1 or Rht semidwarf had the greatest yield in all locations except the least favorable, and Rht_1 Rht_2

dwarf had the lowest yields (McNeal et al., 1972). In this study, low yield potential increased harvest index of tall but not of dwarf wheat. Thus, Rht genes confer a greater increase in harvest index under more favorable conditions.

Differences in harvest index are not due to differences in timing of development of spike and stem, which are slight. In both glasshouse and field experiments, Rht_2 alone in an isogenic background slightly advanced the double ridge stage, but had no effect on anthesis (Brooking and Kirby, 1981). Under controlled conditions with three different temperatures, the double ridge stage was about 2 days earlier for Rht_1Rht_2 dwarf than for tall wheat in two isogenic spring wheat backgrounds, but the date of anthesis did not differ (Bush and Evans, 1988). In this study, there was a single relation between spike weight and weight per unit length of stem, that applied across all environments and reduced height genotypes. This would suggest that when lower stem weight induced by Rht genes is compensated for by increased weight per unit length of stem, it results in greater spike weight and harvest index. Presumably under stressful environments there is less compensation of this sort, and thus less effect of reduced height genes on harvest index.

PARTITIONING AND REDISTRIBUTION OF DRY MATTER

Around the time of anthesis, there is competition for assimilate between the developing spike and the growing stem. During grain filling, assimilate is redistributed from stem to grain (Borrell et al., 1989; Gent, 1994), more so under dry than wet conditions (Bidinger, Musgrave, and Fischer, 1977). Rht genes affect these processes in different ways, and environment influences the effect of dwarfing genes on assimilate partitioning at each stage. In spring wheat grown under controlled conditions and restricted to a single tiller, for 50 days prior to anthesis, dwarf varieties partitioned more dry matter to leaf lamina and less to stem than the tall varieties (Fischer and Stockman, 1986). Partitioning to spikes was higher in dwarf than in tall varieties for 15 days prior to anthesis. However, varieties that differed in height did not differ in the loss of weight in the stem after anthesis (Rawson and Evans, 1971). Among winter wheat varieties in Canada, loss of stem dry matter after anthesis was only weakly related to maximum stem weight or length, and it was not related to weight per unit length (Hunt, 1978).

The fate of ^{14}C after photosynthetic assimilation of $^{14}CO_2$ has been used to track the fate of photosynthetic assimilates in the field and under controlled conditions. Among winter wheat varieties grown in the field in

England, movement from stem to grain of ^{14}C assimilated prior to anthesis was correlated with loss of weight in stem after anthesis (Austin et al., 1977). At maturity, semidwarf varieties lost more stem weight than tall varieties and moved about more ^{14}C to the grain (Thorne, 1982). In winter wheat grown in northeastern USA and labeled 10 days before anthesis, only a small fraction of the total ^{14}C was in the spike at maturity, and the amount in semidwarf varieties was 1.5 fold than that in tall varieties (Gent and Kiyomoto, 1989). There was more ^{14}C in grain at 18 and 58 days after labeling flag leaves at anthesis in a semidwarf than in a tall variety of spring wheat, when grown in wet conditions, but not in dry conditions in Australia (Pheloung and Siddique, 1991). When labeled after anthesis, Rht genes had less effect on ^{14}C distribution. When flag leaves were labeled after anthesis in England, a semidwarf variety had more ^{14}C in the spike 24 hr later than did a tall variety, but differences were rarely significant at maturity (Makunga et al., 1978). When the canopy was labeled after anthesis in northeastern USA, most of the ^{14}C in the plant was in the spike at maturity, and this amount differed little between semidwarf and tall varieties, 0.9 and 0.8, respectively (Gent and Kiyomoto, 1989). Semidwarf wheat retained to harvest a greater fraction of the ^{14}C initially assimilated than did a tall variety. When winter wheat was labeled early or late in grain-fill in northeastern USA, a Rht_1 Rht_2 dwarf isoline moved less ^{14}C to spike than the semidwarf or tall isolines, and retained less of the initial ^{14}C to maturity (Kiyomoto and Gent, 1989b). The tall isoline retained more ^{14}C in the stem when labeled shortly after anthesis, than either semidwarf or dwarf isolines. Thus when measured by ^{14}C label, Rht_1 or Rht_2 semidwarf wheat appears to redistribute more pre-anthesis assimilate to the spike than either tall or dwarf wheat, and retain a greater amount to maturity than dwarf wheat. However, differences in allocation during grain filling are slight, and they are not consistent with loss of dry matter from the stem.

Differences in partitioning of assimilate should also be evident in concentration of nonstructural carbohydrate in the stem and leaves. In spring wheat grown as a single tiller under controlled conditions, stems of dwarf varieties had a greater concentration of carbohydrate at anthesis, but at the time of maximum stem biomass, 15 days post anthesis, the concentration in dwarf and tall wheat were similar (Fischer and Stockman, 1986). In Australia, stem carbohydrate increased after anthesis under wet but not dry conditions, and only under wet conditions, a semidwarf variety could have higher harvest index than a tall variety (Pheloung and Siddique, 1991). Early season application of N fertilizer in the Netherlands depressed stem carbohydrate in a tall variety more than in a semidwarf variety of winter wheat (Spiertz and van de Haar, 1978). In an isogenic winter wheat background in northeastern USA, stem carbohydrate at anthesis was inversely related to stem height (Gent, 1991, unpublished results). After anthesis, the concentration of carbohydrate fell and differences due to Rht_1 and/or Rht_2 were not significant. In all these studies, the carbohydrate concentrations measured in the stem at anthesis were not sufficient to explain quantitatively the loss in weight after anthesis or differences in harvest index at maturity (Gent, 1994).

KERNEL WEIGHT

The competition between stem and spike for carbohydrate should also influence grain size and grain protein. In a comparison of winter wheat varieties in central USA, grain number was correlated to the maximum fructan concentration in the spike, reached 7 to 10 days prior to anthesis (Hendrix et al., 1986). Surprisingly, kernel weight is positively related to plant height, despite apparently greater competition for carbohydrate between stem and developing florets in tall compared to dwarf wheat. Rht_1 and/or Rht_2 genes lowered kernel weight in isogenic backgrounds of soft white wheat in northwestern USA (Allan, 1986). Incorporation of Rht_1 in durum wheat varieties in north central USA resulted in a linear correlation between plant height and kernel weight (McClung et al., 1986). In crosses among spring wheat varieties, grain weight was in the order tall > Rht_1 = Rht_2 > Rht_1Rht_2, independent of plant height, heading date or grains per spike (Pinthus and Levy, 1983). However, in crosses of tall awnletted with semidwarf awned wheat in Canada, awns counteracted the decrease in kernel weight due to Rht_1 (Knott, 1986). The effect of Rht_1 and Rht_2 on grain weight limited yield of widely spaced plants grown in northeastern USA (Keyes and Sorrells, 1989). Grain yield was inversely related to the number of dwarfing genes in hybrids, because the reduction in kernel weight was not offset by other agronomic characteristics. Hybrid vigor increased yield by 0.14, 0.10 and 0.02 times in tall, semidwarf and dwarf hybrids, respectively.

Kernel weight can be altered by changing the source-sink ratio in an artificial manner. For instance, a 0.8 reduction in grains per spike increased kernel weight of older and taller spring wheat varieties by about 0.1 fold, but kernel weight of modern semidwarf varieties increased up to 0.5 (Fischer and HilleRisLambers, 1978). Removal of the outer two grains per spikelet increased weight of the remaining grain by 0.1 fold in tall and 0.4 fold in semidwarf winter wheat varieties in England (Radley and Thorne, 1981). During grain-fill, grain N increased more than grain weight. When assimilate supply was varied by exposing spring wheat plants to 200 or 1000 ppm CO_2, or removing 0.56 part of flag leaf, the range of variation of kernel weight was larger in dwarf, 30 to 47 mg, than in tall, 38 to 42 mg (Winzeler, Monteil, and Nosberger, 1989). When grain number in winter wheat was controlled in the range from 100 to 300 grains per spike, the relation between kernel weight and number, and grain N concentration and number, were the same in Rht_3 and a tall isoline (Flintham and Gale, 1983). Thus, low kernel weight appears to be due to more competition for assimilate during grain filling in dwarf than in tall wheat, rather than to an inherent limit in size of the grain. Although Rht_1 or Rht_2 genes do not limit yield under normal agronomic conditions, these genes may confer a yield and harvest index advantage only under conditions that maintain kernel weight.

DISTRIBUTION OF NITROGEN

Protein concentration of the grain is also correlated with plant height. In north central USA, introduction of Rht_1 into durum wheat varieties reduced grain N from 0.137 to 0.125 g N per gram (McClung et al., 1986). There was a linear correlation between plant height and grain protein concentration. In an isogenic spring wheat background in central USA, the concentration of protein in straw was 0.013, 0.015 and 0.019 in tall, semidwarf and dwarf Centana, respectively, although the tall isoline had more protein in grain (McNeal et al., 1971). However, per unit land area, tall and dwarf isolines did not differ significantly in either straw or grain protein. In another comparison of tall and semidwarf isolines of spring wheat and durum wheat in north central USA, Rht did not affect leaf nitrate, nitrate reductase activity or grain protein. Stem N concentration and N per tiller were the only factors that varied with plant height (Deckard et al., 1977). When spring wheat varieties were grown under controlled conditions and measured at tillering, nitrate reductase activity did not differ consistently due to dwarfing genes (Kulshrestha and Tsunoda, 1981a). In a comparison of fertilizer application in The Netherlands, a tall variety remobilized less N to the grain than a semidwarf winter wheat, except when N application was early in growth (Spiertz and van de Haar, 1978). However, among winter wheat varieties grown in eastern USA, the variation in grain protein concentration was due to a variation in total N uptake and not due to remobilization (Van Sanford and MacKown, 1987). Rht genes do not appear to affect the physiological characters related to uptake and assimilation of N. The higher concentration of grain protein of tall wheat is likely related to the greater amount of N per tiller and per grain at anthesis, most of which is redistributed to the grain at maturity.

STAND ESTABLISHMENT

Beside the influence of environment on the physiology and metabolism in wheat, as described above, stress can also influence the survival and establishment of plants early in the season. When better stand establishment is coupled to better light interception before closure of the canopy, it should give wheat of tall stature a substantial advantage over dwarf wheat. This may be an important factor that results in less yield advantage for dwarf than tall wheat, when grown in stressful environments. Among different genetic backgrounds of winter wheat backgrounds in northwestern USA, emergence rate and final stand differed with Rht_1 or Rht_2 genes (Allan, 1980). Stand establishment averaged 0.39, 0.52 and 0.61 of seeding rate for dwarf, semidwarf and tall, respectively, but this response varied with background. Rht_2 resulted in 0.02 shorter stem length and lower stand establishment than Rht_1, in part due to slower germination (Allan, 1989). In an isogenic Yogo winter wheat background in north central USA, emergence and coleoptile length was positively related to plant height (Allen, Taylor, and Martin, 1986). There was no effect of Rht_1 and/or Rht_2 on winter hardiness. Crossing and back-crossing resulted in reduced height lines with similar winter hardiness to their tall backgrounds in Canada (Gilliland and Fowler, 1988). In a test of winter wheat varieties in central USA, semidwarf varieties were found that had greater yield stability across environments than the tall varieties (Budak et al., 1995).

Rht genes could affect root growth in the same manner as stem growth, and reduce access to water and nutrients in dwarf wheat. In a comparison of semidwarf and tall winter wheat varieties in England, they did not differ in root relative growth rate (Lupton et al., 1974). At depth, semidwarf varieties had more extensive roots and absorbed more ^{31}P from deep soil layers than did tall varieties. In Australia, roots of an old tall variety of spring wheat weighed more than roots of a modern semidwarf, but within a modern background, Rht_1 and/or Rht_2 isolines did not differ in development or root to shoot ratio (Siddique, Belford, and Tennant, 1990). Under controlled conditions, tall isolines of spring wheat germinated faster, and root weights were initially greater, but after stem elongation, root weights were greater in dwarf isolines (Bush and Evans, 1988). In contrast to its action in leaves and stem, the Rht_3 gene does not affect gibberellin metabolism and accumulation in roots (Stoddart, 1984), and Rht genes should affect roots growth much less than of stem and leaves. The simulation of root growth during stem elongation may be due to less competition for assimilate from the tillers.

CONCLUSIONS

This review suggests that incorporating either Rht_1 or Rht_2 genes modulates the morphology and physiology of wheat in a manner that typically increases yield. There are several types of compensation involved in this yield increase. For instance, although Rht genes decrease leaf area, photosynthesis per unit area increases in perfect compensation, so biomass accumulation is rarely altered in Rht_1 or Rht_2 semidwarf compared to tall wheat. Perhaps due to less competition for carbohydrate during stem elongation, there are more fertile florets at anthesis in semidwarf wheat and greater harvest index at maturity. Although tall wheat has a greater weight per kernel at maturity, this does not completely compensate for greater grain number in semidwarf wheat. The compensation that leads to greater yield in semidwarf wheat appears to fail in Rht_1Rht_2 or Rht_3 dwarf wheat. Either specific photosynthesis does not completely compensate for decreased area per leaf, or the reduction in plant height retards canopy closure and efficient interception of solar radiation, resulting in lower biomass and yield of dwarf wheat. This pattern becomes more exaggerated in a dry environment. Although Rht genes increase leaf permeability to water vapor, plant water status changes in compensation to minimize differences in water use efficiency. Drought stress appears to reduce yield and harvest index more in semidwarf than in tall wheat. We speculate that dry conditions that accelerate leaf senescence and reduce photosynthesis would increase competition for carbohydrate between stem and spike at anthesis. This would, in turn, partially negate the greater carbohydrate availability and increase in fertile florets due to Rht genes moderating growth of the stem. Rht genes have insignificant effects on development or winter hardiness, and this broad overview of the effect of Rht genes on physiology appears to apply equally well to winter, spring and durum wheat.

REFERENCES

Allan, R.E. (1980). Influence of semidwarfism and genetic background on stand establishment of wheat. *Crop Science* 20:634-638.

Allan, R.E. (1983). Harvest indexes of backcross derived wheat lines differing in culm height. *Crop Science* 23:1029-1032.

Allan, R.E. (1986). Agronomic comparisons among wheat lines nearly isogenic for three reduced height genes. *Crop Science* 26:707-710.

Allan, R.E. (1989). Agronomic comparisons between Rht_1 and Rht_2 semidwarf genes in winter wheat. *Crop Science* 29:1103-1108.

Allen, S.G., G.A. Taylor and J.M. Martin. (1986). Agronomic characterization of 'Yogo' hard red winter wheat plant height isolines. *Agronomy Journal* 78:63-66.

Appleford, N.J. and J.R. Lenton. (1991). Gibberellins and leaf expansion in near isogenic wheat lines containing Rht_1 and Rht_3 dwarfing alleles. *Planta* 183:229-236.

Austin, R.B., J. Bingham, R.D. Blackwell, L.T. Evans, M.A. Ford, C.L. Morgan and M. Taylor. (1980). Genetic improvement in winter wheat yields since 1900 and associated physiological changes. *Journal of Agricultural Science* (Cambridge) 49:675-689.

Austin, R.B., J.A. Edrich, M.A. Ford, and R.D. Blackwell. (1977). The fate of dry matter, carbohydrates, and ^{14}C lost from leaves and stems during grain filling. *Annals of Botany* 41:1309-1321.

Bidinger, F., R.B. Musgrave and R.A. Fischer. (1977). Contribution of stored pre-anthesis assimilate to grain yield in wheat and barley. *Nature* 270:431-433.

Borner, A., J. Plaschke, V. Korzun and A.J. Worland. (1996). The relationship between dwarfing genes of wheat and rye. *Euphytica* 89:69-75.

Borojevic, S. and W.A. Williams. (1982). Genotype x environment interactions for leaf area parameters and yield components and their effects on wheat yields. *Crop Science* 22:1020-1025.

Borrell, A.K., L.D. Incoll, R.J. Simpson and M.J. Dalling. (1989). Partitioning of dry matter and the deposition and use of stem reserves in a semi-dwarf wheat crop. *Annals of Botany* 63:527-539.

Brooking, I.R. and E.J.M. Kirby. (1981). Interrelationships between stem and ear development in winter wheat: the effects of a Norin 10 dwarfing gene, Gai/Rht_2. *Journal of Agricultural Science* (Cambridge) 97:373-381.

Budak, N., P.S. Baenziger, K.M. Eskridge, D. Baltensperger and N.B. Moreno-Sevilla. (1995). Plant height response of semidwarf and non semidwarf wheats to the environment. *Crop Science* 35:447-451.

Busch, R.H. and T.L. Rauch. (1993). Agronomic performance of tall versus short semidwarf lines of spring wheat. *Crop Science* 33:941-943.

Bush, M.G. and L.T. Evans. (1988). Growth and development in tall and dwarf isogenic lines of spring wheat. *Field Crops Research* 18:243-270.

Cox, T.S., J.P. Shroyer, L. Ben-Hui, R.G. Sears and T.J. Martin. (1988). Genetic improvement in agronomic traits of hard red winter wheat cultivars from 1919 to 1987. *Crop Science* 28:756-760.

Deckard, E.L., K.A. Lucken, L.R. Joppa and J.J. Hammond. (1977). Nitrate reductase activity, nitrogen distribution, grain yield and grain protein of tall and semidwarf near-isogenic lines of *Triticum aestivum* and *T. turgidom*. *Crop Science* 17:293-296.

Ehdaie, B. and J.G. Waines. (1994). Growth and transpiration efficiency of near-isogenic lines for height in a spring wheat. *Crop Science* 34:1443-1451.

Feyerherm, A.M., G.M. Paulsen and J.L. Sebaugh. (1984). Contribution of genetic improvement to recent wheat yield increases in the USA. *Agronomy Journal* 76:985-990.

Fischer, R.A., F. Bidinger, J.R. Syme and P.C. Wall. (1981). Leaf photosynthesis, leaf permeability, crop growth and yield of short spring wheat genotypes under irrigation. *Crop Science* 21:367-373.

Fischer, R.A. and D. HilleRisLambers. (1978). Effect of environment and cultivar on source limitation to grain weight in wheat. *Australian Journal of Agricultural Research* 29:443-458.

Fischer, R.A. and Y.M. Stockman. (1986). Increased kernel number in Norin 10 derived dwarf wheat: evaluation of the cause. *Australian Journal of Plant Physiology* 13:767-784.

Fischer, R.A. and J.T. Wood. (1979). Drought resistance in spring wheat cultivars III yield association with morpho-physiological traits. *Australian Journal of Agricultural Research* 30:1001-1020.

Flintham, J.E. and M.D. Gale. (1983). The Tom Thumb dwarfing gene Rht_3 in wheat. *Theoretical and Applied Genetics* 66:249-256

Gale, M.D., J. Edrich and F.G.H. Lupton. (1974). Photosynthetic rates and the effects of applied gibberellin in some dwarf, semidwarf and tall wheat varieties (*Triticum aestivum*). *Journal of Agricultural Science* (Cambridge) 83:43-46.

Gale, M.D. and S. Youssefian. (1985). Dwarfing genes in wheat. In *Progress in Plant Breeding,* ed. E. Russell, Vol. 1, pp. 1-35.

Gent, M.P.N. (1994). Photosynthate reserves during grain filling in winter wheat. *Agronomy Journal* 86:159-167.

Gent, M.P.N. (1995). Canopy light interception, gas exchange, and biomass in reduced height isolines of winter wheat. *Crop Science* 35:1636-1642.

Gent, M.P.N. and R.K. Kiyomoto. (1985). Comparison of canopy and flag leaf net carbon dioxide exchange of 1920 and 1977 New York winter wheat. *Crop Science* 25:81-86.

Gent, M.P.N. and R.K. Kiyomoto. (1989). Assimilation and distribution of photosynthate in winter wheat cultivars differing in harvest index. *Crop Science* 29:120-125.

Gilliland, D.J. and D.B. Fowler. (1988). Effect of a Rht gene conditioning the semidwarf character on winter hardiness in winter wheat (*Triticum aestivum* L. em Thell*). Canadian Journal of Plant Science* 68:301-309.

Hendrix, J.E., J.C. Linden, D.H. Smith, C.W. Ross and I.K. Park. (1986). Relationship of pre-anthesis fructan metabolism to grain numbers in winter wheat. *Australian Journal of Plant Physiology* 13:391-398.

Ho, T.H.D., R.C. Nolan and D.E. Shute. (1981). Characterization of a gibberellin insensitive dwarf wheat, D6899. *Plant Physiology* 67: 1026-1031.

Holbrook, F.S. and J.R. Welsh. (1980). Soil-water use by semidwarf and tall winter wheat cultivars under dryland field conditions. *Crop Science* 20:244-246.

Hoogendoorn, J., J.M. Rickson and M.D. Gale. (1990). Differences in leaf and stem anatomy related to plant height of tall and dwarf wheat (*Triticum aestivum* L.). *Journal of Plant Physiology* 136:72-77.

Hoshino, T., and H. Seko. (1996). History of wheat breeding for half a century in Japan. *Euphytica* 89:215-221.

Hunt, L.A. (1978). Stem weight changes during grain filling in wheat from diverse sources. *Proceedings of the 5th International Wheat Genetics Symposium*, pp. 923-929.

Keyes, G. and M.E. Sorrells. (1989). Rht_1 and Rht_2 semidwarf genes effect on hybid vigor and agronomic traits of wheat. *Crop Science* 29:1442-1447.

Keyes, G.J., D.J. Paolillo and M.E. Sorrells. (1989). The effects of dwarfing genes Rht_1 and Rht_2 on cellular dimensions and rate of leaf elongation in wheat. *Annals of Botany* 64:683-690.

King, P.W., M.D. Gale and S.A. Quarrie. (1983). Effects of Norin 10 and Tom Thumb dwarfing genes morphology, physiology and abscisic acid production in wheat. *Annals of Botany* 51:201-208.

Kirkham, M.B. and E.L. Smith. (1978). Water relations of tall and short cultivars of winter wheat. *Crop Science* 18:227-230.

Kirkham, M.B. and E.L. Smith. (1979). Water potential gradients of tall and short cultivars of winter wheat. *Plant and Soil* 52:553-559.

Kiyomoto, R.K. and M.P.N. Gent. (1989a). Photosynthetic assimilation of $^{14}CO_2$ and fate of ^{14}C-labeled photosynthate in winter wheat (*Triticum aestivum*) near-isolines differing in alleles at the Rht_1 and Rht_2 reduce height loci. *Annals of Applied Biology* 114:141-148.

Kiyomoto, R.K. and M.P.N. Gent. (1989b). Comparison of flag leaf and canopy net photosynthesis in reduced-height winter wheat isolines. *Photosynthetica* 23:49-54.

Knott, D.R. (1986). Effect of genes for photoperiodism, semidwarfism, and awns on agronomic characters in a wheat crop. *Crop Science* 26:1158.

Kulshrestha, V.P. and S. Tsunoda. (1981a). Nitrate reductase activity in dwarf and tall cultivars of wheat. *Tohoku Journal of Agricultural Research* 32:83-86.

Kulshrestha, V.P. and S. Tsunoda. (1981b). The role of 'Norin 10' dwarfing genes in photosynthetic and respiratory activity of wheat leaves. *Theoretical and Applied Genetics* 60:81-84.

Lecain, D.R., J.A. Morgan and G. Zerbi. (1989). Leaf anatomy and gas exchange in nearly isogenic semidwarf and tall winter wheat. *Crop Science* 29:1246.

Loss, S.P. and K.H.M. Siddique. (1994). Morphological and physiological traits associated with wheat yield increases in Mediterranean environments. *Advances in Agronomy* 52:229-276.

Ludlow, M.M. and R.C. Muchow. (1990). A critical evaluation of traits for improving crop yield in water-limited environments. *Advances in Agronomy* 43:107-153.

Lupton, F.G.H., R.H. Oliver, F.B. Ellis, B.T. Barnes, K.R. Howse, P.J. Welbank and P.J. Taylor. (1974). Root and shoot growth of semidwarf and taller winter wheat. *Annals of Applied Biology* 77:129-144.

Makunga, O.H.D., I. Pearman, S.M. Thomas and G.N. Thorne. (1978). Distribution of photosynthate produced before and after anthesis in tall and semidwarf winter wheat, as affected by nitrogen fertilizer. *Annals of Applied Biology* 88:429-437.

McClung, A.M., R.G. Cantrell, J.S. Quick and R.S. Gregory. (1986). Influence of Rht₁ semidwarf gene on yield, yield components and grain protein of Durum wheat. *Crop Science* 26:1095-1099.

McCullough, D.E. and L.A. Hunt. (1989). Respiration and dry matter accumulation around the time of anthesis in field stands of winter wheat (*Triticum aestivum*). *Annals of Botany* 63:321-329.

McNeal, F.H., M.A. Berg, P.L. Brown and C.F. McGuire. (1971). Productivity and quality response of five spring wheat genotypes, *Triticum aestivum* L., to nitrogen fertilizer. *Agronomy Journal* 63:908-910.

McNeal, F.H., M.A. Berg, V.R. Stewart and D.E. Baldridge. (1972). Agronomic response of three height classes of spring wheat, *Triticum aestivum* L., compared at different yield levels. *Agronomy Journal* 64:362-364.

Montieth, J.L. (1969). Light interception and radiative exchange in crop stands. In *Physiological Aspects of Crop Yield*, ed. J.D. Eastin, F.A. Haskins, C.Y. Sullivan, C.H.M. van Bavel. Madison, WI: American Society of Agronomy, pp. 85-113.

Morgan, J.A., D.R. LeCain and R. Wells. (1990). Semidwarfing genes concentrate photosynthetic machinery and affect leaf gas exchange of wheat. *Crop Science* 30:602-608.

Nizam Uddin, M., and D.R. Marshall. (1989). Effects of dwarfing genes on yield and yield components under irrigated and rainfed conditions in wheat (*Triticum aestivum* L). *Euphytica* 42:127-134.

Paollilo, D.J., and K.J. Niklas. (1996). Effects of Rht dosage on the breaking strength of wheat seedling leaves. *American Journal of Botany* 83:567-572.

Pearman, I., S.M. Thomas and G.N. Thorne. (1979). Effect of nitrogen fertilizer on photosynthesis of several varieties of winter wheat. *Annals of Botany* 43:613-621.

Perry, M.W. and M. D'Antuono. (1989). Yield improvement and associated characteristics of some Australian spring wheat introduced between 1860 and 1982. *Australian Journal of Agricultural Research* 40:457-472.

Pheloung, P.C., and K.H.M. Siddique. (1991). Contribution of stem day matter to grain yield in wheat cultivars. *Australian Journal of Plant Physiology* 18:53-64.

Pinthus, M.J. and M. Abraham. (1996). Effect of light, temperature, gibberellin (GA₃) and their interaction on coleoptile and leaf elongation of tall, semidwarf and dwarf wheat. *Plant Growth Regulation* 18:239-247.

Pinthus, M.J., M.D. Gale, N.E.J. Appleford and J.R. Lenton. (1989). Effect of temperature on gibberellin responsiveness and endogenous GA₁ content of tall and dwarf wheat genotypes. *Plant Physiology* 90:854-859.

Pinthus, M.J. and A.A. Levy. (1983). The relationship between the Rht₁ and Rht₂ dwarfing genes and gram weight in *Triticum aestivum* L. spring wheat. *Theoretical and Applied Genetics* 66:153-156.

Radley, M.E. and G.N. Thorne. (1981). Effects of decreasing the number of grains in ears of cvs Hobbit and Maris Huntsman winter wheat. *Annals of Applied Biology* 98:146-156.

Rawson, H.M. and L.T. Evans. (1971). The contribution of stem reserves to grain development in a range of wheat cultivars of different height. *Australian Journal of Agricultural Research* 22:851-863.

Rawson, H.M., J.H. Hindmarsh, R.A. Fischer and Y.M. Stockman. (1983). Changes in leaf photosynthesis with plant ontogeny and relationships with yield per ear in wheat cultivars and 120 progeny. *Australian Journal of Plant Physiology* 10:503-514.

Siddique, K.H.M, R.H. Belford and D. Tennant. (1990). Root: shoot ratios in old and modern tall and semidwarf wheats in a Mediterranean environment. *Plant and Soil* 121:89-98.

Slafer, G.A., F.H. Andrade and E.H. Sattore. (1990). Genetic improvement effects on preanthesis physiological attributes related to wheat grain yield. *Field Crops Research* 23:255-263.

Spiertz, J.H.J. and H. van de Haar. (1978). Differences in grain growth, crop photosynthesis and distribution of assimilates between a semi-dwarf and a standard cultivar of winter wheat. *Netherlands Journal of Agricultural Science* 26:233-249,

Stoddart, J.L. (1984). Growth and gibberellin-A_1 metabolism in normal and gibberellin insensitive (Rht$_3$) wheat (*Triticum aestivum* L.) seedlings. *Planta* 161:432-438.

Thorne, G.N. (1982). Distribution between parts of the main shoot and the tillers of photosynthate produced before and after anthesis in the top three leaves of main shoots of Hobbit and Maris Huntsman winter wheat. *Annals of Applied Biology* 101:553-559.

Thorne, G.N., I. Pearman, W. Day and A.D. Todd. (1988). Estimation of radiation interception by winter wheat from measurements of leaf area. *Journal of Agricultural Science* (Cambridge) 110:101-108.

Van Sanford, D.A. and C.T. MacKown. (1987). Cultivar differences in nitrogen remobilization during grainfill in soft red winter wheat. *Crop Science* 27:295-300.

Winzeler, M., Ph. Monteil and J. Nosberger. (1989). Grain growth of tall and short spring wheat genotypes at different assimilate supplies. *Crop Science* 29:1487-1491.

Yamada, T. (1990). Classification of GA response, Rht genes and culm length in Japanese varieties and landraces of wheat. *Euphytica* 50:221.

Youssefian, S., E.J.M. Kirby and M.D. Gale. (1992). Pleiotropic effects of the GA insensitive dwarfing genes in wheat. I. Effects of development of ear stem and leaves. *Field Crops Research* 28:191-210.

Yunusa I.A.M., K.H.M. Siddique, R.K. Belford and M.M. Karimi. (1993). Effects of canopy structure on efficiency of radiation interception and use in spring wheat cultivars during the pre-anthesis period in a Mediterranean-type environment. *Field Crops Research* 35:113-122.

SUBMITTED: 06/20/96
ACCEPTED: 12/02/96

Optimization of Vertical Distribution of Canopy Nitrogen: An Alternative Trait to Increase Yield Potential in Winter Cereals

M. F. Dreccer
G. A. Slafer
R. Rabbinge

SUMMARY. This article reviews the literature on the relation between vertical leaf nitrogen distribution and canopy photosynthesis, as a possible route to maximize radiation use efficiency, biomass

M. F. Dreccer and G. A. Slafer, Demonstrator and Adjunct Professor, respectively, Departamento de Producción Vegetal, Facultad de Agronomía, Universidad de Buenos Aires, Av. San Martín 4453, 1417, Buenos Aires, Argentina. R. Rabbinge, Professor, Department of Theoretical Production Ecology, Wageningen Agricultural University, P. O. Box 430, 6700 AK, Wageningen, The Netherlands.

Address correspondence to (current address): M. F. Dreccer, Department of Theoretical Production Ecology, Wageningen Agricultural University, P. O. Box 430, 6700 AK, Wageningen, The Netherlands (E-Mail: Dreccer@students@tpe.wau.nl).

The authors are grateful to V. O. Sadras (CSIRO, Narrabri), J. Goudriaan (Wageningen Agricultural University) and D. Rodriguez (University of Buenos Aires) for comments on early versions of the manuscript. M.F.D. is on leave from the Departamento de Producción Vegetal, Facultad de Agronomía, Universidad de Buenos Aires, and gratefully acknowledges the support given by CONICET (Consejo Nacional de Investigaciones Científicas y Técnicas, Argentina).

[Haworth co-indexing entry note]: "Optimization of Vertical Distribution of Canopy Nitrogen: An Alternative Trait to Increase Yield Potential in Winter Cereals." Dreccer, M. F., G. A. Slafer, and R. Rabbinge. Co-published simultaneously in *Journal of Crop Production* (The Food Products Press, an imprint of The Haworth Press, Inc.) Vol. 1, No. 1 (#1), 1998, pp. 47-77; and: *Crop Sciences: Recent Advances* (ed: Amarjit S. Basra) The Food Products Press, an imprint of The Haworth Press, Inc., 1998, pp. 47-77. Single or multiple copies of this article are available for a fee from The Haworth Document Delivery Service [1-800-342-9678, 9:00 a.m. - 5:00 p.m. (EST). E-mail address: getinfo@haworth.com].

production and yield potential in winter cereals. Main assumptions underlying the prediction of the optimum vertical leaf nitrogen allocation, i.e., the role of light climate within the canopy and total leaf nitrogen content, are revised. Dynamic aspects and possible impact of manipulation of vertical leaf N distribution on yield and quality of winter cereals are discussed and areas for future research highlighted. *[Article copies available for a fee from The Haworth Document Delivery Service: 1-800-342-9678. E-mail address: getinfo@haworth.com]*

KEYWORDS. Leaf nitrogen distribution, optimum leaf nitrogen profile, radiation use efficiency, yield potential, winter cereals

ABBREVIATIONS. A_{max}, light saturated photosynthesis rate, [$mgCO_2$ m^{-2} s^{-1}]; Kl, light extinction coefficient; Kn, nitrogen extinction coefficient; LAI, leaf area index, [m^2 leaf m^{-2} ground]; NTF, Total nitrogen free for mobilization in the leaves, [gN m^{-2} ground]; QE, apparent quantum efficiency, [$mgCO_2$ m^{-2} s^{-1} (J m^{-2} leaf s^{-1})]; RUE, radiation use efficiency, [g MJ^{-1}]; SLN, specific leaf nitrogen, [gN m^{-2} leaf]

INTRODUCTION

General

Breeding efforts, in interaction with improvement in agronomic techniques, have been the routes to increase crop yields since the beginning of agriculture. The distinction between past contribution of each of these factors to increases in grain yield of agricultural crops, though difficult to separate completely, has been attempted by several authors (Jensen, 1978; Slafer and Andrade, 1991). For the most important winter cereal, bread wheat (*Triticum aestivum* L.), it has been shown in different countries that yield increases since the beginning of this century were due equally to genetic gain in yield potential vs. improvement in management practices and gains in genetic resistance (see Slafer, Satorre, and Andrade, 1994). Among the management practices, the increase in nitrogen (N) fertilizer use has been responsible for a great deal of the gain in biomass, yield and quality of cereals (Bell et al., 1995). The fast adoption of this technique was linked to the widespread introgression of major dwarfing genes (Rht) or direct selection for short stature in most winter cereals (Richards, 1992), which reduced the risk of lodging. However, the danger of environmental pollution in conjunction with the trend towards less subsidized agriculture in many cereal growing countries, call for reconsideration of the level of N use. In this context, plant breeding may play a leading role in the search for future increases in grain yields and nitrogen use efficiency.

Crop Yields in Winter Cereals: Past Breeding Achievements and Future Prospects

For most winter cereals, actual prospects for yield improvement contrast widely with those in the beginning of the century. The imposition of organized selection pressure has resulted in a sharp increase in harvest index, the proportion of total biomass allocated into grains (Figure 1) (e.g., oats: Lawes, 1977; Peltonen-Sainio, 1990, 1994; barley: Riggs et al., 1981; Bulman, Mather, and Smith, 1993; bread-wheat: Austin et al., 1980; Austin, Ford, and Morgan, 1989; Siddique et al., 1989; Slafer and Andrade, 1989; Calderini, Dreccer, and Slafer, 1995; durum wheat: Waddington et al., 1987). As harvest index can not be increased beyond certain limits (i.e., 62% as calculated for wheat by Austin et al., 1980), and many crops are already close to this theoretical threshold, the present scope for

FIGURE 1. Changes in harvest index in oats, bread wheat, durum wheat and barley with year of release for different countries.

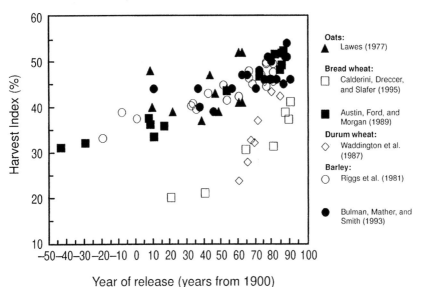

future increases in potential yield through this characteristic is very narrow. Therefore, increasing biomass production must be considered, sooner or later, as the main avenue towards further raising potential yields.

Substantial success has been achieved in increasing yield potential through the empirical selection approach of trial and error directed to yield *per se* (Loss and Siddique, 1994), however, using physiological attributes as selection criteria could accelerate future genetic gains (Shorter, Lawn, and Hammer, 1991). To achieve this target, it will be necessary to identify traits to help in the detection of potential parents and/or the selection of the progeny. Previous experience has indicated that variations in a single characteristic from a relatively simple level of organization, e.g., light saturated leaf photosynthesis rate, have failed to provide cultivars of higher dry matter production (Evans, 1990). This may be indicating that biomass increases will presumably require changes in several aspects of photosynthesis biochemistry at a time (Lawlor, 1995), although it may also point that selection has to focus on a particular trait provided it is represented, as yield, at the crop level of organization. In the following sections, we will analyze crop level physiological components of final biomass and try to recognize those which could be taken into account in future breeding aimed to further increase yield potential of winter cereals.

Components of Biomass Generation and Possibilities for Their Genetic Manipulation

Crop biomass production can be analyzed as the product of two major components: the amount of accumulated intercepted radiation and its efficiency of conversion into new dry matter or radiation use efficiency (RUE) (Monteith, 1977; Gallagher and Biscoe, 1978). The amount of accumulated intercepted radiation depends on the level of incident radiation, the proportion of that intercepted by photosynthetically active surfaces of the crop and the length of the growing season. Plant breeders can potentially modify the last two components of accumulated intercepted radiation. The proportion of incident radiation that is intercepted by the crop depends on its architecture, which in fact, is quite responsive to management practices (e.g., plant density). Then, it may prove very inefficient to increase this attribute by breeding. On the other hand, genetic manipulations of phenological responses to environmental factors that determine the final length of the growing season have been successfully done, so that anthesis can be adjusted to the optimum time for a particular location (Flood and Halloran, 1986). For most temperate, wheat growing areas, breeders have already made available genotypes that fully exploit the growing season as delimited by frost immediately before anthesis and relatively high temperatures during the grain filling period. Therefore, prospects for further increasing yield will not likely result from manipulation of the growing season; although there is room for speculation about improving the relative allocation of time to different phenophases (Slafer, Calderini, and Miralles, 1996). In this context, it is possible that RUE could be more efficiently affected by genetic improvement than radiation interception.

In attempting to increase RUE, attention must be paid to avoid unwanted reductions in biomass quality as, the lower the energetic content of the biomass, the higher the RUE. If biomass composition is not changed, increases in RUE will depend on genetic manipulation of the overall photosynthetic output of the canopy. The hypothesis that increasing the level of the basic process responsible for the gain of dry matter, leaf photosynthesis, would result in increased biomass has often been considered (Austin, Ford, and Morgan, 1989; Carver and Nevo, 1990; Austin, 1992). However, a striking lack of association between leaf photosynthesis and biomass production has been frequently found, particularly when comparing old vs. new varieties or genotypes with different levels of ploidy (Austin et al., 1982; Johnson et al., 1987; Carver, Johnson, and Rayburn, 1989). One of the reasons might be that the rate of photosynthesis and leaf size are so often negatively correlated that pleiotropic effects of genes for high photosynthetic rate on reduced leaf area have been suggested (Bhagsari and Brown, 1986; Austin, 1989). Another limitation of the approach could be that the impact of a trait from a low level of organization (organ) was expected to be additive on scaling to a higher level of organization (crop). However, the basic idea of improving leaf photosynthesis might still be useful, provided the potential for photosynthesis at different leaf layers in the canopy is considered.

Radiation Use Efficiency: Routes for Maximization

Because leaf photosynthesis responds to incident radiation and leaf N content, different possibilities are open for maximization of photosynthesis at the canopy level and ultimately, RUE (Field and Mooney, 1983; Loomis, 1993). The nature of the photosynthesis response to irradiance indicates that the efficiency (slope) is reduced as the irradiance level is increased (Figure 2). The two main parameters describing this relation are the maximum efficiency at low levels of irradiance or apparent quantum efficiency (QE) and the rate of photosynthesis at saturating irradiance (the asymptote, A_{max}). Then, to increase RUE at the canopy level, the inefficiencies occurring in the uppermost layers exposed to high irradiance must be considered (see zone 1 in Figure 2). The most frequently discussed and

FIGURE 2. Leaf photosynthesis rate as a function of irradiance. See text for explanation of numbers.

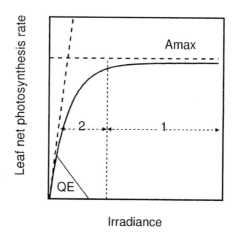

Irradiance

experimentally explored alternative has been to improve the distribution of radiation within the canopy through varying leaf angles along the plant. The ideal genotype would have more erect leaves in the top leaf layers (Duncan, 1971). As a consequence, the amount of radiation intercepted at saturating irradiance would be reduced and light penetration improved. Furthermore, the sunlit area in the bottom of the canopy would be increased, i.e., where radiation levels are lower and photosynthesis is more efficient (see zone 2 in Figure 2). As a result of increasing leaf erectness, the coefficient of light attenuation (Kl) would be reduced (Figure 3a).

Carvalho and Qualset (1978) have found genes having a major effect on flag leaf angle. Moreover, genetic variability for canopy photosynthesis or RUE in wheat associated with different patterns of radiation distribution within the canopy have been reported (Austin et al., 1976; Rasmusson, 1987; Aikman, 1989; Green, 1989), as well as in other crops (Kiniry et al., 1989). However, the impact of changes in leaf angle on winter cereals yields have been inconclusive. For example, Innes and Blackwell (1983) in wheat and Angus, Jones, and Wilson (1972) in barley have reported that crops with erect upper leaves produced higher yields than those with predominantly horizontal leaves. But, Austin et al. (1976) and Tungland et al. (1987) found little evidence that erect leaf angle enhanced yield in wheat and in barley, respectively. The apparent contradiction may stem from the fact that the advantage of leaf erectness can be better expressed only at high leaf area index (LAI) values and under high irradiance (Duncan, 1971; Goudriaan, 1988). This may be indirectly confirmed by the fact

FIGURE 3. Changes in (a) light intensity for different light extinction coefficients and (b) SLN allocation for different SLN distribution models, as a function of relative cumulative leaf area (0 and 1 are the top and bottom of the canopy).

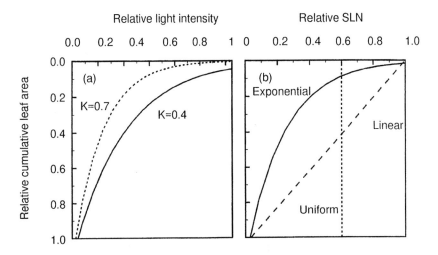

that in summer crops, such as rice, a positive effect of leaf erectness on yield has been more consistently observed (e.g., Chang and Tagumpay, 1970; Trenbath and Angus, 1975; Peng, Kuhsh, and Cassman, 1994). Then, the room for impact of leaf erectness on RUE of winter cereals grown under potential conditions may be limited to the grain filling period in crops of temperate regions, i.e., when the angle of incident irradiance is highest, or to crops grown at low latitudes. Araus, Reynolds, and Acevedo (1993) have put a cautionary note on the role of leaf erectness on dry matter production, discussing the possible existence of allometric relationships among this character and smaller leaves, spikes and stems, which may hamper breeding progress through this trait.

An alternative approach to increase RUE would be the maximization of growth through the improvement of the vertical distribution of N among leaves. Leaf N content strongly determines the maximum photosynthetic rate at high irradiances (Field and Mooney, 1986; van Keulen, Goudriaan, and Seligman, 1989; Evans, 1989a). Therefore, theoretical studies have suggested that canopy photosynthesis would be maximized if N is preferentially allocated to the more illuminated leaves (Field, 1983; Hirose and

Werger, 1987b). In this scenario, for a fixed amount of leaf N in the canopy, the possibility of increasing RUE would rely on the redistribution of canopy N from bottom to uppermost leaf layers (Figure 3b). In closed canopies, leaf N gradients are frequently observed, and have been interpreted as an adaptative response to the light environment that allows a higher canopy productivity than that expected from the uniform distribution, i.e., when the N content of every leaf equals the mean N content of the canopy (Mooney and Gulmon, 1979; Mooney et al., 1981).

The leaf N vertical distribution or profile that yields maximum canopy carbon gain has been termed 'optimum profile' and can be calculated with canopy photosynthesis models that take light and leaf N gradients into consideration (Hirose and Werger, 1987a; Anten, Schieving, and Werger, 1995; Goudriaan, 1995). Application of optimization theory and models have been formerly used to test if the naturally uneven distribution of canopy N is required to maximize growth, focusing on perennial herbs and native shrubs (Hirose and Werger, 1987a; Schieving, Werger, and Hirose, 1992a). This theoretical interpretation may also be helpful for assessing the impact of leaf N distribution on growth rate of agricultural crops, which are subjected to changes in canopy structure and N availability during the growing season and through management practices. Aspects of the issue whether leaf N distribution limits crop canopy photosynthesis in extensive crops have been addressed in summer crops (peanut: Hammer and Wright, 1994; Wright and Hammer, 1994; soybean: Sinclair and Shiraiwa, 1993; sunflower: Sadras, Hall and Connor, 1993; Connor, Sadras and Hall, 1995; Giménez, Connor, and Rueda, 1994). For winter cereals grown in temperate areas, the topic has not been methodically addressed by any study. The reason might be that these crops are mostly exposed to relatively low levels of radiation. However, the fact that their photosynthetic rate is saturated at a lower irradiance threshold could make the hypothesis of N redistribution as contributing to maximize RUE equally applicable for winter cereals particularly if we consider that a critical growth phase for biomass accumulation and yield formation in these cereals, i.e., the period of kernel number determination (Fischer, 1985), occurs with increasing irradiance during spring.

Objective

The objective of this article is to review the literature on the relation between vertical leaf N distribution and canopy photosynthesis, and to speculate on the possibilities for manipulation of canopy N profiles in order to maximize growth in winter cereals. For this purpose, we will initially address the influence of leaf N content on photosynthesis and

RUE and analyze the expected impact of a gradient in leaf N distribution. The mechanisms behind leaf N distribution in vegetative canopies will be briefly considered. A following section is focused on the revision of main assumptions of models that have been used to calculate the N profile that maximizes canopy photosynthesis. Then, major dynamic aspects that should be included in studies of optimization of N distribution for crop growth are identified. Finally, possible applications of the topic in different winter cereals are discussed and areas for future research highlighted.

LEAF NITROGEN, PHOTOSYNTHETIC CAPACITY AND RADIATION USE EFFICIENCY

Photosynthetic response to irradiance is largely determined by leaf N content (Field and Mooney, 1986; Evans, 1989a). Since approximately 40-70 % of the soluble protein in the leaf is concentrated in the carboxylation enzymes (Evans, 1983, 1989a; Terashima and Evans, 1988), the relation between photosynthesis and leaf N is not surprising. The response of photosynthesis to N can be analyzed through the effect on A_{max} and QE. In many crop plants, A_{max} increases asymptotically with leaf N content per unit leaf area (SLN, specific leaf nitrogen) (wheat: Evans, 1983; soybean: Sinclair and Horie, 1989; potato: Marshall and Vos, 1991; sunflower: Connor, Hall, and Sadras, 1993; peanut: Sinclair, Bennet, and Boote, 1993; Muchow and Sinclair, 1994) (see examples in Figure 4). Among the factors contributing to the degree of curvature of the A_{max} − SLN relation, the saturating relation between SLN and the content of carboxylating enzyme and the increasing irradiance threshold needed to reach light saturation in leaves with higher SLN, have been mentioned (Evans, 1989a). Linear increases in A_{max} with SLN have also been reported (Anten, Schieving, and Werger, 1995; Peng, Cassman, and Kropff, 1995).

At the canopy level of organization, RUE response to N availability has been documented in many crops (Green, 1987; Muchow and Sinclair, 1994; Hall, Connor, and Sadras, 1995) and investigated with models that integrate leaf carbon assimilation over canopy architecture and environmental gradients. Based on such a model, Sinclair and Horie (1989) theoretically developed the relation between RUE and SLN as hyperbolic, highly sensitive to low leaf N and species-dependent. Their initial calculations were performed under the assumption that SLN was equal at any canopy height, or uniformly distributed. The observation of uneven N distribution in natural canopies posed a question: to what extent could SLN distribution affect canopy assimilation rate or RUE? Evans (1993) calculated that daily gains in canopy photosynthesis with actual SLN

FIGURE 4. Light saturated leaf photosynthesis rate as a function of leaf SLN content in different crops: maize-sorghum (Muchow and Sinclair, 1994); rice and soybean (Sinclair and Horie, 1989), sunflower (Connor, Sadras and Hall, 1993), peanut (Sinclair, Bennet, and Boote, 1993).

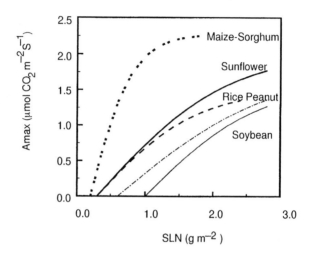

profiles can be 1-36% higher than those under uniform SLN distribution. Whereas, the comparison between canopy photosynthesis calculated with actual vs. optimal profiles yielded up to 7% gain (Schieving, Werger, and Hirose, 1992b; Connor, Sadras, and Hall, 1995). Furthermore, it has been reported that including a SLN gradient in the calculation of RUE could increase its value from 1-20% in peanut (Hammer and Wright, 1994) or even more in soybean (Sinclair and Shiraiwa, 1993).

The above-mentioned conclusions about estimated quantitative effect of N profiles on canopy photosynthesis must be taken cautiously since they were reached accepting assumptions that may not yet be sufficiently tested. One of them is that the SLN content does not generally affect QE. Consequently the increase in A_{max} due to higher SLN in the upper layers of the canopy during the period of the day of saturating irradiance would not be accompanied by a decreasing QE in the lower layers, and by an increased QE in the upper layers during the rest of the day. This certainly deserves a rigorous testing before the conclusions (and particularly those quantitative) could be fully accepted and confidently extrapolated. Not only little is known on the effects of SLN on QE, but also the few studies in which this relationship was investigated yielded variable results. QE has been linearly (Hirose and Werger, 1987a; Dingkuhn and Kropff, 1996) or hyperbolically (Pons et al., 1989) related to SLN in some species, while these traits appear to be independent in others (Connor, Hall, and Sadras, 1993; Anten, Schieving, and Werger, 1995). These considerations are probably of consequence for a winter cereal crop, in which a significant part of the growing cycle is exposed to low radiation and a large fraction of leaf area is shaded during most of the season.

To explore the sensitivity of daily total gross photosynthesis to the presence of a gradient in A_{max} or QE in the canopy, a simulation model was built following the techniques proposed by Goudriaan and van Laar (1994). Parameters for wheat were derived from Goudriaan and van Laar (1994) and van Heemst (1988). The model calculates daily total photosynthesis by integrating canopy photosynthesis obtained three times a day; each time, canopy photosynthesis is the integral of assimilation rate of sunlit and shaded leaves at five different LAI values corresponding to different canopy depths. Both integrations, over LAI and time, are performed at gaussian intervals (Goudriaan and van Laar, 1994). Total incident radiation is separated in direct and diffuse flux. The proportion of diffuse flux is calculated on basis of the atmospheric transmisivity, which changes during the year according to the sine of the solar elevation and the solar constant (Goudriaan and van Laar, 1994). The gradient in photosynthesis parameters for the present analysis was arbitrarily chosen within the ranges observed in the literature: A_{max} and QE decrease 25%, linearly, from top to bottom canopy layers. Calculations were performed for a crop with a LAI = 5, grown in a temperate region in the southern hemisphere (e.g., Rolling Pampas, Argentina, at 35° lat.), at different values of daily total radiation, for a day in the end of July and mid November. Those dates correspond roughly to the stages of tillering and mid grain filling of a crop sown under current practices. The relative decrease in canopy gross photosynthesis as a percentage of that of a canopy with a constant value of A_{max} and QE over canopy depth are presented in Table 1. The outcome indicates that wheat canopy photosynthesis is likely to be more sensitive to changes in QE than in A_{max}, but the impact is more similar at high radiation levels. As expected, the risk of overestimating photosynthesis by a constant A_{max} value increases with increasing radiation, while the trend is inverse for QE. In conclusion, canopy photosynthesis in wheat is sensitive to changes in QE, and therefore, those factors affecting QE, such as possibly SLN, merit further attention. Other authors have calculated that, in temperate regions, wheat dry matter production would be fairly insensitive to increases in A_{max} and suggested that genetically manipulating QE would

TABLE 1. Relative decrease in daily total gross photosynthesis (%) by introducing a 25% linear decrease in A_{max} or QE from top to bottom canopy layers with respect to a unique A_{max} or QE (see text for details).

Daily Total Radiation [MJ m^{-2} d^{-1}]	A_{max} reduction	QE reduction
31 July, 35° lat S		
7	5.3	21.5
14	11.3	16.4
21	12.5	15.0
15 November, 35° lat S		
7	4.3	22.4
14	7.3	20.3
21	10.1	18.1
28	12.5	15.6

yield high returns on canopy photosynthesis (Day and Chalabi, 1988; Ort and Baker, 1988).

Leaf N Distribution: Effect of Light Climate Within the Canopy and Leaf Age

The main proposed factors controlling leaf N allocation in dense vegetative canopies have been light gradient (Hirose et al., 1988) and leaf age (Mooney et al., 1981; Field, 1983) and both may operate simultaneously. Several authors have tried to manipulate them independently to evaluate the importance of each in the definition of SLN gradients, using different experimental approaches. For instance, Schieving, Werger, and Hirose (1992b), growing a monocot herb which develops new leaves from a basal meristem, and hence places them in the shade, still observed a steep decline in SLN with depth. Burkey and Wells (1991) observed in dense soybean canopies that the SLN gradient was largely reversible when the stand was thinned. Hikosaka, Terashima, and Katoh (1994) examined the effects of irradiance level and age by creating two types of shade gradients on a horizontally grown vine. In one of the gradients, younger leaves received more shade, the other simulated a canopy-type shading. They concluded that the effect of radiation level on SLN was more significant than that of age. Important quantitative effects of radiation level over age in the definition of the SLN gradients have been also observed by Woledge (1986).

The overall conclusion of the above-mentioned experiments, that the light gradient has a preeminent role in the definition of SLN has been included as the backbone criteria of optimization of N distribution for canopy photosynthesis. Theoretically, the closer the patterns of light and N distribution, the smaller the gap between actual and maximum capacity for biomass production, assuming that the effect of SLN on QE is small or negligible.

Optimization of Leaf N Distribution: Revision of Model Assumptions

In most of the studies where photosynthetic production of the canopy under actual and optimal N allocation patterns was compared, the optimal SLN profile was steeper than the actual, i.e., SLN was overestimated in top and underestimated in bottom leaf layers (Hirose and Werger, 1987b; Pons et al., 1989; Evans, 1993; Giménez, Connor, and Rueda, 1994; Anten, Schieving and Werger, 1995). Since the bias was independent of the magnitude of the predicted gain in canopy photosynthesis under optimum SLN distribution (see Giménez, Connor, and Rueda, 1994 as an example), it is possible to suspect that one or more of the criteria used to dictate the optimum N distribution are not robust enough, or their domain of applicability is restricted to certain conditions not yet sufficiently defined. To draw solid conclusions about possible gains in biomass production and efficient use of light and N, the basis for prediction of the optimum SLN profile has to be reliable. Particularly, when the benefits need to be assessed under changing N availability, canopy structure and environmental conditions as is the case of agricultural crops. For this reason, we revised two main assumptions in models that predict canopy photosynthesis under optimum N-allocation by contrasting them with experimental results.

The Pattern of SLN Distribution Is Determined by Light Distribution Within the Canopy

The core of optimization models of SLN distribution is that the optimum solution is determined by radiation distribution. The shape of the

SLN profile is usually described with the coefficient of SLN allocation, Kn. The actual Kn values are calculated by regression of SLN and LAI at different canopy heights. The value of Kn equals zero when the SLN of every leaf equals the mean canopy SLN, uniform profile, and increases as more N is partitioned to upper leaf layers. In studies performed in native perennial species, SLN distribution was fitted to exponential models (Hirose and Werger, 1987b; Pons et al., 1989), and inferred that canopy photosynthesis would be maximized when the SLN allocation mimicked that of radiation distribution within the canopy (Kn = Kl). For crop canopies, few and contrasting descriptions of SLN profiles are available to date. In soybean (Shiraiwa and Sinclair, 1993) and peanut (Wright and Hammer, 1994) a linear decline in SLN with cumulative LAI has been reported, whereas an exponential decline has been observed in a grain legume and forage sorghum (Charles-Edwards et al., 1987) and sunflower (Sadras, Hall, and Connor, 1993). The linear decline is, a priori, not related to the extinction of radiation and the mechanisms behind its development and possible advantages for crop growth merit further research. Therefore, the following analysis has been restricted to those situations where an exponential decline in SLN could be fitted.

If the radiation distribution dictates the SLN profile, then changes in structural characteristics of the canopy are expected to influence Kn (Hirose et al., 1988; Anten, Schieving, and Werger, 1995). When differences in Kl among species were pictured against Kn, no clear relation emerged (Figure 5a). The effect of canopy structure on SLN profiles was also tested by manipulation of stand density by Hirose et al. (1988), who found that the distribution of SLN was markedly more non-uniform in the dense than in the open stand. In their experiment, the LAI was the main source of variation in canopy structure, and hence associated with the change in Kn. Sadras, Hall, and Connor (1993) pictured the relation between LAI and Kn including data from different species during vegetative growth (see data and regression line in upper part of Figure 5b), and suggested that a departure from this trend could be accounted for by differences in Kl among species. However, data of four species from the study of Anten, Schieving, and Werger (1995) did not fit in this regression though the Kls of the species used were in the range of those presented by Sadras, Hall, and Connor (1993) (Figures 5a and b). Other studies have failed to find any effect of plant density on the steepness of SLN gradients. For instance, no consistent differences were found in Kn of the perennial herb *Solidago altissima,* even though 53% vs. 7-10% of total incident PFD reached the lowest branches in open vs. dense stands, respectively (Werger and Hirose, 1988). Similar results were reported for a range of population

FIGURE 5. Relation between the coefficient of leaf nitrogen allocation (Kn) and (a) KI, and (b) LAI for different species: *Solidago altissima* (Hirose and Werger, 1987b), *Lysimachia vulgaris* (open or dense canopy: Hirose et al., 1988), *Medicago sativa* (Lucerne: Lewmaire et al., 1991); *Helianthus annus* (sunflower: Sadras, Hall, and Connor, 1993); *Oryza sativa, Glycine max, Sorghum bicolor* and *Amaranthus cruentus* (Anten, Schieving, and Werger, 1995). Regression in (b) r^2 = 0.95, P < 0.005 (Sadras, Hall, and Connor, 1993).

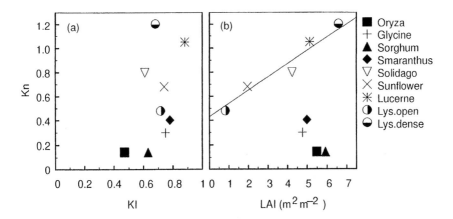

densities of sunflower (Sadras, Hall, and Connor, 1993) and soybean (Shiraiwa and Sinclair, 1993).

If, as observed by detailed experimentation (cf. previous section), the light gradient has a regulatory role on SLN distribution, why is Kn not always associated with the characteristics that determine the light climate within the canopy? Several reasons can be given. On one hand, the ability to respond or not to light distribution could be species-dependent. Aerts and de Caluwe (1994) have suggested that low-productivity species have a low phenotypic plasticity to shape the vertical SLN distribution. However, this concept can not be easily extrapolated to crops, which are usually selected under potential growing conditions and have high growth rate. Another issue is that most of the studies have been performed on closed or nearly closed canopies (Table 2). If the lag period between the imposition of a certain light environment and the corresponding SLN gradient is several days long (Pons and Pearcy, 1994), actual SLN profiles would be more uniform than expected. This effect could be clarified by performing a sequence of harvests and assessing optimality on a period of several days. The fact that SLN profiles are usually more uniform than expected,

TABLE 2. Coefficient of N allocation (Kn) and intercepted radiation at the moment of Kn determination.

Species	Kn	Intercepted radiation	Source
–Oryza HN[1]	0.14	0.92	Anten (1995); Anten, Schieving and Werger (1995)
Oryza LN	0.17	0.87	
Glycine HN	0.30	0.97	
Sorghum HN	0.14	0.97	
Sorghum LN	0.22	0.94	
Amaranthus HN	0.40	0.97	
Amaranthus LN	0.49	0.81	
–Solidago	0.79	0.92	Hirose and Werger (1987b)
–Sunflower	0.68	0.76	Sadras, Hall and Connor (1993)
–Lucerne	1.05	0.99	Lemaire et al. (1991)
–Lysimachia open	0.48	0.45	Hirose et al. (1988)
Lysimachia dense	1.20	0.99	

[1] Legends as in Figures 5 and 6.

could also originate in the limited chance to do a precise determination of the response of QE and respiration to SLN with measurements available from portable photosynthetic systems, since the error of the measurement usually increases as the CO_2 exchange decreases. If the linear response of both respiration and QE to SLN are underestimated, optimization procedures will tend to allocate N in upper layers at a very low cost. Finally, although light is an important factor controlling the distribution of SLN, the fact that its influence is only detected under certain conditions may indicate that it is not the only regulatory factor.

Does Kn Increase with Total Leaf N Content in the Canopy?

In most models of optimization of N allocation among leaves, the leaf N content is divided in two functional pools, labile metabolic N or free for mobilization vs. structural N, as proposed by Caloin and Yu (1984). Hirose and Werger (1987b), by means of a numerical model, derived that as canopy N content increased, so did the Kn necessary to maximize canopy photosynthesis. Their explanation was that more N becomes available for translocation to more illuminated micro-sites. This notion was incorporated in the analytical calculation of the optimum N distribution developed by Anten, Schieving, and Werger (1995), where Kn is proportional to the amount of leaf free SLN integrated over leaf area (NTF). Despite the theoretical demonstrations, actual Kn values do not meet this presumption straightforward, questioning the realism of the assumption (see data from Aerts and de Caluwe, 1994; Hikosaka, Terashima, and Katoh, 1994; Anten, 1995; Anten, Schieving, and Werger, 1995). In Figure 6, Kn is pictured as a function of NTF, to allow comparison among species and growing conditions without the confounding effect of the proportion of structural N. The data presented belong to crops grown at two levels of N from Anten (1995). By contrast to model expectations, a slight negative relation between Kn and NTF is apparent (Figure 6). DeJong, Day, and Johnson (1989) reported that in peach, a major influence of N fertilizer was to increase the photosynthetic capacity of partially shaded leaves but not the A_{max} of highly exposed leaves. Both the data of Anten (1995) and

FIGURE 6. Relation between the coefficient of leaf nitrogen allocation (Kn) and total leaf free nitrogen (NTF) in the canopy of *Oryza sativa, Glycine max, Sorghum bicolor* and *Amaranthus cruentus* at high and low nitrogen availability (HN and LN respectively) (Anten, 1995).

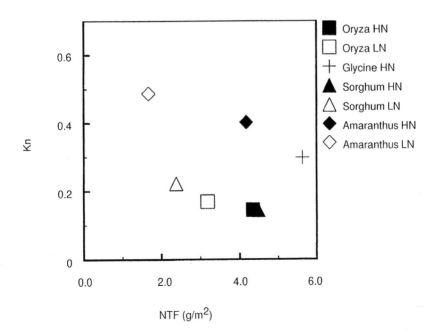

the evidence of DeJong, Day, and Johnson (1989) suggest that, as total canopy N content in the leaves increases, canopy photosynthesis is fostered by keeping a more uniform SLN profile, i.e., retaining more N in the lower leaves. These observations, agree with sensitivity studies on the introduction of SLN gradients on RUE, which have shown that the most significant impact could be expected in canopies with low SLN (Sinclair and Shiraiwa, 1993; Hammer and Wright, 1994).

INTRODUCING DYNAMIC ASPECTS IN SLN OPTIMIZATION STUDIES

Changes Along Crop Ontogeny

Studies on optimization of N allocation have generally relied on short term observations. However, crops are exposed to a fluctuating environment and changes in source-sink relationships during development. The whole growth cycle should be considered to adequately describe the dynamics of the relationship between net photosynthesis and SLN. This should be taken as an essential condition if the aim is to identify those phenological stages where optimization of N distribution may have an impact on the growth rate of winter crops. The main features of the crop that should be considered in relation to its phenology are changes in light distribution and in source-sink ratios with advancing development.

The Light Environment Changes During the Growth Cycle

Due to the phases of leaf expansion and senescence, the growth of reproductive organs and the evolution of total radiation during the season, the radiation environment within the canopy changes dramatically during crop ontogeny. Kl may vary with canopy depth within a developmental stage and between stages as well (Russell, Jarvis, and Monteith, 1989; Baldocchi and Collineau, 1994; Meinke, 1996). However, the role of Kl in determining biomass production of winter cereals or dictating the SLN profiles in vegetative canopies after canopy closure does not seem conclusive (see Figure 5a). Instead, Meinke (1996) has alerted that ignoring the higher efficiency of light capture of wheat canopies during early development, by using a constant Kl, could lead to serious underestimation of biomass production at anthesis when maximum LAI is low. As pointed before, only sporadic observations about the development of SLN profiles before canopy closure are available in the literature, and therefore its implications for early biomass production can hardly be discussed.

As the growing season advances, the proportion of senescent tissue increases and the reproductive structures grow. The role of the inflorescence in intercepting light has been minimized in natural herbaceous canopies (Werger and Hirose, 1991). However, in agricultural crops, it can exert a major difference in light climate during grain filling by reducing the incident radiation on leaves (Rosenthal, Arkin, and Howell, 1985; Yunusa et al., 1993). An extreme case would be that of a flowering canopy of oilseed rape, which reflects most of the radiation, thereby reducing light available for leaves and immature pods in a critical stage for yield formation (Yates and Steven, 1987). A final remark is that the photosynthesis of the inflorescence can make an important contribution to grain filling in several species (Rood, Major, and Charnetski, 1984; Rosenthal, Arkin, and Howell, 1985). All these aspects should be considered when analyzing the potential impact of SLN profiles development during the reproductive phase.

Source-Sink Relations During the Reproductive Stage

The idea that the optimum SLN profile, as dictated by the pattern of radiation attenuation, could change during ontogeny has been addressed by comparing late vegetative and reproductive stages. Lack of difference (Werger and Hirose, 1988; Schieving, Werger, and Hirose, 1992a) or ample contrast (Sadras, Hall, and Connor, 1993) between SLN profiles corresponding to those stages have been documented. In this context, the possibility that fruits with contrasting N requirements for seed formation, as oilseed rape and wheat, exert a differential influence on the optimum distribution of SLN during grain filling has not been explored yet. If sink capacity plays a role in the regulation of photosynthetic rate (Lawlor, 1995), considering its influence will be an essential step in order to set the limits for manipulation of SLN profiles on crop growth during grain filling.

Reproductive canopies offer an interesting opportunity for the study of the impact of SLN profiles on biomass formation since it is possible to combine large LAI with low SLN contents. Penning de Vries, van Keulen, and Alagos (1988), for rice, have calculated with a simulation model that N redistribution from leaves to grains during grain filling could be responsible for a yield loss of up to 10% at high yield levels (ca. 10 Mg ha^{-1}). The authors proposed that yield reduction could be decreased if stems instead of the leaves provided a larger share of the N allocated to grains. However, N storage in organs of non-legume crops is mostly packaged in photosynthesis-related pigments and proteins, while the evidence about the role of storage proteins in cereals reported so far has been variable and highly dependent on the species (Williams, Farrar, and Pollock, 1989; MacKown, Van Sanford, and Zhang, 1992). Therefore, the development of differently shaped N profiles could be regarded as a promising alternative to increase total canopy photosynthesis during grain fill for a given crop N status.

POSSIBLE APPLICATIONS OF CANOPY SLN DISTRIBUTION TO INCREASE GRAIN YIELD AND QUALITY IN WINTER CEREALS

The Case of Bread Wheat and Malting Barley

Bread wheat and barley for malting are two widespread, often fertilized, winter cereals that contrast in the quality standards required for industrialization, specifically, high vs. low protein percentage in the grains for wheat and barley, respectively. Therefore, they offer an interesting case to analyze comparatively the impact of the development of SLN profiles on yield and quality formation. The analysis proposed here is focused on the grain filling period.

In Figure 7, a diagrammatic model of the changes in SLN profiles from anthesis to maturity is presented. During grain filling, massive leaf senescence and N translocation to the grains take place, therefore, total canopy N content in the leaves declines. The idea that genotypic differences in the pattern of N vertical distribution during this stage could have different consequences on yield and protein percentage is explored with genotypes A and B. Several assumptions have been made for this analysis. Among them, that the genotypes do not differ in the response of photosynthesis parameters to SLN, that both crops have been grown under sufficient N

FIGURE 7. Changes in SLN with depth in the canopy from anthesis to maturity for genotype A and B. See text for explanation.

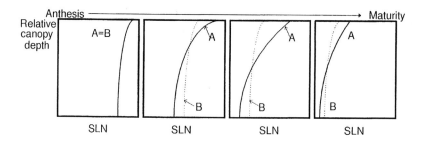

provision and that they have equal total N uptake during grain filling. The main difference between genotypes A and B is their capacity to develop SLN profiles.

Genotype A can develop more non-uniform SLN profiles than genotype B at a given total amount of leaf N in the canopy (Figure 7). Thus, in A, leaf N is depleted to a greater extent from lower and intermediate leaf layers than in genotype B, while SLN is kept higher at the upper, light-saturated leaf layers. Consequently, leaf loss in the bottom of the canopy will take place earlier in genotype A than B, and canopy photosynthesis will be concentrated on the upper leaf layers. By contrast, in genotype B, the amount of N lost is similar among leaf layers for a considerable part of the grain filling period. The steepness of the SLN profile in B is only slightly altered towards maturity. Eventually, both genotypes will present a uniform profile at maturity, with the SLN content approaching the content of structural N.

The relative advantage of one genotype over the other in canopy photosynthesis will certainly depend on the magnitude and shape of the response of A_{max} and QE to SLN. For instance, shortly after anthesis, SLN content in upper leaf layers is still high and probably comprised in the plateau region of the A_{max}-SLN response for both genotypes (Figure 2). If the response of QE to SLN were linear, the decline in canopy photosynthesis would be explained by the drop in QE due to N mobilization from the light-non-saturated leaf layers. Under those conditions, genotype B may outyield canopy photosynthesis of genotype A around anthesis and early grain filling. But, as grain filling proceeds, SLN content of light-saturated leaf layers will inevitably fall in the linear region of the response, then, the maintenance of higher SLN in upper leaf layers of genotype A may help to compensate for the general loss in potential canopy photosynthesis. Another interesting possibility is open if the response of A_{max} to SLN were not only steeper than that of QE to SLN, but also linear. Then, the genotypes that can easily develop steep SLN profiles would have an advantage in growth rate.

One of the assumptions of the previous analysis was that both genotypes, at any moment during grain filling, presented the same total amount of N in the leaves. However, the assumption that changes in Kn have no relation with the capacity of the crop for N translocation to the grains, may be a rather simplistic view. It is possible that genotype A also has a greater efficiency for N translocation, i.e., total leaf N in the canopy decreases to a greater extent than in genotype B. If that is the case, both genotypes may reach similar total dry matter production during grain filling with different strategies, A through a more steep SLN profile and B keeping a higher total canopy N content in the leaves. The different strategies may have consequences on the definition of the grain protein content. In genotype B, the trend towards slow changes in the SLN gradient may be associated with a lower rate of leaf N turnover and mobilization. Then, in genotype B, an increase in N availability during heading and grain filling will affect relatively more grain dry matter accumulation than protein yield, thus protein concentration will be kept low. Therefore, ideotypes like B could be searched for in barley breeding. By contrast, genotype A could be more suitable as a bread wheat ideotype. If canopy photosynthesis during grain filling is the main source of assimilates for the growing grains, genotype A would reach the maximum thousand grain weight, maximizing protein yield and percentage in the grain.

CONCLUDING REMARKS AND FUTURE RESEARCH

In the past, breeding and selection successfully managed to improve yield potential of winter cereals through the increase in dry matter allocation to the grains, without major changes in total biomass production. As the calculated limit in harvest index is being approached, increases in biomass production are likely to be the route towards future higher yields. The expected increase will probably rely on the development of new insights into physiological processes at the crop level more than on empirical selection. Crop growth models may help in the achievement of this search and the definition of ideotypes. When the total amount of radiation intercepted is maximized through exploitation of the whole length of the growing season, increases in biomass will be based on gains in RUE. Such increase may probably not be expected by the improvement of a single physiological attribute but to the addition of marginal contributions achieved by modifications in several traits at a time. Among them a reduction in photorespiration and maintenance respiration and an increase in canopy gross assimilation capacity have been mentioned (Slafer, Satorre, and Andrade, 1994). In this paper we have focused on one of the possibilities to increase RUE and yield (if harvest index is maintained) in winter cereals, by examining the role of optimization of vertical distribution of canopy N.

During the review and analysis of literature on the topic, several gaps in understanding and information became apparent. Firstly, a limited amount of experimentation has been carried out in crops, particularly winter crops. Secondly, the criteria accepted to rule N redistribution performed rather poorly when examined in the light of experimental results. More work will be needed at the interface between processes and mechanisms regulating

N mobilization in order to have sound optimization criteria with a defined domain of applicability. For instance, when the division of N pools in the leaf is considered, all the non-structural N is assumed as equally ready for mobilization, though there are reports that the different proteins involved in photosynthesis may not decrease in parallel during senescence (Makino, Mae, and Ohira, 1983). In addition, for a given leaf N content, photosynthetic capacity may vary with the level of irradiance during growth (Evans, 1989b).

A third point is that, to define the importance of SLN profiles to increase biomass and yield potential, the whole growth cycle should be studied, since the impact may change among growth stages. Then, windows of opportunity (that may differ among crops) could be detected, where an increase in crop growth rate has a direct bearing on yield potential. In wheat, this is likely to be the case of the period of grain number definition around anthesis, as established by Fischer (1985). In addition, since the possibilities of development of SLN profiles and their impact may vary with growing conditions (Figure 6), experimentation should include monitoring of the SLN profiles under combination of factors such as nitrogen availability and density. Furthermore, the assessment should be made frequently during the stages under study.

To answer the question whether the development of different patterns of leaf N allocation within the canopy can effectively impact biomass production, research will be needed in certain basic areas. The relation between photosynthesis parameters and SLN has been documented to change in shape and magnitude in many crops (e.g., rice: Peng, Khush, and Cassman, 1994). This variation, has often been associated to the growing conditions, such as exposure to high vs. low radiation (DeJong, Day, and Johnson, 1989) and needs to be appraised in order to define the potential of photosynthesis parameters at different canopy depths. Implications of such changes are expected to be important for the definition of the optimum SLN profiles. For instance, increasing investment in SLN in upper leaf layers may have different profit when the A_{max}-SLN relation is asymptotic or linear. In addition, the importance of photosynthesis at low radiation levels for total canopy assimilation has been highlighted and thus, the impact of SLN as potentially limiting QE merits further research.

An interesting area which remains open for research is whether the impact of SLN profiles on biomass production will be different under elevated CO_2 concentration environments, expected as a result of global change. We hope that some of the elements given in this paper may help to stimulate discussion on this topic. Finally, it is necessary to remark that we are aware of the shortcomings of optimization theory, e.g., time constraints, oversimplifications and lack of insights as discussed by Chen et al. (1993). Nevertheless, it offers a quite simple framework to analyze exploitation of scarce resources or environmental features in agriculture (Loomis, 1993). The study of solutions originated in such analysis, followed by appropriate design of phenotypes, is possibly a valid tool to help move upwards the actual barrier of yield potential in winter cereals.

REFERENCES

Aerts, R. and H. de Caluwe. (1994). Effects of nitrogen supply on canopy structure and leaf nitrogen distribution in *Carex* species. *Ecology* 75: 1482-1490.

Aikman, D.P. (1989). Potential increase in photosynthetic efficiency from the redistribution of solar radiation in a crop. *Journal of Experimental Botany* 40: 855-864.

Angus, J.F., R. Jones and J.H. Wilson. (1972). A comparison of barley cultivars with different leaf inclination. *Australian Journal of Agricultural Research* 23: 945-957.

Anten, N.P.R. (1995). Canopy structure and patterns of leaf nitrogen distribution in relation to carbon gain. PhD Thesis, University of Utrecht, The Netherlands.

Anten, N.P.R., F. Schieving, and M.J.A. Werger. (1995). Patterns of light and nitrogen distribution in relation to whole canopy carbon gain in C3 and C4 mono- and dicotyledonous species. *Oecologia* 101: 504-513.

Araus, J.L., M.P. Reynolds, and E. Acevedo. (1993). Leaf posture, growth, leaf structure, and carbon isotope discrimination in wheat. *Crop Science* 33: 1273-1279.

Austin, R.B. (1989). Genetic variation in photosynthesis. *Journal of Agricultural Science* 112: 287-294.

Austin, R.B. (1992). Can we improve on nature? Abstracts of the First International Crop Science Congress, Ames, IA, p. 29.

Austin, R.B., J. Bingham, R.D. Blackwell, L.T. Evans, M.A. Ford, C.L. Morgan, and M. Taylor. (1980). Genetic improvement in winter wheat yield since 1900 and associated physiological changes. *Journal of Agricultural Science* 94: 675-689.

Austin, R.B., M.A. Ford, J.A. Edrich, and B.E. Hooper. (1976). Some effects of leaf posture on photosynthesis and yield in wheat. *Annals of Applied Biology* 83: 425-446.

Austin, R.B., M.A. Ford and C.L. Morgan. (1989). Genetic improvement in the yield of winter wheat: A further evaluation. *Journal of Agricultural Science:* 112-295.

Austin, R.B., C.L. Morgan, M.A. Ford, and S.C. Bhagwat. (1982). Flag leaf photosynthesis of *Triticum aestivum* and related diploid and tetraploid species. *Annals of Botany* 49: 177-189.

Baldocchi, D. and S. Collineau. (1994). The physical nature of solar radiation in heterogeneous canopies: spatial and temporal attributes. In *Exploitation of Environmental Heterogeneity by Plants: Ecophysiological Processes Above and Belowground,* eds. M.M. Caldwell and R.W. Pearcy. San Diego, CA: Academic Press, pp. 21-71.

Bhagsari, A.S. and R.H. Brown. (1986). Leaf photosynthesis and its correlation with leaf area. *Crop Science* 26: 127-131.

Bell, M.A., R.A. Fischer, D. Byerlee, and K. Sayre. (1995). Genetic and agronomic contributions to yield gains: a case study for wheat. *Field Crops Research* 44: 55-65.

Bulman, P., D.E. Mather, and D.L. Smith. (1993). Genetic improvement of spring barley cultivars grown in eastern Canada from 1910 to 1988. *Euphytica* 71: 35-48.

Burkey, K.O. and R. Wells. (1991). Response of soybean photosynthesis and chloroplast membrane function to canopy development and mutual shading. *Plant Physiology* 97: 245-252.

Calderini, D.F., M.F. Dreccer, and G.A. Slafer. (1995). Genetic improvement in wheat yield and associated traits. A re-examination of previous results and latest trends. *Plant Breeding* 114:108-112.

Caloin, M. and O. Yu. (1984). Analysis of the time course of change in nitrogen content in *Dactylis glomerata* L. using a model of plant growth. *Annals of Botany* 44: 523-535.

Carvalho, F.I.F. and C.O. Qualset. (1978). Genetic variation for canopy architecture and its use in wheat breeding. *Crop Science* 18: 561-567.

Carver, B.F., R.C. Johnson, and A.L. Rayburn. (1989). Genetic analysis of photosynthetic diversity in hexaploid and tetraploid wheat and their interspecific hybrids. *Photosynthesis Research* 20: 105-118.

Carver, B.F. and E. Nevo. (1990). Genetic diversity of photosynthetic characters in native populations of *Triticum dicoccoides*. *Photosynthesis Research* 25: 119-128.

Chang, T.T. and O. Tagumpay. (1970). Genotypic association between grain yield and six agronomic traits in a cross between rice varieties of contrasting plant types. *Euphytica* 19: 356-363.

Charles-Edwards, D.A., H. Stutzel, R. Ferraris, and D.F. Beech. (1987). An analysis of spatial variation in the nitrogen content of leaves from different horizons within a canopy. *Annals of Botany* 60: 421-426.

Chen, J.L., J.F. Reynolds, P.C. Harley, and J.D. Tenhunen. 1993. Coordination theory of leaf nitrogen distribution in a canopy. *Oecologia* 93: 63-69.

Connor, D.J., A.J. Hall, and V.O. Sadras. (1993). Effect of nitrogen content on the photosynthetic characteristics of sunflower leaves. *Australian Journal of Plant Physiology* 20: 251-263.

Connor, D.J., V.O. Sadras, and A.J. Hall. (1995). Canopy nitrogen distribution and the photosynthetic performance of sunflower crops during grain filling–a quantitative analysis. *Oecologia* 101: 274-281.

Day, W., Z.S. Chalabi. (1988). Use of models to investigate the link between the modification of photosynthetic characteristics and improved crop yields. *Plant Physiology and Biochemistry* 26: 511-517.

DeJong, T.M., K.R. Day, R.S. Johnson. (1989). Partitioning of leaf nitrogen with respect to within canopy light exposure and nitrogen availability in peach (*Prunus persica*). *Trees* 3: 89-95.

Dingkuhn, M. and M. Kropff. (1996). Rice. In *Photoassimilate Distribution in Plants and Crops. Source-Sink Relationships,* eds. E. Zamski and A.A. Schaffer. New York, NY: Marcel Dekker Inc., pp. 519-547.

Duncan, W.G. (1971). Leaf angles, leaf area, and canopy photosynthesis. *Crop Science* 11: 482-485.

Evans, J.R. (1983). Nitrogen and photosynthesis in the flag leaf of wheat (*Triticum aestivum* L). *Plant Physiology* 72: 297-302.

Evans, J.R. (1989a). Photosynthesis and nitrogen relationships in leaves of C$_3$ plants. *Oecologia* 78: 9-19.

Evans, J.R. (1989b). Photosynthesis–The dependence on nitrogen partitioning. In *Causes and Consequences of Variations in Growth Rate and Productivity of Higher Plants,* eds. H. Lambers, M.L. Cambridge, H. Konings and T.L. Pons. The Hague: SPB Academic Publishing, pp. 159-174.

Evans, J.R. (1993). Photosynthetic acclimation and nitrogen partitioning within a lucerne canopy. II Stability through time and comparison with a theoretical optimum. *Australian Journal of Plant Physiology* 20: 55-67.

Evans, L.T. (1990). Assimilation, allocation, explanation, extrapolation. In *Theoretical Production Ecology: Reflections and Prospects,* eds. R. Rabbinge, J. Goudriaan, H. van Keulen, F.W.T. Penning de Vries, and H.H. van Laar. Wageningen: PUDOC, pp. 77-87.

Field, C. (1983). Allocating leaf nitrogen for the maximization of carbon gain: leaf age as a control on the allocation program. *Oecologia* 56: 341-347.

Field, C. and H.A. Mooney. (1983). Leaf age and seasonal effects on light, water, and nitrogen use efficiency in a California shrub. *Oecologia* 56: 348-355.

Field, C. and H.A. Mooney. (1986). The photosynthesis-nitrogen relationship in wild plants. In *On the Economy of Plant Form and Function,* ed. T.J. Givnish. Cambridge: Cambridge University Press, pp. 25-55.

Fischer, R.A. (1985). Number of kernels in wheat crops and the influence of solar radiation and temperature. *Journal of Agricultural Science* 105: 447-461.

Flood, R.G. and G.M. Halloran. (1986). The influence of genes for vernalisation response on development and growth in wheat. *Annals of Botany* 58: 505-508.

Gallagher, J.N. and P.V. Biscoe. (1978). Radiation absorption, growth and yield of cereals. *Journal of Agricultural Science* 105: 447-461.

Giménez, C., D.J. Connor, and F. Rueda. (1994). Canopy development, photosynthesis and radiation use efficiency in sunflower in response to nitrogen. *Field Crops Research* 38: 15-27.

Goudriaan, J. (1988). The bare bones of leaf-angle distribution in radiation models for canopy photosynthesis and energy exchange. *Agricultural and Forest Meteorology.* 43: 155-169.

Goudriaan, J. (1995). Optimization of nitrogen distribution and of leaf area index for maximum canopy assimilation rate. In *SARP Research Proceedings: Nitrogen Management Studies in Irrigated Rice,* eds. T.M. Thiyagarajan, H.F.M. ten Berge, and M.C.S. Wopereis. Wageningen: AB-DLO, TPE-WAU, and Los Baños: IRRI, pp. 85-97.

Goudriaan, J. and H.H. van Laar. (1994). *Modelling Potential Crop Growth Processes.* Dordrecht: Kluwer Academic Press.

Green, C.F. (1987). Nitrogen nutrition and wheat in growth in relation to absorbed solar radiation. *Agricultural and Forest Meteorology* 41: 207-248.

Green, C.F. (1989). Genotypic differences in the growth of *Triticum aestivum* in relation to absorbed solar radiation. *Field Crops Research* 19: 285-295.

Hall, A.J, D.J. Connor, and V.O. Sadras. (1995). Radiation use efficiency of sunflower crops: effects of specific leaf nitrogen and ontogeny. *Field Crops Research* 41: 65-77.

Hammer, G.L. and G.C. Wright. (1994). A theoretical analysis of nitrogen and radiation effects on radiation use efficiency in peanut. *Australian Journal of Agricultural Research* 45: 575-589.

Hikosaka, K., I. Terashima, and S. Katoh (1994). Effects of leaf age, nitrogen nutrition and photo flux density on the distribution of nitrogen among leaves of a vine (*Ipomoea tricolor* Cav.). grown horizontally to avoid mutual shading of leaves. *Oecologia* 97: 451-457.

Hirose, T. and M.J.A. Werger. (1987a). Nitrogen use efficiency in instantaneous and daily photosynthesis of leaves in the canopy of *Solidago altissima* stand. *Physiologia Plantarum* 70: 215-222.

Hirose, T. and M.J.A. Werger. (1987b). Maximizing daily canopy photosynthesis with respect to the leaf nitrogen allocation pattern in the canopy. *Oecologia* 72: 520-526.

Hirose, T., M.J.A. Werger, T.L. Pons and J.W.A. van Rheenen. (1988). Canopy structure and leaf nitrogen distribution in a stand of *Lysimachia vulgaris* L. as influenced by stand density. *Oecologia* 77: 145-150.

Innes, P. and R.D. Blackwell. (1983). Some effects of leaf posture on the yield and water economy of winter wheat. *Journal of Agricultural Science* 101: 367-376.

Jensen, N.F. (1978). Limits to growth in world food production. Ceilings for wheat yields are coming in developed countries. *Science* 201: 317-320.

Johnson, R.C., H. Kebede, D.W. Mornhinweg, B.F. Carver, A.L. Rayburn, H.T. Nguyen. (1987). Photosynthetic differences among *Triticum* accessions at tillering. *Crop Science* 27: 1046-1050.

Kiniry, J.R., C.A. Jones, J.C. O'Toole, R. Blanchet, M. Cabelguenne, D.A. Spanel. (1989). Radiation-use efficiency in biomass accumulation prior to grain-filling for five grain-crop species. *Field Crops Research* 20: 51-64.

Lawes, D.A. (1977). Yield improvement in spring oats. *Journal of Agricultural Science* 89: 751-757.

Lawlor, D.W. (1995). Photosynthesis, productivity and environment. *Journal of Experimental Botany* 46: 1449-1461.

Lemaire, G., B. Onillon, G. Gosse, M. Chartier, J.M. Allirand. (1991). Nitrogen distribution within a lucerne canopy during regrowth: relation with light distribution. *Annals of Botany* 68: 483-488.

Loomis, R.S. (1993). Optimization theory and crop improvement. In *International Crop Science I,* Madison, WI: Crop Science Society of America, pp. 583-588.

Loss, S.P., K.H.M. Siddique. (1994). Morphological and physiological traits associated with wheat yield increases in Mediterranean environments. *Advances in Agronomy* 52: 229-276.

MacKown, C.T., D.A. Van Sanford, and N. Zhang. (1992). Wheat vegetative nitrogen compositional changes in response to reduced reproductive sink strength. *Plant Physiology* 99: 1469-1474.

Makino, A., T. Mae and K. Ohira. (1983). Photosynthesis and ribulose 1,5-bisphosphate carboxylase in rice leaves. *Plant Physiology* 73: 1002-1007.

Marshall, B. and J. Vos. (1991). The relation between the nitrogen concentration and photosynthetic capacity of potato (*Solanum tuberosum* L.) leaves. *Annals of Botany* 68: 33-39.

Meinke, H. (1996). Improving wheat simulation capabilities in Australia from a cropping systems perspective. PhD Thesis, Wageningen Agricultural University, The Netherlands.

Monteith, J.L. (1977). Climate and the efficiency of crop production in Britain. *Philosophical Transactions of Royal Society of London,* B, 281: 277-294.

Mooney, H.A., C. Field, S.L. Gulmon, and F.A. Bazzaz. (1981). Photosynthetic capacity in relation to leaf position in desert versus old-field annuals. *Oecologia* 50: 109-112.

Mooney, H.A., and S.L. Gulmon. (1979). Environmental and evolutionary constraints on the photosynthetic characteristics of higher plants. In *Topics in Plant Population Biology,* ed. O.T. Solbrig, S. Jain, G.B. Johnson, and P.H. Raven. New York, NY: Columbia University Press, pp. 316-337.

Muchow, R.C. and T.R. Sinclair. (1994). Nitrogen response of leaf photosynthesis and canopy radiation use efficiency in field grown maize and sorghum. *Crop Science* 34: 721-727.

Ort, D.R., N.R. Baker. (1988). Consideration of photosynthetic efficiency at low light as a major determinant of crop photosynthetic performance. *Plant Physiology and Biochemistry* 26: 555-565.

Peltonen-Sainio, P. (1990). Genetic improvements in the structure of oat plants in northern growing conditions during this century. *Plant Breeding* 104: 340-345.

Peltonen-Sainio, P. (1994). Productivity of oats: genetic gains and associated physiological changes. In *Genetic Improvement of Field Crops,* ed. G.A. Slafer. New York, NY: Marcel Dekker Inc, pp. 69-94.

Peng, S., K.G. Cassman, and M.J. Kropff. (1995). Relationship between leaf photosynthesis and nitrogen content of field-grown rice in the tropics. *Crop Science* 35: 1627-1630.

Peng, S., G.S. Khush, K.G. Cassman. (1994). Evolution of the new plant ideotype for increased yield potential. In *Breaking the Yield Barrier: Proceedings of a Workshop on Rice Potential in Favourable Environments,* ed. K.G. Cassman. Los Baños: IRRI, pp. 5-20.

Penning de Vries, F.W.T., H. van Keulen, and J.C. Alagos. (1988). Nitrogen redistribution and potential production in rice. In *Proceedings of the International Congress of Plant Physiology*, New Delhi, India, February 15-20, pp. 511-520.

Pons, T.L., and R.W. Pearcy. (1994). Nitrogen reallocation and photosynthetic acclimation in response to partial shading in soybean plants. *Physiologia Plantarum* 92: 636-644.

Pons, T.L., F. Schieving, T. Hirose, and M.J.A. Werger. (1989). Optimization of leaf nitrogen allocation for canopy photosynthesis in *Lysimachia vulgaris*. In *Causes and Consequences of Variations in Growth Rate and Productivity of Higher Plants,* ed. H. Lambers, M.L. Cambridge, H. Konings and T.L. Pons. The Hague: SPB Academic Publishing, pp. 173-186.

Rasmusson, D.C. (1987). An evaluation of ideotype breeding. *Crop Science* 27: 1140-1146.

Richards, R. (1992). The effect of dwarfing genes in spring wheat in dry environments. I. Agronomic characteristics. *Australian Journal of Agricultural Research* 43: 517-523.

Riggs, T.J., P.R. Hanson, N.D. Start, D.M. Miles, C.L. Morgan, and M.A. Ford. (1981). Comparison of spring barley varieties grown in England and Wales between 1880 and 1980. *Journal of Agricultural Science* 97: 599-610.

Rood, S.B., D.J. Major, W.A. Charnetski. (1984). Seasonal changes in $^{14}CO_2$ assimilation and ^{14}C translocation in oilseed rape. *Field Crops Research* 8: 341-348.

Rosenthal, W.D., G.F. Arkin, and T.A. Howell. (1985). Transmitted and absorbed photosynthetically active radiation in grain sorghum. *Agronomy Journal* 77: 841-845.

Russell, G., P.G. Jarvis, and J.L. Monteith. (1989). Absorption of radiation by canopies and stand growth. In *Plant Canopies: Their Growth, Form and Function,* ed. G. Russell, L.B. Marshall and P.G. Jarvis. Cambridge: Cambridge University Press, pp. 21-39.

Sadras, V.O., A.J. Hall, and D.J. Connor. (1993). Light-associated nitrogen distribution profile in flowering canopies of sunflower (*Helianthus annuus* L.) altered during grain growth. *Oecologia* 95: 488-494.

Schieving, F., M.J.A. Werger, and T. Hirose. (1992a). Canopy structure, nitrogen distribution and whole canopy photosynthetic carbon gain in growing and flowering stands of tall herbs. *Vegetatio* 102: 173-181.

Schieving, F., M.J.A. Werger, and T. Hirose. (1992b). The vertical distribution of nitrogen and photosynthetic activity at different plant densities in *Carex acutiformis*. *Plant and Soil* 14: 9-17.

Shiraiwa, T. and T.R. Sinclair. (1993). Distribution of nitrogen among leaves in soybean canopies. *Crop Science* 33: 804-808.

Shorter, R., R.J. Lawn, and G.L. Hammer. (1991). Improving genotypic adaptation in crops–a role for breeders, physiologists and modelers. *Experimental Agriculture* 271: 155-175.

Siddique, K.H.M., R.K. Belford, M.W. Perry, D. Tennant (1989). Growth, development and light interception of old and modern wheat cultivars in a Mediterranean-type environment. *Australian Journal of Agricultural Research* 40: 473-487.

Sinclair, T.R., J.M. Bennet, K.J. Boote. (1993). Leaf nitrogen content, photosynthesis and radiation use efficiency in peanut. *Peanut Science* 20: 40-43.

Sinclair, T.R. and T. Horie. (1989). Leaf nitrogen, photosynthesis and crop radiation use efficiency: A review. *Crop Science* 29: 90-98.

Sinclair, T.R. and T. Shiraiwa. (1993). Soybean radiation use efficiency as influenced by non-uniform specific leaf nitrogen distribution and diffuse radiation. *Crop Science* 33: 808-812.

Slafer, G.A., and F.H. Andrade. (1989). Genetic improvement in bread wheat (*Triticum aestivum* L.) yield in Argentina. *Field Crops Research* 21: 351-367.

Slafer, G.A. and F.H. Andrade. (1991). Changes in physiological attributes of the dry matter economy of bread wheat (*Triticum aestivum*) through genetic improvement of grain yield potential at different regions of the world. A review. *Euphytica* 58: 37-49.

Slafer, G.A., D.F. Calderini and D.J. Miralles. (1996). Generation of yield components and compensation in wheat: opportunities for further increasing yield potential. In: *Increasing the Yield Potential in Wheat: Breaking the Barriers,* eds. S. Rajaram and M. Reynolds. Mexico: CYMMYT, in press.

Slafer, G.A., E.H. Satorre, and F.H. Andrade. (1994). Increases in grain yield in bread wheat from breeding and associated physiological changes. In *Genetic Improvement of Field Crops,* ed. G.A. Slafer: New York, NY: Marcel Dekker Inc., pp. 1-68.

Terashima, I. and J.R. Evans. (1988). Effects of light and nitrogen nutrition on the organization of the photosynthetic apparatus in spinach. *Plant and Cell Physiology* 29: 143-155.

Trenbath, B.R. and J.F. Angus. (1975). Leaf inclination and crop production. *Field Crop Abstracts* 28: 231-244.

Tungland, L., L.B. Chapko, J.V. Wiersma and D.C. Rasmusson. (1987). Effect of leaf angle on grain yield in barley. Crop Science 27: 37-40.

van Heemst, H.D.J. (1988). *Plant Data Values Required for Simple Crop Growth Simulation Models: Review and Bibliography.* Wageningen: Simulation Report CABO-TT.

van Keulen, H., J. Goudriaan, N.G. Seligman. (1989). Modelling the effects of nitrogen on canopy development and crop growth. In *Plant Canopies: Their Growth, Form and Function,* eds. G. Russell, L.B. Marshall and P.G. Jarvis. Cambridge: Cambridge University Press, 83-104.

Waddington, S.R., M. Ormanzai, M. Yoshida, and J.K. Ransom. (1987). The yield of durum wheats released in Mexico between 1960 and 1984. *Journal of Agricultural Science* 108: 469-477.

Werger, M.J.A. and T. Hirose. (1988). Effects of light climate and nitrogen partitioning on the canopy structure of stands of a dicotyledonous, herbaceous vegetation. In *Plant Form and Vegetation Structure,* eds. M.J.A. Werger and T. Hirose. The Hague: SPB Academic Publishing bv, pp. 171-181.

Werger, M.J.A. and T. Hirose. (1991). Leaf nitrogen distribution and whole canopy photosynthetic carbon gain in herbaceous stands. *Vegetatio* 97: 11-20.

Williams, M.L., J.F. Farrar, C.J. Pollock. (1989). Cell specialization within the parenchymatous bundle sheath of barley. *Plant Cell Environment* 12: 909-918.

Woledge, J. (1986). The effect of age and shade on the photosynthesis of white clover leaves. *Annals of Botany* 57: 257-262.

Wright, G.C. and G.L. Hammer. (1994). Distribution of nitrogen and radiation use efficiency in peanut canopies. *Australian Journal of Agricultural Research* 45: 565-574.

Yates, D.J. and M.D. Steven. (1987). Reflection and absorption of solar radiation by flowering canopies of oilseed rape (*Brassica napus* L.). *Journal of Agricultural Science* 109: 409-502.

Yunusa, I.A.M., K.H.M. Siddique, R.K. Belford, M.M. Karimi. (1993). Effect of canopy structure on efficiency of radiation interception and use in spring wheat cultivars during the pre-anthesis period in a mediterranean environment. *Field Crops Research* 35: 113-122.

SUBMITTED: 11/11/96
ACCEPTED: 01/01/97

Genetic Diversity
and Phylogenetic Relationships in Cotton
Based on Isozyme Markers

Sukumar Saha
Allan Zipf

SUMMARY. Phylogenetic relationships among 16 diploid species, including at least one species for each of the A, C, D, E and G genomic groups, and for the natural AD tetraploid species and one synthetic AD tetraploid species were investigated using starch gel electrophoretic techniques for isozyme detection in conjunction with multivariate analysis. The species were polymorphic for phosphoglucomutase (PGM), malate dehydrogenase (MDH), shikimate dehydrogenase (SKDH), aconitase (ACO) and isozymes, but monomorphic for phosphoglucoisomerase (PGI). Similar isozyme arrays indicated close or perhaps conspecific relationships between natural tetra-

Sukumar Saha, formerly Research Assistant Professor, and Allan Zipf, Research Assistant Professor, Department of Plant and Soil Science, Alabama A&M University, Normal, AL 35762, USA.

Address correspondence to (current address): Sukumar Saha, USDA-ARS, Crop Science Research Laboratory, P. O. Box 5367, Mississippi State, MS 39762, USA (E-mail: Saha@ra.msstate.edu).

The authors would like to gratefully acknowledge Dr. David Stelly, Crop and Soil Science Department, Texas A&M University, College Station, TX, for his direct support in conducting this study. The authors also recognize Cotton Incorporated, USAID and the CSREES Capacity Building Program of the USDA for their support in manuscript preparation. The authors would also like to thank Dr. Govind Sharma (AAMU) and Dr. Alex Diner (USFS) for critical reading of the manuscript.

[Haworth co-indexing entry note]: "Genetic Diversity and Phylogenetic Relationships in Cotton Based on Isozyme Markers." Saha, Sukumar, and Allan Zipf. Co-published simultaneously in *Journal of Crop Production* (The Food Products Press, an imprint of The Haworth Press, Inc.) Vol. 1, No. 1 (#1), 1998, pp. 79-93; and: *Crop Sciences: Recent Advances* (ed: Amarjit S. Basra) The Food Products Press, an imprint of The Haworth Press, Inc., 1998, pp. 79-93. Single or multiple copies of this article are available for a fee from The Haworth Document Delivery Service [1-800-342-9678, 9:00 a.m. - 5:00 p.m. (EST). E-mail address: getinfo@haworth.com].

ploids *G. hirsutum* and *G. lanceolatum* and between *G. barbadense* and *G. darwinii*. Natural tetraploid *G. mustelinum* was found to be relatively distinct from the other tetraploid species. The synthetic AD tetraploid was intermediate between the natural AD tetraploid and the A and D diploid clusters. The two A genome species, *G. aboreum* and *G. herbaceum* were closely related to each other. These results supported the hypothesis that the A and D genome species, or their closely related progenitors, were the ancestors of the tetraploid species. The C $_1$ and E2 genome diploid species differed extensively from other diploid and tetraploid species, suggesting that they are very distantly related to the diploid and tetraploid species. These genome types would be very useful resources for germplasm introgression and improvement of the cultivated diploid and tetraploid species because of their significant genetic differences. *[Article copies available for a fee from The Haworth Document Delivery Service: 1-800-342-9678. E-mail address: getinfo@haworth.com]*

KEYWORDS. *Gossypium,* phylogenetic, isozyme, tetraploid, diploid, species, taxonomy, evolution

COTTON TAXONOMY

Cotton, the genus *Gossypium,* includes 32 diploid species ($2n = 26$) and six disomic tetraploid species ($2n = 4x = 52$) (Fryxell, 1979). The currently described diploids include seven genomes, namely A, B, C, D, E, F, G; the tetraploids are disomic and carry both A and D genomes (Endrizzi, Turcotte and Kohel, 1985). Taxonomic classification of the genus *Gossypium* has been based primarily on karyotype, meiotic chromosome affinity, ability to hybridize, geographic distribution, and morphology.

There is divergence of opinion regarding the taxonomic treatment of the *Gossypium* species (Watt, 1907; Hutchinson, Silow and Stephens, 1947; Fryxell, 1979; Brubaker and Wendel, 1993). Factors that have contributed to the complexity of the genus classification include: (1) the genus is distributed in a wide range of geographical and ecological habitats; (2) several species were domesticated long ago in different parts of the world; (3) introgression has occurred between wild species and cultivated germplasm; (4) there have been relatively few genetic markers useful for identification; and (5) diversification of the genus occurred about 1.1 million years ago (Wendel, 1989). There have been some reports on *Gossypium* systematics and evolution based on biochemical or molecular marker studies (Cherry and Katterman, 1971; Johnson, 1975; Hancock,

1982; Wendel, 1989; Percy and Wendel, 1990; Brubaker and Wendel, 1994; Reinisch et al., 1994; Wendel, Schnabel and Seelanan, 1995). Biochemical or molecular markers have advantages over morphological markers in quantifying genetic relationships, due to the former's complete penetrance, co-dominant gene action, lack of epistasis, absence of environmental effects and non-deleterious phenotypes.

A phylogenetic classification of the genus that combines geographical, morphological and molecular data would assist more accurate classification and efficacious utilization of germplasm resources. The objectives of this paper were to utilize isozyme markers in: (1) evaluating the phylogenetic relationships among the tetraploid species of *Gossypium*, (2) studying the relationships between diploid and tetraploid species of *Gossypium*, especially between the tetraploid AD genomes and diploid A and D genomes, and (3) illustrating the origin of the tetraploid species of *Gossypium*.

ISOZYME ANALYSIS

Seeds were obtained from the USDA National Germplasm Center and the Texas Agricultural Experiment Station cotton cytogenetics project for 26 accessions of 16 diploid and tetraploid species of *Gossypium* including AD, A, D, C, E, and G genomes (Table 1). We had to use, in some cases, one or two accessions as representative of a species because many of the wild species are very poor in germination. The cotyledon samples were analyzed based on the banding patterns of five isozymes, namely phosphoglucomutase (PGM), malate dehydrogenase (MDH), shikimate dehydrogenase (SKDH), aconitase (ACO) and phosphoglucoisomerase (PGI). The data were based on 67 different electromorphs of the isozymes and were treated as 67 different binary variables because the mode of genetic inheritance of some of the isozymes were not known. Two multivariate statistical approaches, hierarchical cluster analysis and nonmetric multidimensional scaling analysis, were used to analyze the data. The group average method was applied to a cluster analysis on the basis of an association matrix from simple matching as recommended by Sneath and Sokal (1973). Data were also subjected to a nonmetric multidimensional scaling analysis (MDS) following principal coordinate analysis (PCORD Biostat II program, SigmaSoft™, 1989). Rolf (1972) and Pimental (1981) illustrated that MDS performs appropriately in systematic applications. The three eigenvectors (E1, E2, and E3) calculated for each accession served as 3-D coordinates to construct the three-dimensional graphic plot (Figure 2) using SYGRAPH 5.0 (Pimental and Smith, 1986).

TABLE 1. Species sampled to study phylogenetic relationships in *Gossypium*

Species	Genome	Accession	Sample No.[1]	Geographic Origin
G. hirsutum L.	$(AD)_1$	TM1	13	Central America
G. hirsutum L.	$(AD)_1$	TX-2114	13	Central America
G. barbadense L.	$(AD)_2$	Pima 3-79	15	South, Central America
G. barbadense L.	$(AD)_2$	K-80	15	South, Central America
G. barbadense L.	$(AD)_5$	K-100	15	South, Central America
G. darwinii Watt.	$(AD)_5$	P1499720	16	Galapagos Island
G. darwinii Watt.	$(AD)_5$	P1499724	16	Galapagos Island
G. darwinii Watt.	$(AD)_5$	CB3120	17	Galapagos Island
G. tomentosum Nutt. ex Seem.	$(AD)_3$	Beasley Lab[2]	18	Hawaii
G. mustelinum Miers ex Watt.	$(AD)_4$	Athens #57	19	Brazil
G. mustelinum Miers ex Watt.	$(AD)_4$	Beasley Lab[2]	19	Brazil
G. lanceolatum Tod.	(AD)	TX-1	14	Mexico
Synthetic tetraploid	$2(A_2D_3)$	Beasley Lab[2]	20	USA
G. herbaceum L.	A_1	A_1-46	1	People's Republic of China
G. herbaceum L.	A_1	A_1-18	2	Afghanistan
G. herbaceum L.	A_1	A_1-V797	12	India
G. arboreum L.	A_2	v. indicum	3	India
G. arboreum L.	A_2	A_2-100	3	Mexico
G. thurberi Tod.	D_1	D_1-4	4	Mexico
G. davidsonii Kell.	D_3-d	D_3-d2	5	Mexico
G. raimondii Vlbr.	D_5	D_5-2	6	Peru
G. laxum Phillips	D_8	D_8-2	7	Mexico
G. somalense (Anrke) Hutch	E_2	E_2-1	8	Arabia
G. somalense (Anrke) Hutch	E_2	E_2-2	9	Arabia
G. sturtianum J. H. Wills	C_1	C_1-1	10	Australia
G. bickii Prokh.	G_1	G_1-1	11	Australia

[1]Numbers correspond to sample number used in the nonmetric multidimensional scaling analysis (Figure 3).
[2]This particular line was available from Dr. David Stelly, Texas A&M University, College Station, TX, without an accession number.

Electrophoretic analysis of 26 accessions revealed polymorphisms in four of the five isozyme systems among the diploid and tetraploid species of *Gossypium*. The isozyme banding pattern among seeds within an accession were identical except that variation existed in one electromorph within certain seed samples of *G. lanceolatum* (2%) and one accession of *G. hirsutum* (4%). This variation of the genotypes might be due to seed mixture or germplasm introgression. Four of the five isozymes (80%) were polymorphic among the diploid and tetraploid species of *Gossypium*. Bourdon (1986) observed that seven of the nine isozyme systems studied (77%) were polymorphic in two cultivated tetraploid species of cotton.

PGM Isozymes

PGM isozyme profiles of the diploid and tetraploid species of cotton exhibited four to eight bands. The xymogram of PGM isozymes could be divided into two distinct zones. The most cathodal region included two distinct bands, whereas the slightly anodal zone included four to six bands, depending on the species (Figure 1). Tetraploid species *G. hirsutum* and *G. lanceolatum* exhibited identical banding patterns for PGM isozymes (Figure 1). Saha and Stelly (1994) reported that two distinct cathodal bands were coded for by the *PGM-7* locus which was polymorphic between *G. hirsutum* and *G. barbadense*. They also reported that this locus was located on the long arm of chromosome 12. Polymorphisms were detected among the three accessions of *G. darwinii*, two of which exhibited a banding pattern identical to that of *G. barbadense*. The most cathodal major PGM band from *G. mustelinum* was unique in mobility when compared to the other species (Figure 1). The PGM banding pattern in the most cathodal zone of *G. tomentosum* differed from those of the other tetraploid species but the individual bands towards the anodal region exhibited mobilities similar to specific bands of *G. hirsutum* and *G. barbadense* (Figure 1).

PGM bands of the cathodal zones of the natural tetraploid species (4X) were found to be absent in the diploid species (2X) and the synthetic tetraploid species (4X), indicating that the duplicate PGM loci of the A and D genomes of the natural tetraploid species have changed structurally in the course of evolution. Another possibility is that chromatin has been lost from the synthetic $2(A_2D_3)$ hybrid. Selection for loss of chromatin may have resulted due to the D_3 hybrid lethality system (Lee, 1981). Further insight could be gained by screening the isozyme banding patterns of the specific diploid parental accessions and the corresponding synthetic AD tetraploid.

Polymorphism for PGM isozymes was observed among the four

FIGURE 1. Banding patterns of PGM and SKDH isozymes of the tetraploid *Gossypium* species. Germinating seeds were homogenized in a chilled mortar and pestle using a 1:1 ratio (w/v) of tissue to extraction buffer containing 0.1 M Tris-HCl, 7% (w/v) sucrose, 5% (w/v) polyvinylpyrrolidone, 10 mM mercaptoethanol, 5.8 mM diethyldithiocarbamate, 5 mM ascorbic acid, 4.8 mM sodium bisulfite and 6.5 mM dithioerythritol, pH 7.5 (Saha, 1989). The slurries were centrifuged at 1500 × g at 4°C and the supernatants were subjected to horizontal starch gel electrophoresis using histidine-citrate (Kirkpatrick, Decker and Wilson, 1985) and Tris-citrate (Chenicek and Hart, 1987) buffer systems. Following electrophoresis, the histidine-citrate gels were sliced and then stained for phosphoglucoisomerase (PGI), phosphoglucomutase (PGM), and malate dehydrogenase (MDH) activities. Triscitrate gels were stained for aconitase (ACO), shikimate dehydrogenase (SKDH) and for phosphoglucomutase (PGM), and malate dehydrogenase (MDH) activities. 1a.) PGM isozyme banding patterns of the following tetraploid *Gossypium* species: Lane 1, *G. darwinii*; Lane 2, *G. tomentosum*; Lane 3, *G. mustelinum*; Lanes 4-6 *G. hirsutum*; Lanes 7-9, *G. barbadense*; Lanes 10-11, *G. hirsutum* (TM1). 1b.) SKDH isozyme banding patterns of the following tetraploid *Gossypium* species: Lane 1, *G. mustelinum*; Lane 2, *G. barbadense*; Lane 3, *G. lanceolatum*; Lane 4, *G. hirsutum*.

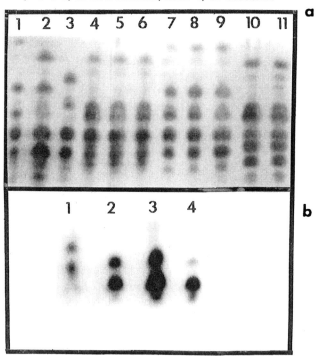

D-genome species tested, *G. thurberi, G. davidsonii, G. raimondii* and *G. laxum.* Little variation was observed between the two A genome species. Out of the five A genome accessions, which included both A_1 and A_2 genomes, four exhibited identical banding patterns; while one accession differed only in the mobility of one PGM band. Only two of the six PGM bands among the two accessions of *G. somalense* (E genome) were identical in position, indicating that significant variation existed between these two accessions. *G. sturtianum* (C_1 genome) exhibited a PGM banding pattern different from those of the other diploid species including *G. bickii* (G_1 genome), the only other Australian species tested. Our results indicated that considerable variation existed with respect to PGM isozymes within the diploid and tetraploid species of *Gossypium.*

MDH Isozymes

MDH isozymes of diploid and tetraploid species were polymorphic except that three accessions of *G. barbadense,* the synthetic $2(A_2D_3)$ tetraploid, and one accession of *G. darwinii* exhibited identical banding patterns. *G. hirsutum* and *G. tomentosum* accessions exhibited a similar banding pattern that differed from that of *G. lanceolatum* in the mobility of only one band. The banding pattern of *G. mustelinum* was different from the other tetraploid species. Banding patterns of the A genome and D genome accessions were very similar to each other, but were very different from the banding patterns of the other diploid species screened (C_1, E_1, E_2, and G_1 genomes).

SKDH Isozymes

Polymorphisms existed within the diploid and tetraploid species of *Gossypium* for SKDH isozymes. *G. hirsutum, G. barbadense, G. lanceolatum, G. tomentosum,* and two accessions of *G. darwinii* exhibited a common SKDH banding pattern, whereas the other tetraploid species, *G. mustelinum,* exhibited a unique SKDH banding pattern (Figure 1). Diploid species exhibited single or double SKDH bands, which suggests the possible presence of duplicated loci, given the monomeric nature of SKDH isozymes. Polymorphisms existed among the D genome species and the banding patterns of SKDH for the D genome and A genome species were more similar to each other than of the other diploid species.

ACO Isozymes

Polymorphism for ACO isozymes was detected among the diploid and tetraploid species of *Gossypium.* Six or seven bands were observed in samples from each tetraploid species and three or four bands were present in each sample of the diploid species. In one accession of *G. hirsutum* and in *G. lanceolatum*, a few seed samples exhibited an ACO banding pattern similar to *G. barbadense*, indicating that heterogeneity exists within the respective accessions. Only one ACO band was polymorphic in one of the five accessions of the two A genome species, while polymorphisms were more prevalent among the D genome species.

PGI Isozymes

Two PGI isozyme bands were detected in *Gossypium*. The banding patterns were identical for all tested diploid and tetraploid species of *Gossypium*. This result may be due to the conservative nature of the genetic system for this isozyme, or some inherent problem in the electrophoretic system used to separate the isozyme molecules. Suiter (1988) observed two zones of enzyme activity for PGI isozymes in the diploid species of *Gossypium*. However, his results revealed a polymorphism for one PGI locus between the two A genome species.

Statistical Analysis

Results from the two methods of multivariate statistical analysis, namely cluster analysis (Figure 2), and principal coordinate analysis (PCORD) followed by non-metric multidimensional scaling (MDS) analysis (Figure 3) were essentially concordant about the species relationships of the accessions. PCORD analysis was used in an attempt to resolve the origin of *G. hirsutum* using restriction fragment length polymorphisms (RFLPs) (Brubaker and Wendel, 1994). Although a cluster analysis dendogram can summarize well the relationships among species, it provides only a one-dimensional summary of the relationships among multidimensionally distributed objects. The combination of PCORD and MDS analyses and three-dimensional plotting was used to overcome this limitation. Calculation in MDS depends on the rank order of similarities rather than absolute agreement, so the relationships are measured in more than one dimension on the basis of relative agreement, rather than absolute compliance. MDS also takes into account non-linear monotone relationships between individuals (Rolf, 1972).

RELATIONSHIPS AMONG THE TETRAPLOID SPECIES

The results indicated that *G. hirsutum* and *G. lanceolatum* were very closely related. The two species differed in only one of the sixty-seven

FIGURE 2. Dendrogram derived from cluster analysis based on simple matching of isozyme banding patterns showing relationships of the species of *Gossypium*. The data were based on 67 different electromorphs of the isozymes and were treated as 67 different binary variables. The group average method was applied to a cluster analysis on the basis of an association matrix from simple matching as recommended by Sneath and Sokal (1973).

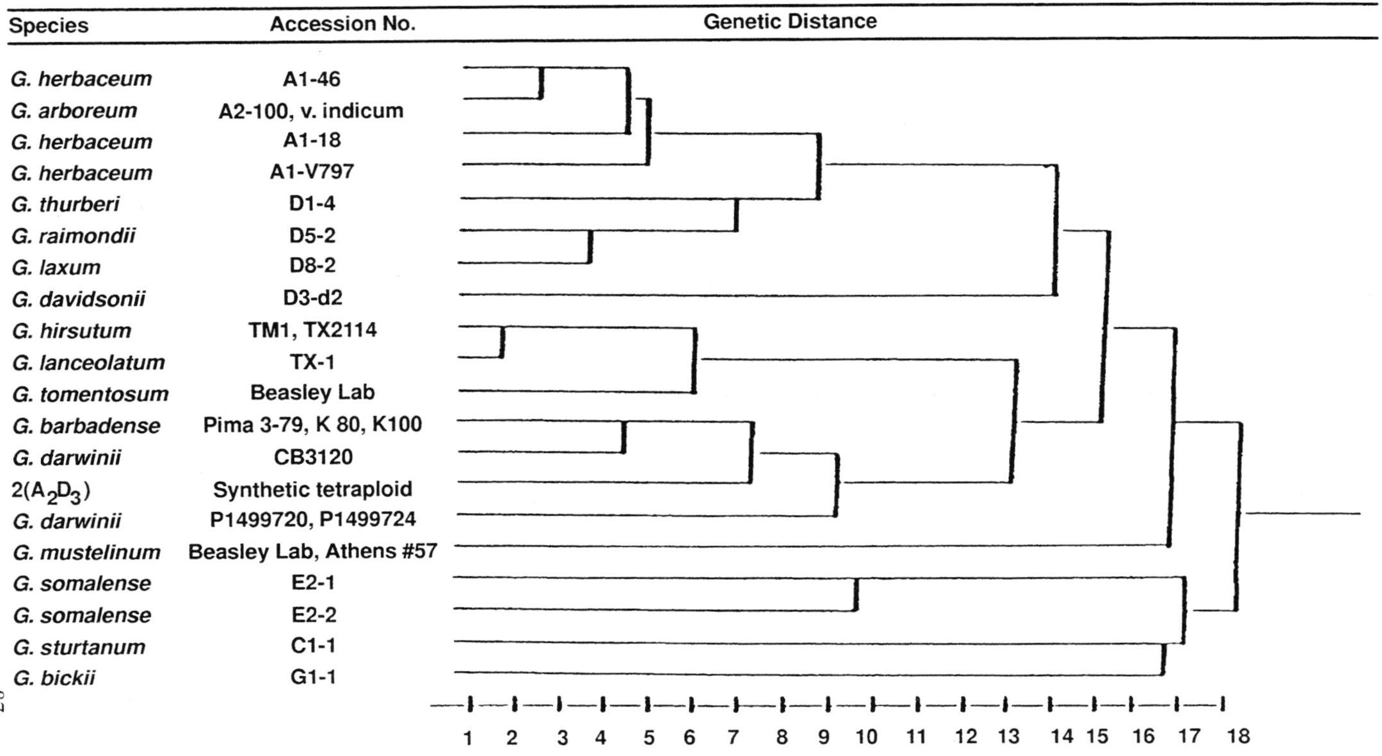

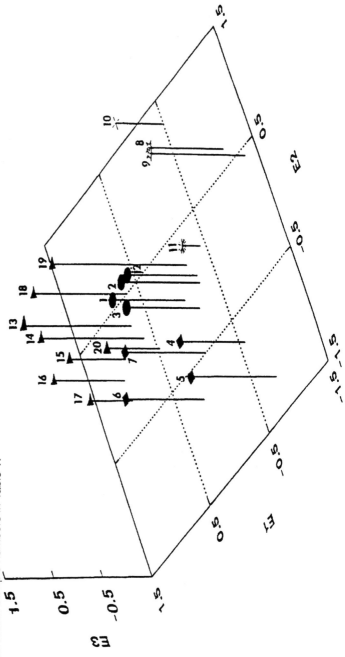

FIGURE 3. Phylogenetic relationship among the species of *Gossypium* on the basis of nonmetric multidimensional scaling analysis. The data were based on 67 different electromorphs of the isozymes and were treated as 67 different binary variables. Data were subjected to a nonmetric multidimensional scaling analysis (MDS) following principal coordinate analysis (PCORD Biostat II program). The MDS was calculated for three and four dimensions, but the stress values were "fair" for just three dimensions (Johnson and Wichern, 1982), i.e., sufficient for representative graphics. The three eigenvectors (E1, E2, and E3) calculated for each accession served as 3-D coordinates to construct the three-dimensional graphic plot using SYGRAPH 5.0. Plot numbers refer to accession sample numbers in Table 1.

electromorphs. Although these two species are morphologically very different, they are sympatric in habitat (Fryxell, 1979), and *G. lanceolatum* was at one time considered as *G. hirsutum* var. *palmeri* (Hutchinson, 1959). However, Johnson (1975) concluded, on the basis of seed protein banding patterns, that the tetraploid species *G. hirsutum, G. barbadense,* and *G. tomentosum* were remarkably uniform, except for *G. hirsutum* var. *palmeri*. Based on the results of Johnson (1975), Fryxell (1979) tentatively treated *G. hirsutum* var. *palmeri* as a separate species, *G. lanceolatum*. On the other hand, *G. lanceolatum* was not resolved from *G. hirsutum* in analysis of chloroplast and ribosomal DNA restriction site variations (DeJoode and Wendel, 1992).

The results also indicated that considerable variation existed among the tetraploid species *G. hirsutum, G. barbadense, G. tomentosum* and *G. mustelinum* (Figures 2 and 3). Percy and Wendel (1990) reported that the level of genetic variation in *G. barbadense* was moderate based on the allozyme analysis. *G. mustelinum,* a native to northeastern Brazil, exhibited isozyme banding patterns very distinct from the other tetraploid species (Figure 2). That *G. mustelinum* is distinctive was also shown during ribosomal DNA analysis where this species apparently demonstrates concerted evolution toward an A genome repeat type whereas all the other allopolyploid species concert to a D genome repeat type (Wendel, Schnabel and Seelanan, 1995). Two accessions of *G. darwinii* were found to be very closely related to *G. barbadense* (Figures 2 and 3). These two species interbreed freely. The results support the earlier theory (Hutchinson, Silow and Stephens, 1947) that taxonomic distinction between these two species should be at the sub-species level.

RELATIONSHIPS AMONG THE DIPLOID SPECIES

All of the A genome accessions were very closely allied with each other as were the D genome species. Johnson (1975) reported that the two A genome species exhibited identical seed protein electrophoretic banding patterns. Wendel (1989) could not detect any mutational differences in chloroplast DNA between the two A genome species. Our results indicated that the two A genome species were very similar to each other, although a single reciprocal translocation has been reported to distinguish *G. arboreum* from *G. herbaceum* (Gerstel, 1953). Sequence data from internal transcribed spacer regions also showed little heterogeneity between these two species (Wendel, Schnabel and Seelanan, 1995). Hutchinson, Silow and Stephens (1947) proposed that *G. arboreum* originated from *G. herbaceum* early in the history of diploid cotton cultivation.

Overall results attested to the fact that the C and E genome species were far removed from the A genome, D genome and AD tetraploid species (Figure 3). Similar results were also found during analysis of chloroplast DNA restriction site data (Wendel and Albert, 1992). In general, our results are concordant with the belief that genetic diversification in *Gossypium* occurred within the backdrop of the natural geographic distribution.

ORIGIN OF THE TETRAPLOID SPECIES

The results indicated (Figures 2 and 3) that the A genome and D genome diploid species (Table 1) were more closely allied with the tetraploid species than with the other diploid species. This supports the view that the progenitors of the current A genome and D genome species are closely related to the ancestors of the modern tetraploid species (Endrizzi, Turcotte and Kohel, 1984; Wendel and Albert, 1992; Wendel, Schnabel and Seelanan, 1995). Both the dendrogram and the 3-D graphic plot revealed specific clustering patterns within the diploid and tetraploid species of cotton (Figures 2 and 3). Among the tetraploid species, *G. hirsutum, G. lanceolatum,* and *G. tomentosum* were found to be most closely allied while *G. barbadense* and *G. darwinii* constituted a second tetraploid group.

Analysis indicated that *G. mustelinum* was far removed from the other natural tetraploids, which suggests that it may be an important source for new germplasm. Given its geographic isolation, it was surprising that *G. tomentosum* was not more distant than *G. mustelinum* from the other tetraploids. However, DeJoode and Wendel (1992) proposed that *G. tomentosum* was sister to *G. hirsutum* based on allozyme and chloroplast and ribosomal DNA restriction site analyses. The distinct grouping pattern among the tetraploid species (Figure 3) may support the polyphyletic origin theory (Johnson, 1975), with the idea that each tetraploid grouping may have originated from a different ancestral source. However, one can not exclude the possibility of diversification while at the tetraploid level. Among the D genome species, *G. raimondii* (D_5 genome) was found to be the most closely related to the tetraploid species, thus supporting the theory that the D_5, or a closely related ancestral genome, contributed to the origin of the natural tetraploid cottons. Wendel (1989) reported that the cytoplasm donor for all of the tetraploid species was an A genome diploid that was similar to *G. arboreum* or *G. herbaceum* based on RFLP analysis. He also suggested that the tetraploid cotton originated relatively recently, within the last 1 to 2 million years, with subsequent rapid evolution and diversification throughout the New World tropics. Further detailed RFLP analysis (Reinisch et al., 1994) and chloroplast restriction site data (Wendel and Albert, 1992) also supported such a recent origin.

In conclusion, the results from our data (Figures 2 and 3) indicated that *G. hirsutum* and *G. lanceolatum* were very closely related to each other, thus supporting the earlier treatment of *G. lanceolatum* as *G. hirsutum* var *palmeri* (Hutchinson, Silow and Stephens, 1947). Isozyme banding patterns also revealed two subgroups of tetraploids, *G. hirsutum* and *G. lanceolatum,* and *G. barbadense* and *G. darwinii. G. mustelinum* was found to be very distinct from the other tetraploid species. The A and D genome species were found to be the more closely related to the tetraploid species than were the other diploid genomes, which corroborates the designation of A and D genomes as comprising the tetraploid genomes. The results suggested that the two A genome species were very similar to each other. Furthermore, the results also supported the hypothesis that D_5 (*G. raimondii*) or its progenitor contributed significantly to the origin of the tetraploid species (Endrizzi, 1991). Wendel and Albert (1992) found that the G and C genomes formed a monophyletic clade corresponding to the Australian continent based on chloroplast DNA restriction site analysis. It is curious that our results indicated that *G. bickii* (G genome) and *G. sturtianum* (C genome), both native to Australia, may not be much more closely related to each other than they are to the Arabian diploid species, *G. somalense* (E genome).

The results advocate the hypothesis that the ancestors of the tetraploid species were the A and D genome species, or their closely related progenitors. The C_1 and E_2 diploid genomes differed significantly from the other diploid and tetraploid species based on our isozyme analysis. We also observed that *G. mustelinum* differed significantly from the other tetraploid species. Species of these genome types would be very useful resources for germplasm introgression and enhancement of the cultivated species because of their wide genetic divergence.

REFERENCES

Bourdon, C. (1986). Enzymatic polymorphism and genetic organization of two cotton tetraploid cultivated species, *G. hirsutum* and *G. barbadense. Cotton et Fibres Tropicales* 44:191-210.

Brubaker, C. and F. Wendel. (1993). On the specific status of *Gossypium lanceolatum* Todaro. *Genetic Resources and Crop Evolution* 40:165-170.

Brubaker, C. and F. Wendel. (1994). Reevaluating the origin of domesticated cotton (*Gossypium hirsutum*; Malvaceae) using nuclear restriction fragment length polymorphisms (RFLPs). *American Journal of Botany* 81:1309-1326.

Cherry, J. and F. Katterman. (1971). Nonspecific esterase isozyme polymorphism in natural populations of *Gossypium thurberi. Phytochemistry* 10:141-145.

DeJoode, D. and J. Wendel. (1992). Genetic diversity and origin of the Hawaiian Islands cotton, *Gossypium tomentosum*. *American Journal of Botany* 79:1311-1319.

Endrizzi, J. (1991). The origin of the allotetraploid species of *Gossypium*. In *Chromosome Engineering in Plants: Genetics, Breeding, Evolution*, Part B, eds. T. Tsuchiya and P. Gupta. Amsterdam: Elsevier, pp. 449-482.

Endrizzi, J., E. Turcotte and R. Kohel. (1984). Quantitative genetics, cytology and cytogenetics. In *Cotton*, eds. R. Kohel and C. Lewis. Madison, WI: American Society of Agronomy, pp. 81-129.

Endrizzi, J., E. Turcotte and R. Kohel. (1985). Genetics, cytology and evolution of *Gossypium*. *Advances in Genetics* 23:271-375.

Fryxell, P. (1979). *The Natural History of the Cotton Tribe*. College Station, TX: Texas A&M University Press.

Gerstel, D. (1953). Chromosome translocations in interspecific hybrids of the genus *Gossypium*. *Evolution* 7:234-244.

Hancock, J. (1982). Alcohol dehydrogenase isozymes in *Gossypium hirsutum* and its putative diploid progenitors. *Plant Systematics and Evolution* 140:141-149.

Hutchinson, J. (1959). *The Application of Genetics to Cotton Improvement*. London: Cambridge University Press.

Hutchinson, J., R. Silow and S. Stephens. (1947). *The Evolution of Gossypium*. London: Oxford University Press.

Johnson, B. (1975). *Gossypium palmeri* and a polyphyletic origin of the New World cottons. *Bulletin of the Torrey Botanical Club* 102:340-349.

Lee, J. (1981). Genetics of D_3 complementary lethality in *Gossypium hirsutum* and *Gossypium barbadense*. *Journal of Heredity* 72:299-300.

Percy, R. and J. Wendel. (1990). Allozyme evidence for the origin and diversification of *Gossypium barbadense* L. *Theoretical and Applied Genetics* 79:529-542.

Pimental, R. (1981). A comparative study of data and ordination techniques based on a hybrid swarm of sand verbenas (Abronia Juss.). *Systematic Zoology* 30:250-267.

Pimental, R. and J. Smith. (1986). *BioSTAT II, a Multivariate Statistical Toolbox*. Placentia, CA: SigmaSoft™.

Reinisch, A., J. Dong, C. Brubaker, D. Stelly, J. Wendel and A. Paterson. (1994). A detailed RFLP map of cotton, *Gossypium hirsutum* X *Gossypium barbadense*: Chromosome organization and evolution in a disomic polyploid genome. *Genetics* 138:829-847.

Rolf, F. (1972). An empirical comparison of three ordination techniques in numerical taxonomy. *Systematic Zoology* 21:271-280.

Saha, S. and D. Stelly. (1994). Chromosomal location of *Phosphoglucomutase7* locus in *Gossypium hirsutum*. *Journal of Heredity* 85:35-39.

Sneath, P. and R. Sokal. (1973). *Numerical Taxonomy*. San Francisco: W. H. Freeman and Co.

Suiter, K. (1988). Genetics of allozyme variation in *Gossypium arboreum* L. and *Gossypium herbaceum* L. (Malvaceae). *Theoretical and Applied Genetics* 75:259-271.

Watt, G. (1907). *The Wild and Cultivated Cotton Plants of the World.* London: Longmans, Green and Co.

Wendel, J. (1989). New World tetraploid cottons contain Old World cytoplasm. *Proceedings of the National Academy of Sciences, USA* 86:4132-4136.

Wendel, J. and V. Albert. (1992). Phylogenetics of the cotton genus (*Gossypium*): Character-state weighted parsimony analysis of chloroplast-DNA restriction site data and its systematic and biogeographic implications. *Systematic Botany* 17:115-143.

Wendel, J., A. Schnabel and T. Seelanan. (1995). Bidirectional interlocus concerted evolution following allopolyploid speciation in cotton (*Gossypium*). *Proceedings of the National Academy of Sciences, USA* 92:280-284.

SUBMITTED: 12/15/96
ACCEPTED: 01/08/97

Mechanisms of Heterosis in Crop Plants

A. S. Tsaftaris
M. Kafka

SUMMARY. Although the biological basis of heterosis remains unknown, plant breeders have made wide use of this phenomenon and it is generally believed that the understanding of the mechanisms underlying heterosis will enhance our ability to form new genotypes which may be used directly as F_1 hybrids or form the basis for future selection programmes. The original data concerning the phenomenon came out of studies at the phenotypic level. They were followed by physiological and later by biochemical data with the advent of electrophoresis and the easy accumulation of data related to isozyme variability. More recently, efforts have been made at the molecular level and the results suggest the significance of both the regulatory proteins (and their encoding genes) and the mechanisms of regulation of gene activity in manifestation of complicated phenomena such as heterosis. In this article emphasis is given to the molecular mechanisms of heterosis. Recent data on quantifying gene expression at the protein or mRNA level in different parental inbreds and F_1 hybrids, and some regulatory mechanisms involved such as transcription factor protein and genomic DNA methylation are discussed. *[Article copies available for a fee from The Haworth Document Delivery Service: 1-800-342-9678. E-mail address: getinfo@haworth.com]*

A. S. Tsaftaris, Professor and M. Kafka, Graduate Student, Department of Genetics and Plant Breeding, Aristoteles University of Thessaloniki, Thessaloniki, Greece.

Address correspondence to: A. S. Tsaftaris, Department of Genetics and Plant Breeding, Aristoteles University of Thessaloniki, Thessaloniki, Greece (E-mail: TSAFTARIS@olymp.ccf.auth.gr).

Part of the work presented in this article from the authors' laboratory is financed by E.U., P.T.P. roigect AMICA and the Greek Secretariat for Research and Technology.

[Haworth co-indexing entry note]: "Mechanisms of Heterosis in Crop Plants." Tsaftaris, A. S., and M. Kafka. Co-published simultaneously in *Journal of Crop Production* (The Food Products Press, an imprint of The Haworth Press, Inc.) Vol. 1, No. 1 (#1), 1998, pp. 95-111; and: *Crop Sciences: Recent Advances* (ed: Amarjit S. Basra) The Food Products Press, an imprint of The Haworth Press, Inc., 1998, pp. 95-111. Single or multiple copies of this article are available for a fee from The Haworth Document Delivery Service [1-800-342-9678, 9:00 a.m. - 5:00 p.m. (EST). E-mail address: getinfo@haworth.com].

KEYWORDS. Heterosis, hybrid vigor, molecular heterosis, *Z. mays*, quantitative trait loci, restriction fragment length polymorphism

ABBREVIATIONS. RFLP: Restriction Fragment Length Polymorphism; GA: Gibberelic acid; PAP: Protein Amount Polymorhism; QTL: Quantitative Trait Loci; HHF1: High Heterosis F1; LHF1: Low Heterosis F1; RAP: RNA Amount Polymorphism

INTRODUCTION

While agronomists have been utilizing heterosis as a means of improving crop productivity, the biological basis of heterosis and the mechanisms involved remain largely unknown. Understanding heterosis will facilitate not only the easier creation of new and more productive F_1 hybrids but will help the initial component of all plant breeding work, i.e., the creation of useful genetic variability prior to the second part, that of evaluation of the different genotypes (through their phenotypes) and selection of the superior ones. It is commonly accepted that crosses giving rise to heterotic F_1s and superior F_2s are those giving the most desirable new genetic variability for attaining faster progress through selection (Fasoulas, 1988).

Investigations of the phenomenon of heterosis have followed two separate courses. The first has been largely descriptive and concerned with the frequency and magnitude of the observed effects, the plant characters involved, etc. Speculations as to the underlying causes began to appear early in this observational period. These speculations became phrased in more specific terms after the rediscovery of Mendelism. The second course, on the other hand, centered primarily on genetic, biochemical and physiological studies, on the nature of processes that presumably caused heterosis. But, the causal factors at the biochemical and molecular level remain virtually as obscure as they were 60 years ago.

In this article, we will briefly consider genetic hypotheses that have been advanced as explanations for heterosis and the data accumulated over the years from studies made at the morphological, physiological and biochemical levels. Emphasis will be given to the most current studies, particularly at the molecular level, which have been conducted to elucidate the mechanisms behind this important biological phenomenon.

GENETIC DISTANCE AND HETEROSIS

Since the earliest attempts to explain hybrid vigor in Mendelian terms, there have been two principal hypotheses to explain the genetic basis of heterosis: the dominance and overdominace hypotheses. Because of the observed correlation between recessiveness and detrimental effect, the dominance hypothesis attributes the increased vigor of heterozygosity to their dominant alleles. The overdominance hypothesis assumes that heterozygosity, *per se,* is important, i.e., there exist loci albeit relatively rare, at which the heterozygote is superior to either homozygote.

These two hypotheses hold important differences for the practical plant breeders, since they propose two different methods for attaining maximum improvement. Heterosis due to dominance should, in principle, be fixable by inbreeding. The dominance hypothesis suggests that it should be possible to circumvent inbreeding depression by recombination among inbred lines to derive an inbred line with all the desirable dominant alleles. However, in practice, heterosis due to dominance is fixable only if the number of dispersed complementary genes is small. If the number of favorable and unfavorable allele pairs differentiating the parents is large, the probability of recovering an F_2 segregant with all the desired favorable dominant alleles is extremely small. Heterosis due to overdominance or pseudo-overdominance (i.e., tightly linked complementary genes) is not fixable by inbreeding.

Early data on heterosis derived from studies of quantitative genetics or different plant breeding methodologies (for review see Sprague, 1983). Thus, Fasoulas and Allard (1962) devised a system to compare a series of homozygous and heterozygous loci in a group of isogenic barley lines. The technique, utilized by Russell and Eberhart (1970) and by Russell (1971), provided clear evidence for overdominance and also for epistasis.

Sprague and Miller (1950) suggested a breeding scheme designed to provide information on the relative importance of dominance and overdominance in yield ability of hybrids. The scheme involved use of two maize populations and a single inbred tester parent. Selection was to be practiced in each population based on test cross performance. Under either hypothesis, successive cycles of selection should affect gene frequency changes within each population. According to overdominance hypothesis, gene changes should be such as to give an increasing degree of heterozygosity relative to the tested parent and a corresponding increase in genetic similarity of the two parental populations. Two sets of experiments were initiated to test this hypothesis. Reports were published on each set after the completion of five cycles of selection. Russell et al. (1973) presented data on one set which involved an open-pollinated variety Alph and the F_2 of a single cross (WF9 × B7) with the common tester parent, B14. As a result of the selection, yield increased in each population from cycle 0 through cycle 5. Furthermore, the population crosses C5 × C5 gave significantly higher yield than the Co × Co crosses. This is the type of response which would be expected under the dominance hypothesis and the reverse of the expected if overdominance was the major type of gene action. Data summarizing five cycles in the second set have been reported by Walejco and Russell (1977). The data from both sets of experiments indicate that the additive genetic effects were primarily responsible for the increase in heterosis.

Instead of assuming different genetic diversity in populations from designed breeding programmes, as previously described, different investigators used markers to assess directly the genetic diversity of parental genotypes. Initial studies employed morphological mutations as genetic markers. Sax (1923) first reported association of a simply inherited genetic marker with a quantitative trait in plants when he observed segregation for seed size associated with segregation for a seed coat color marker in beans (*Phaseolus vulgaris* L.). Concerning the correlation between the morphological resemblance of parents with heterosis of their respective F_1 hybrid, only Moll et al. (1965) working with maize found a positive result, but no correlation was found in experiments with other species.

Major limitations of studies using morphological markers included the limited number of available markers and the undesirable effects on phenotype of many of the morphological markers. However, these studies provided a background of theory and observation for more recent work with biochemical and molecular markers.

With the advent of electropohoresis and the consequent ease of accumulation of data related to isozyme variability, a number of attempts have been made for genes controlling isozymes as a criterion of genetic relatedness between plants (for a review see Tsaftaris, 1987). Several investigators have estimated the correlations between isozyme allelic diversity and grain yield of single-cross hybrids derived from commercially used lines of maize (Hunter and Kannenberg, 1971; Heidrich-Sobrinho and Cordeiro, 1975; Gonella and Peterson, 1978; Hadjinov et al., 1982; Tsaftaris and Efthimiadis, 1987). These studies used 11 or fewer isozyme loci and 15 or fewer inbred parental lines. Estimated correlations between isozyme diversity and specific combining ability were low and non-significant. When we analyzed 16 inbreds for 47 isozymic loci, a significant but relatively small correlation ($r = 0.23$) was recorded. Even in a study which used more than 100 maize hybrids derived from 37 elite inbred lines in which associations were evaluated with 30 isozyme loci, Smith and Smith (1989) reported an r^2 value of only 0.36 when F_1 yield was plotted against isozyme allele diversity.

From the above studies, it became obvious that isozyme diversity provides limited value in the prediction of hybrid performance. A number of different factors have been postulated to attribute to this somewhat disappointing conclusion (Tsaftaris and Efthimiadis, 1987; Tsaftaris, 1995), the small number of loci examined being the most significant limitation.

To overcome this limitation, other techniques for protein analysis such as two-dimensional gel electrophoresis were employed (Damerval, Hebert, and De Vienne, 1987; Leonardi, Damerval, and De Vienne, 1987, 1988; Tsaftaris, 1990; Leonardi et al., 1991). But this kind of analysis with two dimensional separation of proteins is limited to a very small number of individual inbreds. Fortunately, the recent development of molecular markers has made it possible to markedly increase the number of markers in large numbers of inbreds.

Many investigators (Lee et al., 1989; Godshalk, Lee, and Lamkey, 1990; Melchinger et al., 1990, 1991, 1992; Dudley, Saghai-Maroof, and Rufener, 1991; Boppenmaier et al., 1992) used RFLPs to examine the relationship between RFLP-based distance and single-cross grain yield in maize. Correlations of genetic distance with F_1 yield heterosis were positive but relatively small for prediction purposes. In all the above studies, the number of lines analyzed using RFLP were relatively small (< 20). The number of markers used, even though greater than the number of morphological or isozyme markers, was still relatively small.

In a study where the number of inbreds analyzed was increased to 230 lines and again for a small number of 76 marker loci (9 isozymic and 67 RFLPs), a highly significant relationship ($r = 0.68$) was found between parental heterozygosity and hybrid yield (Stuber et al., 1992).

When both these parameters, namely, the number of genotypes analyzed and the number of markers, were increased simultaneously, the results were strikingly different. In the most thorough study reported to date by Smith et al. (1990), associations of grain yield with diversity of RFLP genotypes were measured using more than 100 hybrids derived from 37 elite maize inbred lines. When F_1 grain yield was plotted against RFLP diversity, based on 230 enzyme-probe markers the r^2 value was 0.87. This represents a striking difference compared with most of the previous studies and possibly indicates that when the analysis is carried out in a large number of inbreds and, most importantly, when a large number of markers are used, there will be a positive relation between parental genetic distance and F_1 performance.

PHYSIOLOGICAL AND METABOLIC MECHANISMS

Efforts to understand the physiological basis of heterosis have focused on the study of physiological parameters judged by their expression to the intermediate characters associated with yield (e.g., growth rate of shoots or roots, etc.) or the underlying physiological processes (e.g., absorption of minerals, production of energy in the mitochondria, ability to fix CO_2, respiration, etc.).

A number of studies, for example, of early growth during the first 2 weeks of germination demonstrated that the heterotic F_1 seedling possessed higher growth rates than the inbreds (Luckwill, 1937; Srivastava, 1981). The basic physiological activity during the early seedling growth is the formation of enzyme patterns, translocation, transformation and utilization of stored food material in the seed and then in the building-up of the active protoplasmic base for further physiological activity. At this stage, the heterotic effect is generally apparent and is visualized in the acceleration of growth rate and in the building-up of differential level of metabolism which ultimately results in quantitative differences in size, vigor, and yield. Cherry, Lund, and Earley (1960) reported increased amount of embryonic nucleotides and RNA in the hybrids and suggested that this perhaps may be the reason for faster growth and development of the hybrid. Better and profuse root systems in hybrids leading to improved absorption of mineral nutrients such as N, P and K in contrast to inbreds have also been implicated (Kiesslbach and Weightling, 1935).

Sarkissian and Huffaker (1962) reported that a barley hybrid derived from essentially isogenic parents exhibited heterosis in seedling growth and ability to fix CO_2. The ability of chloroplasts of hybrids to carry out the Hill reaction and non-cyclic phosphorylation was found to be intermediate between the activities of the parents (Miflin and Hageman, 1966). Studies with maize seedlings have shown that heterotic hybrids possess a greater rate of respiration and higher amount of reducing sugars (Sarkissian, Kessinger, and Harris, 1964) than do their parents.

Following the original demonstration by Hanson, Hageman, and Fisher (1960) that heterosis exhibited by hybrid maize seedlings is related to mitochondrial activity, a number of reports appeared (McDaniel and Sarkissian, 1966; Sarkissian and Srivastava, 1969; Hobson, 1971; McDaniel, 1972, 1973, 1974) confirming that energy production is higher in the F_1 hybrids compared with their parents due to mitochondrial complementation in the hybrid.

Hormonal basis of heterosis in an F_1 maize hybrid and its inbred parental lines has been investigated (Rood et al., 1983; Rood, Blake, and Pharis, 1983). Under controlled environmental conditions or field conditions, vegetative growth (height, internode length, leaf area, shoot dry weight and grain yield) was greater in this hybrid than in either parent. It was noted that endogenous gibberelic acid (GA)-like substances in apical meristem cylinders were higher in the hybrid than in either inbred parent. No qualitative differences in GA-like substances were apparent and no consistent differences in abscisic acid levels among the three genotypes were observed. Complementing their observations with studies involving the response of the three genotypes to exogenous GA_3, these investigators concluded that GA plays a physiological role in the expression of heterosis in maize.

Despite the progress of physiological understanding of heterosis and the possible practical significance of these observations, the genetic mechanism involved still remains obscure. The genetic questions concerning the two hypotheses have been simply translocated from the phenotypic to the physiological level (e.g., why is the hybrid fixing more CO_2 or synthesizing more hormones than the inbreds?).

One very promising concept, that of the metabolic balance, originally formulated by Mangelsdorf (1952) and elaborated later by Schrader (1985) and Rhodes et al. (1992) is analogous to the concept of limiting factors or physiological bottlenecks, where bottlenecks may reside at different loci (and hence possibly in different metabolic pathways) in different inbreds. The concept predicts that for each parental inbred should be one major bottleneck locus and potentially a series of less serious minor loci and for the hybrid to exhibit heterosis the two parental inbreds must have different bottlenecks. Measuring such metabolic distances (e.g., amino acid profiles of the parental inbreds) Rhodes et al. (1992) reported significant relationship of this parameter and hybrid yield.

MOLECULAR MECHANISMS

It has been suggested that molecular foundations of phenotype diversity reside in the variability of genome expression. Some authors have even proposed that the evolution of patterns of gene expression during development lies at the heart of organismal evolution (for a review see Atchley, 1989). The question then arises as to whether or not the variability of genetic expression also plays a role in the phenotypic diversity of the different genotypes of a plant species and even its relation to variation in different physiological processes, and finally to heterotic manifestations in crosses between such genotypes. Variability in gene expression can be assessed through the polymorphism of individual protein amounts (PAP) which is consistently detected when comparing genotypes by two-dimensional PAGE of denatured proteins, and/or through the polymorphism of individual RNA amounts (RAP), which is consistently detected when comparing genotypes by Northern analysis of RNA, dot or slot blot hybridizations of RNA and lately with reverse PCR.

The advantages of PAP have been enumerated using 5 maize lines where PAP was related to a Mahalanobis distance computed from general combining abilities of 14 morphological and developmental traits (Damerval, Hebert, and De Vienne, 1987; Leonardi et al., 1991). A nonadditive inheritance for protein amounts in hybrids of maize was also reported (Leonardi, Damerval, and De Vienne, 1987, 1988). More recently, in a study of their group, 28 maize single-cross hybrids of a diallel design between eight parental lines, were characterized for agromorphological performance in 4 different environments (Leonardi et al., 1991). PAP analysis was made from coleoptiles of the parental lines and for the 10 hybrids between 5 of the parental lines. Numerous significant correlations between PAP indices and hybrid vigor for agronomical traits were found. The correlations were mostly significant or highly significant for the height and yield characters for experiments over three years in three locations. The significant correlation between PAP and hybrid values led these researchers to conclude that loci controlling protein would themselves be quantitative trait loci (QTLs). Moreover, the PAP index is correlated to the number of non-additivity situations, which appeared also to be related to hybrid values. This correlation can not, in any case, be explained in terms of linkage disequilibrium.

Concerning the polymorphism of individual RNA amounts (RAP), the results obtained were similar to those with PAP. In one such RAP analysis (Tsaftaris and Polidoros, 1993), the expression of 35 genes in different maize tissues and in 4 developmental stages was measured. Three maize parental inbreds (B73, H108 and H109) and two of their hybrids: the highly heterotic F_1 H109 × B73 (HHF$_1$) and the low heterotic F_1 H109 × H108 (LHF$_1$) were used. The analysis was carried out by comparing the intensity of the hybridization signals, resulted after hybridizing immobilized RNA from each material/developmental stage with 35 single-copy cloned genes of maize used as probes. In accordance with the protein data, the qualitative analysis with RAP gave similar results, i.e., no large genetic dissimilarities were detected among the 5 genotypes for the number of active genes and most importantly none of these differences were correlated to agronomic performance of the different genotypes. The numbers of active genes, for a specific developmental stage, among the 3 inbreds or between HHF$_1$ and the LHF$_1$ and most importantly, between inbreds and hybrids, were not significantly different. The highest number of active genes was observed early in development, some of them becoming gradually silent at later stages. On the contrary, significant quantitative differences in the amounts of the individual RNAs for the 35 tested genes were found, among the 5 genotypes, in every developmental stage (Figure 1). This quantitative variation in the amounts of individual RNA was signifi-

FIGURE 1. The genome activity of parental lines (B73, H109, H108) and their hybrids, the high heterotic and the low heterotic ones, in 4 developmental stages, calculated after pooling the quantitative measurements of the individual mRNAs for the 35 tested gene loci for every genotype in each of 4 developmental stages.

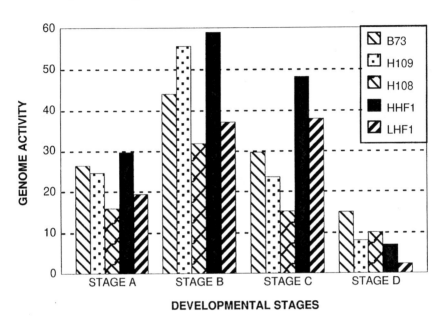

cantly correlated with agronomic performance. The genome activity, estimated as the average of the mRNA quantities for 35 gene loci, was higher for the HHF1 compared with that of the LHF₁ and the parental lines at the first three developmental stages. Similarly, significant differences were observed after comparing male parents of the two half-sib hybrids. In a more detailed analysis, the deviation in quantity of each RNA in the two hybrids from the quantity of the same mRNA in the two parental inbreds, at the same developmental stage, was examined. The heterotic hybrid had a significant number of genes expressed over the quantity of the better parent at the first three stages. Genes belonging to this category were 30% of the tested genes at the first stage and reached 63% at the third stage. The non-heterotic hybrid had only a few genes expressed over the better parent at the first and the second stage, representing no more than 15% of the tested genes, and reaching 57% at the third stage. In LHF₁ hybrid also, a significant number of genes (28% of the tested genes) expressed below the level of the inferior parent. Thus, genes of the two hybrids are differentially expressed during development.

These results are in accordance with the data of Romagnoli et al. (1990), who also conducted a RAP analysis in one heterotic hybrid in comparison with its parental inbreds. They showed differential expression of many genes, since approximately 33% of the major proteins of the hybrid, translated *in vitro* from isolated mRNA, were more abundant or possibly new, in relation to parental mRNA. They also suggested that increased expression of certain loci may be important in manifestation of vigor.

The above data with PAP and RAP indicate that variations in the expression of certain loci may be important in vigor manifestation, and underlines the significance of regulatory mechanisms of gene expression, in manifestation of heterosis. It remains then, for the regulatory mechanisms responsible for this extended variation in the amounts of mRNAs or proteins to be analyzed, their mode of action to be defined, as well as their relation to the QTLs to be followed through gene mapping approaches.

In an effort to decipher the involvement of such mechanisms underlying the regulation of gene expression and heterosis, we have analyzed different transcription factors in maize parental inbreds and their hybrids as per the hypothesis of Tsaftaris (1995) that some of the QTLs could code for regulatory proteins and are able to control a vast array of other structural genes, the products of which are necessary for the expression of complicated characters such as yield and heterosis for yield. A few of these identified proteins are all multimeric proteins with the heteropolymers exhibiting significantly different activities in comparison with the homopolymers, which is in compliance with the clear overdominance manifestation of the few QTLs analyzed so far. Our analysis involved electrophoretic separation of maize proteins and Western blotting quantitation of individual transcription factors (after denaturing SDS electrophoresis) or their multimeric complexes (after non-denaturing native electrophoresis) and making use of specific antibodies against the individual transcription factors. As shown in Figure 2, significant variation exists for the amounts of the transcription factors among the parental inbreds and the hybrids as well (E. Galani, E. Hasiotou and A.S. Tsaftaris, in preparation). In the example shown we see a negative regulation of the stress related ABA transcription factor Rab21 (Mundy and Chua, 1988), which is highly concentrated in one parental line but is present in very low amounts in the other parental line as well as in the F_1 hybrid.

Finally, in order to obtain some information concerning the nature of the possible regulatory mechanisms involved, we examined the methyla-

FIGURE 2. Western blot analysis of individual transcription factors or their multimeric complexes, after **A**. Denaturing SDS-PAGE electrophoresis of proteins extracted from dry seeds and **B**. Non-denaturing native electrophoresis of proteins extracted from dry seeds, in two parental lines and their respective hybrid, using specific antibodies.
1: Parental line 1
2: Hybrid between parent 1 and 2
3: Parental line 2

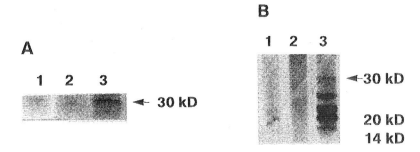

FIGURE 3. Linear regression of genome activity estimated as average mRNA quantity for the 35 cloned genes tested, on total DNA genome methylation, for 5 maize genotypes (3 parental inbreds and 2 F_1 hybrids) in 3 developmental stages.

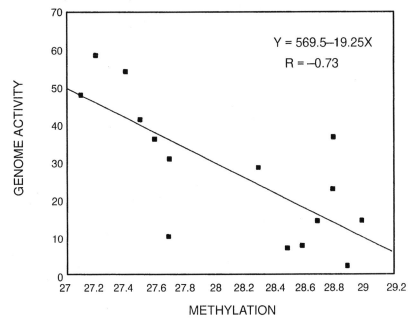

FIGURE 4. Total DNA genome methylation of maize genotypes grown under spaced and dense planting in the field. The methylation estimated by the 5-methylcytosine (5-mC) content expressed as 5-mC/(5-mC+C) × 100. **A.** At the first stage in this experiment in field-grown plants 44 days after planting **B.** At the second stage, 66 days after planting.

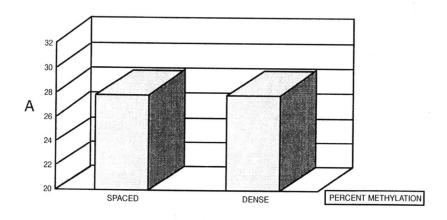

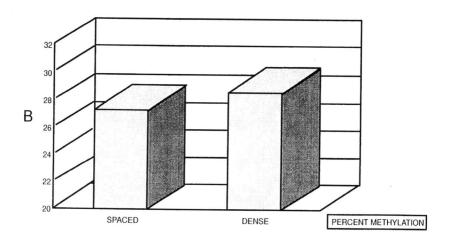

tion of DNA in the above genetic material and in different developmental stages. DNA methylation has been found in many studies to be involved in the quantitative regulation of gene expression (Yisraeli and Szyf, 1984). Tsaftaris and Polidoros (1993) found significant variation in total DNA methylation among different maize genotypes and a significant negative correlation (r = −0.739) was recorded between genome activity and DNA methylation (Figure 3). It is worth noting that DNA methylation apart from being tissue- and developmental stage dependent, was also significantly affected by the environmental conditions under which plants grow and develop (Tsaftaris and Polidoros, 1993). The environmental parameter taken into account in this work was the stress caused by the degree of density between maize plants. At 66 days after planting, there were significant differences in the level of methylation since maize plants almost always show a higher methylation content under dense planting (Figure 4). At 44 days after planting the stress of density has not been expressed yet, therefore genotypes show little or no differences at both conditions of planting (M. Kafka, A. Polidoros, I. Niopas and A.S. Tsaftaris, in preparation).

REFERENCES

Atchley, W.R. (1989). Introduction to the symposium "Evolution of Developmental Systems." *American Naturalist* 134: 437-512.

Boppenmaier, J., A.E. Melchinger, E. Brunklaus-June, H.H. Geiger and R.G. Herrmann. (1992). Genetic diversity for RFLPs in European maize inbreds. I. Relation to performance of flint × dent crosses for forage traits. *Crop Science* 32: 895-902

Cherry, J., H.A. Lund and E.B. Earley. (1960). Effect of gibberellic acid on growth and yield of corn. *Agronomy Journal* 52: 167-170

Damerval, C., Y. Hebert and D. De Vienne. (1987). Is the polymorphism of protein amounts related to phenotypic variability? A comparison of two-dimensional electrophoresis data with morphological traits in maize. *Theoretical and Applied Genetics* 74: 194-202.

Dudley J.W., M.A. Saghai-Maroof and G.K. Rufener. (1991). Molecular markers and grouping of parents in corn breeding programs. *Crop Science* 31: 718-722.

Fasoulas, A.C. ed. (1988). *The Honeycomb Methodology of Plant Breeding*. University of Thessaloniki, Greece.

Fasoulas, A.C. and R.W. Allard. (1962). Nonallelic gene interaction in the inheritance of quantitative characters in barley. *Genetics* 47: 899-907

Godshalk, E.B., M. Lee and K.R. Lamkey. (1990). Relationship of restriction fragment length polymorphisms to single-cross hybrid performance of maize. *Theoretical and Applied Genetics* 80: 273-280.

Gonella, J.A. and P.A. Peterson. (1978). Isozyme relatedness of inbred lines of maize and performance of their hybrids. *Maydica* 27: 135-149.

Hadjinov, M.I., V.S. Scherbok, N.I. Benko, V.P. Gusev, T.B. Sukhorzhevokaya and L.P. Voronova. (1982). Interrelationships between isozymic diversity and combining ability in maize lines. *Maydica* 27: 135-149.

Hanson, J.B., R.H. Hageman and M.E. Fisher. (1960). The association of carbohydrates with the mitochondria of corn scutellum. *Agronomy Journal* 52: 49-52.

Heidrich-Sobrinho, E. and A.R. Cordeiro. (1975). Codominant isozymic alleles as markers of genetic diversity correlated with heterosis in maize (*Zea mays*). *Theoretical and Applied Genetics* 46: 197-199.

Hobson G.E. (1971). A study of mitochondrial complementation in wheat. *Biochemical Journal, Proceedings of Biochemical Society* 515th Meeting, Bangkok, p. 10.

Hunter, R.B. and L.W. Kannenberg. (1971). Isozyme characterization of corn (*Zea mays*) inbreds and its relationship to single cross hybrid performance. *Canadian Journal of Genetics and Cytology* 13: 649-655.

Kiesselbach, T.A. and R.M. Weightling. (1935). The comparative root development of selfed lines of corn and their F_1 and F_2 hybrids. *Journal of the American Society of Agronomy* 27: 538-541.

Lee, M., E.B. Godshalk, K.R. Lamkey and W.W. Woodman. (1989). Association of restriction fragment length polymorphisms among maize inbreds with agronomic performance of their crosses. *Crop Science* 29: 1067-1071.

Leonardi, A., C. Damerval and D. De Vienne. (1987). Inheritance of protein amounts: comparison of two-dimensional electrophoresis patterns of leaf sheath of two maize lines (*Zea mays* L.) and their hybrids. *Genetical Research* 50: 1-5.

Leonardi, A., C. Damerval and D. De Vienne. (1988). Organ-specific variability and inheritance of maize proteins revealed by two-dimensional electrophoresis. *Genetical Research* 52: 97-103.

Leonardi, A., C. Damerval, Y. Hebert, A. Gallais and D. De Vienne. (1991). Association of protein amount polymorphism (PAP) among maize lines with performances of their hybrids. *Theoretical and Applied Genetics* 82: 552-560.

Luckwill, L.C. (1937). Studies in the inheritance of physiological characters. IV. Hybrid vigor in the tomato, part 2. Manifestation of hybrid vigor during the flowering period. *Annals of Botany New Series* (London) 1: 379-408.

Mangelsdorf, A.J. (1952). Gene interaction in heterosis. In *Heterosis,* ed. J.W. Gowen, Ames, IA: Iowa State College Press, pp. 320-329.

McDaniel, R.G. (1972). Mitochondrial heterosis and complementation as biochemical measures of yield. *Nature, New Biologist* 236: 190-191.

McDaniel, R.G. (1973). Genetic factors influencing seed vigor: Biochemistry of heterosis. *Seed Science and Technology* 1: 25-50.

McDaniel, R.G. (1974). Mitochondrial complementation: A plant-breeding tool for estimating combining ability of wheats and Triticale. *Proceedings of 4th International Wheat Genetics Symposium,* Columbia 1973, pp. 541-546.

McDaniel, R.G. and I.V. Sarkissian. (1966). Heterosis: Complementation by mitochondria. *Science* 152: 1640-1642.

Melchinger, A.E., J. Boppenmaier, B.S. Dhillon, W.G. Pollmer and R.G. Herrmann. (1992). Genetic diversity for RLFPs in European maize inbreds: II. Relation to performance of hybrids within versus between heterotic groups for forage traits. *Theoretical and Applied Genetics* 84: 672-681.

Melchinger, A.E., M. Lee, K.R. Lamkey and W.L. Woodman. (1990). Genetic diversity for restriction fragment length polymorphisms: Relation to estimated genetic effects in maize inbreds. *Crop Science* 30: 1033-1040.

Melchinger, A.E., M.M. Messmer, M. Lee, W.L. Woodman and K.R. Lamkey. (1991). Diversity and relationships among U.S. maize inbreds revealed by restriction fragment length polymorphism. *Crop Science* 31: 669-678.

Miflin, B.J. and R.H Hageman. (1966). Activity of chloroplasts isolated from maize inbreds and their F_1 hybrids. *Crop Science* 6: 185-187.

Moll, R.H., J.H. Lonquist, J. Velez and E. Johnson. (1965). The relationship of heterosis and genetic divergence in maize. *Genetics* 52: 139-144.

Mundy, J. and N.H. Chua. (1988). Absisic acid and water stress induce the expression of a novel rice gene. *The EMBO Journal* 7: 2279-2286.

Rhodes, D., G.C. Ju, W. Yang and Y. Samaras. (1992). Plant metabolism and heterosis. *Plant Breeding Reviews* 10:53-91.

Romagnoli, S., M. Maddaloni, C. Livini and M. Motto. (1990). Relationship between gene expression and hybrid vigor in primary root tips of young maize (*Zea mays* L.) plantlets. *Theoretical and Applied Genetics* 80: 767-775.

Rood, S.B., T.J. Blake and R.P. Pharis. (1983). Gibberellin and heterosis in maize. II. Response to gibberellic acid and metabolism of ^3H gibberellin A_{20}. *Plant Physiology* 71: 645-651.

Rood, S.B., R.P. Pharis, M. Koshiola and D.J. Major. (1983). Gibberellin and heterosis in maize. I. Endogenous gibberellin-like substances. *Plant Physiology* 71: 639-644.

Russell, W.A. (1971). Types of gene action at three gene loci in sub-lines of a maize inbred line. *Canadian Journal of Genetics and Cytology* 13: 322-334.

Russell, W.A. and S.A. Eberhart. (1970). Effects of three gene loci in the inheritance of quantitative characters in maize. *Crop Science* 10: 165-169.

Russell, W.A., S.A. Eberhart, A. Urbano and O. Vega. (1973). Recurrent selection for specific combining ability for yield in two maize populations. *Crop Science* 13: 257-261.

Sarkissian, V.I. and R.C. Huffaker. (1962). Depression and stimulation by chloramphenicol of development of carboxylating enzyme activity in inbred and hybrid barley. *Proceedings of the National Academy of Sciences USA* 48: 735-743.

Sarkissian, V.I., M.A. Kessinger and W. Harris. (1964). Differential rates of development of heterotic and non-heterotic young seedlings. I. Correlation of different morphological development with physiological differences in germinating seeds. *Proceedings of the National Academy of Sciences USA* 51: 212-218.

Sarkissian, V.I. and H.K. Srivastava. (1969). High efficiency, heterosis and homeostasis in mitochondria of wheat. *Proceedings of the National Academy of Sciences USA* 63: 302-309.

Sax, K. (1923). The association of size differences with seed coat pattern and pigmentation in *Phaseolus vulgaris. Genetics* 8: 552-560.

Schrader, L.E. (1985). Selection for metabolic balance in maize. In *Exploitation of Physiological and Genetic Variability to Enhance Crop Productivity,* eds. J.E. Harper, L.E. Schrader and R.W. Howell. Baltimore, MD: Waverly Press, pp. 79-89.

Smith, J.S.C. and O.S. Smith. (1989). The description and assessment of distance between inbred lines of maize. II. The utility of morphological, biochemical, and genetic descriptors and a scheme for the testing of distinctiveness between inbred lines. *Maydica* 34: 151-161.

Smith, O.S., J.S.C. Smith, S.L. Bowen, R.A. Tenborg and S.J. Wall. (1990). Similarities among a group of elite maize inbreds as measured by pedigree, F_1 grain yield, grain yield heterosis and RFLPs. *Theoretical and Applied Genetics* 80: 833-840.

Sprague, G.F. (1983). Heterosis in maize: Theory and practice. In *Heterosis: Reappraisal of Theory and Practice,* ed. R. Frankel. Berlin: Springer-Verlag.

Sprague, G.F. and P.A. Miller. (1950). A suggestion for evaluating current concepts of the genetic mechanisms of heterosis in corn. *Agronomy Journal* 42: 161-162.

Srivastava, H.K. (1981). Intergenomic interaction, heterosis and improvement of crop yield. *Advances in Agronomy* 34: 117-195.

Stuber, C.W., S.E. Lincoln, D.W. Wolff, T. Helentjaris and E.S. Lander. (1992). Identification of genetic factors contributing to heterosis in a hybrid from two elite maize inbred lines using molecular markers. *Genetics* 132: 823-839.

Tsaftaris, A.S. (1987). Isozymes in plant breeding. In *Isozymes: Current Topics in Biological and Medical Research,* Vol. 13, eds. M.C. Rattazi, J.C. Scandalios and G.S. Whitt. New York: Alan R. Liss. pp. 103-119.

Tsaftaris, A.S. (1990). Biochemical analyses of inbreds and their heterotic hybrids in maize. In *Isozymes: Structure, Function and Use in Biology and Medicine,* eds. A. Ogita and C. Markert. New York: Wiley-Liss. pp. 639-664.

Tsaftaris, A.S. (1995). Molecular aspects of heterosis in plants. *Physiologia Plantarum* 94: 362-370.

Tsaftaris, A.S. and P. Efthimiadis. (1987). F_1 heterosis and heterozygosity for isozymic structural loci in maize. In *Isozymes: Current Topics in Biological and Medical Research,* Vol. 16, eds. M.C. Rattazi, J.C. Scandalios, and G.S. Whitt. New York: Alan R. Liss. pp. 157-174.

Tsaftaris, A.S. and A.N. Polidoros. (1993). Studying the expression of genes in maize parental inbreds and their heterotic and non-heterotic hybrids. In *Proceedings of XVI Eucarpia Maize and Sorghum Conference,* eds. A. Bianci, E. Lupotto and M. Motto, Bergamo: Italy, pp. 283-292.

Walejco, R.N. and W.A. Russell. (1977). Evaluation of recurrent selection for specific combining ability in two open-pollinated maize cultivars. *Crop Science* 17: 647-651.

Yisraeli, J. and M. Szyf. (1984). Gene methylation patterns and expression. In *DNA Methylation. Biochemistry and Biological Significance,* eds. A. Razin, H. Cedar and A.D. Riggs. New York: Springer-Verlag. pp. 353-378.

SUBMITTED: 09/23/96
ACCEPTED: 12/22/96

Microsatellite Markers for Molecular Breeding

Kurt Weising
Peter Winter
Bruno Hüttel
Günter Kahl

SUMMARY. Microsatellites are tandem repeats of short sequence motifs that occur ubiquitously in eukaryotic genomes. A key feature of this class of repetitive DNA is an extraordinarily high level of variation among taxa, mainly expressed as a variable copy number of tandem repeats. A multitude of techniques were described that exploit microsatellite variability as molecular markers. Basically, these approaches can be classified into four different experimental strategies. (1) Oligonucleotides complementary to microsatellites are used as hybridization probes for multilocus RFLP fingerprinting. (2) Microsatellite-complementary oligonucleotides serve as PCR primers, either alone

Kurt Weising, Peter Winter, Bruno Hüttel, and Günter Kahl, Professor, Plant Molecular Biology, Department of Biology, Biozentrum, University of Frankfurt am Main, D-60439 Frankfurt, Germany.

Address correspondence to: Günter Kahl, Plant Molecular Biology, Department of Biology, Biozentrum, University of Frankfurt am Main, D-60439 Frankfurt, Germany (E-mail: Kahl@em.uni-frankfurt.de).

The authors dedicate this article to their colleagues at ICARDA (Aleppo, Syria). The invitation to contribute this review to the inaugural issue of the *Journal of Crop Production* is acknowledged.

Research of the authors is supported by grants from German Ministry of Technical Cooperation (BMZ grant 89.7860.3-01.130), German Research Council (DFG grants Ka 332/14.1, 14-2, 14-3) and the European Union.

[Haworth co-indexing entry note]: "Microsatellite Markers for Molecular Breeding." Weising et al. Co-published simultaneously in *Journal of Crop Production* (The Food Products Press, an imprint of The Haworth Press, Inc.) Vol. 1, No. 1 (#1), 1998, pp. 113-143; and: *Crop Sciences: Recent Advances* (ed: Amarjit S. Basra) The Food Products Press, an imprint of The Haworth Press, Inc., 1998, pp. 113-143. Single or multiple copies of this article are available for a fee from The Haworth Document Delivery Service [1-800-342-9678, 9:00 a.m. - 5:00 p.m. (EST). E-mail address: getinfo@haworth.com].

or in combination with arbitrary primers, to amplify certain regions of genomic DNA. (3) (Non)radioactively labelled microsatellite motifs are hybridized to electrophoretically resolved RAPD fragments, resulting in new and unexpected banding patterns on the autoradiograms. (4) Length variation of individual microsatellite loci is analyzed by PCR with a pair of locus-specific flanking primers. In the present review, we will summarize the principles, advantages and limitations of the different microsatellite-based marker techniques developed so far. Appropriate application areas are discussed for each type of marker, and typical results are exemplified by our own research on chickpea (*Cicer arietinum*), an important grain legume of the Eastern Mediterranean region and the Indian subcontinent. Locus-specific microsatellite analysis is currently the method of choice for creating almost optimal molecular markers, and is discussed in more detail. *[Article copies available for a fee from The Haworth Document Delivery Service: 1-800-342-9678. E-mail address: getinfo@haworth.com]*

KEYWORDS. Microsatellites, simple tandem repeats, molecular markers, DNA fingerprinting, marker technology, plant breeding, chickpea

INTRODUCTION

In the past decade, the design of molecular techniques and the development of molecular markers brought a new dimension into the traditional area of plant breeding. Especially markers based on selectively neutral DNA sequence polymorphisms made their way into the breeding business, and did so with great success. Molecular markers not only allowed the easy and reliable identification of clones, breeding lines, hybrids and cultivars, but also facilitated the monitoring of introgression, and the estimation of genetic diversity and relatedness among germplasm used in breeding programs. All the more, in breath-taking speed, advanced high-density genetic linkage maps for a series of economically important crops have been established using molecular markers. These maps in turn provide a basis for *marker-assisted selection* of agronomically useful traits, for *pyramiding of resistance genes,* and the isolation of these and other important genes by *map-based cloning* strategies (see Tanksley, Ganal, and Martin, 1995 for a review). It is safe to say that molecular marker techniques will gain more and more influence on plant breeding in the future, and will speed up breeding processes considerably.

The repertoire of molecular marker techniques has become abundant, so that a review on all aspects of marker technology would certainly be beyond the scope of this article. Instead, we would like to portray a special and highly successful marker system, the microsatellites, and list some of the benefits the breeder and plant pathologist can derive from these sequences. We will not only outline the principles, advantages and application areas, but also critically discuss the limitations of currently used microsatellite-based marker techniques. Typical results obtained by various approaches are exemplified by our own studies on chickpea (*Cicer arietinum* L.) and related species.

MICROSATELLITE-BASED MARKERS: A BRIEF SURVEY

A molecular marker can be derived from any kind of molecular data which provides a screenable polymorphism between two organisms that are to be compared. Various techniques to visualize such polymorphisms have been or are being developed (reviewed by Winter and Kahl, 1995). An ideal marker system would have to meet a number of criteria (Table 1). Though no markers are yet available which fulfill all of these criteria, one can already choose between a variety of marker systems each of which combines at least some of the properties summarized in Table 1.

In recent years, microsatellites have become the molecular marker target sequences of choice for a wide range of applications in genetic mapping and genome analysis. Microsatellites, also called "simple sequence repeats (SSRs)," "simple tandem repeats (STRs)" or "simple sequences" consist of head-to-tail tandem arrays of short DNA motifs (usually 1-5 bases). They are a common component of eukaryotic genomes but are

TABLE 1. Desirable Properties of Molecular Markers

(1)   High level of polymorphism
(2)   Codominant inheritance (discrimination of homo- and heterozygotic states)
(3)   Unambiguous designation of alleles
(4)   Frequent occurrence in the genome
(5)   Even distribution throughout the genome
(6)   Selectively neutral behaviour (no pleiotropic effects)
(7)   Easy access (no cloning)
(8)   Easy and fast assay (e.g. by procedures amenable to automation)
(9)   High reproducibility
(10) Easy exchange of data between laboratories
(11) Development at reasonable cost

almost absent from prokaryotes. Several properties make microsatellites exceptionally useful as molecular markers (for recent reviews see Weising et al., 1995a; Gupta et al., 1996; Powell, Machray, and Provan, 1996). First, microsatellites are highly variable among taxa, mainly due to a variable *n*umber of *t*andem *r*epeat (*VNTR*)-type of polymorphism. Second, microsatellites occur ubiquitously and in large numbers in all eukaryotic genomes tested so far. Third, microsatellites are more or less evenly dispersed throughout the genome. Fourth, most microsatellites are probably nonfunctional, and therefore selectively neutral.

A number of strategies have been developed to exploit microsatellite polymorphisms as molecular markers in eukaryotes. The very first approach relied on *hybridization*: as early as 1986, oligonucleotides complementary to microsatellites (e.g., [GATA]₄) were introduced as end-labelled probes for multilocus *RFLP fingerprinting* in humans by Epplen and colleagues (Ali, Müller, and Epplen, 1986). Nowadays, most microsatellite-based marker techniques rather rely on *amplification* by the polymerase chain reaction (PCR). The different principles underlying these approaches are schematically outlined in Figure 1.

In 1989, several groups introduced microsatellite amplification by locus-specific PCR (Litt and Luty, 1989; Smeets et al., 1989; Tautz, 1989; Weber and May 1989). This strategy, later called *STMS analysis* ("sequence-tagged microsatellite sites"; Beckmann and Soller, 1990) involves screening a genomic library for microsatellite repeats, sequencing of positive clones, designing primers which flank the repeats, and amplifying genomic DNA with these locus-specifc primers. Since STMS analysis involves cost and labour-intensive cloning and sequencing work, alternatives which exploit the variability of microsatellites *without* cloning steps were also looked for. In 1993, microsatellite-complementary oligonucleotides were introduced as single primers in a PCR with genomic DNA (Meyer et al., 1993). This "microsatellite-primed PCR" (*MP-PCR*) strategy yielded multilocus banding patterns resembling those obtained by the RAPD (random amplified polymorphic DNA) approach (Williams et al., 1990; Gupta et al., 1994; Weising, Atkinson, and Gardner, 1995b). In a closely related technique (*"Inter-SSR-PCR"*), Zietkiewicz, Rafalski, and Labuda (1994) used 5'-or 3'-anchored microsatellites of the dinucleotide repeat type as PCR primers (e.g., GG[CA]₇). Still another variant of the approach combined a 5'-anchored microsatellite primer with an arbitrary primer, resulting in so-called "random amplified microsatellite polymorphisms" (*RAMPs;* Wu et al., 1994). Finally, several groups found that blotting and hybridization of RAPDs to microsatellite-complementary probes resulted in unexpected, polymorphic banding patterns. This new

FIGURE 1. A schematic survey of PCR-based microsatellite marker methodologies. The different techniques mainly differ in the position and type of primer(s) used for amplification. See text for details.

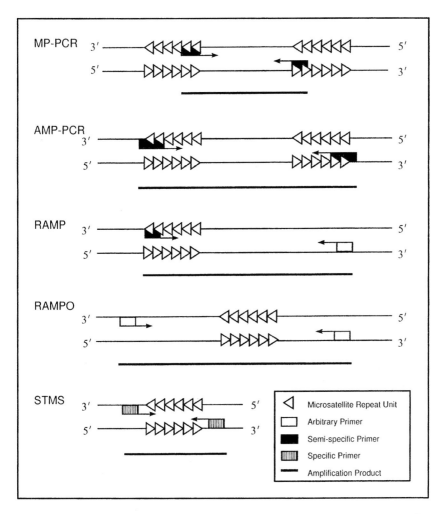

and so far untapped marker source was coined *RAMPO* (randomly amplified microsatellite polymorphisms; Richardson et al., 1995), *RAHM* (randomly amplified hybridizing microsatellites; Cifarelli, Gallitelli, and Cellini, 1995) or *RAMS* (randomly amplified microsatellites; Ender et al., 1996).

Occurrence and Frequency of Microsatellites in Plant Genomes

Before discussing the different techniques and their advantages and limitations in more detail, we will briefly review our knowledge about the distribution of microsatellite motifs in plants. Though the presence of simple repeats in wheat and barley was recognized as early as 1980 (Dennis, Gerlach, and Peacock, 1980), the following years experienced only few studies on this class of repetitive DNA sequences in plants. This situation changed remarkably, when the potential of microsatellite-complementary probes to yield highly polymorphic plant DNA fingerprints was recognized in 1989 (Weising et al., 1989), and CA- and GA-repeats were first cloned from several tropical tree species (Condit and Hubbell, 1991). Since then, the ubiquitous presence and polymorphism of simple tandem repeats in the plant genome has been amply demonstrated (see e.g., Panaud, Chen, and McCouch, 1995; Sharma et al., 1995).

Database studies revealed the interesting fact that the frequency and distribution of microsatellites in plants and animals differ considerably (Beckmann and Weber, 1992; Lagercrantz, Ellegren, and Andersson, 1993; Morgante and Olivieri, 1993; Wang et al., 1994). This is not only true for the total number of microsatellites (plant genomes are estimated to contain on an average 10 times less microsatellites than the human genome; Powell et al., 1996), but also for the relative abundance of each motif. For example, the $[CA]_n$ repeat is comparatively rare in plants while it is very frequent in animals. In plants, $[AT]_n$, $[GA]_n$, $[GAA]_n$ and $[TAA]_n$ repeats are the most common types of di- and trinucleotide repeats, respectively. Stretches of poly[A] occur in large numbers in both plant and animal genomes, probably as a consequence of retroposon activity.

Cloning experiments showed that relative and absolute microsatellite frequencies also vary within the plant kingdom, and range from one per 17-100 kb for $[GA]_n$ in several tropical tree species (Condit and Hubbell, 1991) to one per 430 kb for $[CA]_n$ in *Arabidopsis thaliana* (Bell and Ecker, 1994). Significant differences were observed between monocots, dicots and algae (Wang et al., 1994). Defining a minimum length of 20 bp, one microsatellite (of any kind) per 65 kb was found in monocotyledons, whereas one per 21 kb occurs in dicotyledons, and one per 19 kb in algae. There was no correlation with genome size observed. Microsatellites are generally rare in chloroplast and mitochondrial genomes, as might be expected from the putative prokaryotic origin of these organelles. Only four microsatellites exceeding a size of 20 bp were found in the database entries, all of which were located in chloroplast DNA (Wang et al., 1994). This frequency equals one microsatellite per 423 kb of sequenced organellar DNA. Only one type of motif occurs quite commonly in chloroplast genomes, i.e., short- to medium-sized stretches of poly[A] (Powell et al., 1995a,b).

Of course, calculations of microsatellite frequencies have to be treated cautiously, since they depend on various parameters. One important variable is the lower threshold of repeat copy number which defines a microsatellite. Moreover, the exact hybridization and washing conditions applied in cloning experiments exert a strong influence on the number of "positive" clones (Panaud, Chen, and McCouch, 1995). Frequencies of AT- and A-repeats are especially hard to estimate, since both types of repeats are refractory to cloning. Finally, calculations based on database entries may be biased by an over-representation of coding sequences. Nevertheless, the abundance of microsatellites in plant genomes is certainly high enough to provide an almost infinite number of polymorphic markers for plant breeding.

Oligonucleotide Fingerprinting

Historically, the first microsatellite-based marker technique was a derivative of RFLP analysis using microsatellite-complementary oligonucleotides as probes (Ali, Müller, and Epplen, 1986). This procedure, called "oligonucleotide fingerprinting," comprises several steps: (1) Isolation of high molecular weight genomic DNA; (2) complete digestion of the genomic DNA with an appropriate restriction enzyme; (3) electrophoretic separation of the restriction fragments in agarose gels; (4) denaturation and immobilization of the separated DNA fragments within the gel (or, alternatively, blotting onto a membrane); (5) hybridization of the dried gel (or the membrane) to (non)radioactively labelled, microsatellite-complementary oligonucleotide probes; (6) detection of hybridizing fragments (i.e., fingerprints) by autoradiography or by chemiluminescence, and documentation by photographing (see Epplen, 1992; Weising et al., 1995a; Weising and Kahl, 1996, for a detailed description of the methodology).

Oligonucleotide fingerprinting has been applied to a large number of plant species, and for many different purposes (reviewed by Weising et al., 1995a). The effects of different combinations of probe, enzyme and target DNA have been most comprehensively documented in cultivated chickpea, *Cicer arietinum* (Weising et al., 1992; Sharma et al., 1995) and in *Arabidopsis thaliana* (Depeiges et al., 1995). A total of 14 different restriction enzymes and 38 oligonucleotide probes were used to analyze four accessions of chickpea. Among 38 tested probes, 35 yielded detectable hybridization signals. The abundance and level of polymorphism of the target sequences varied considerably (see Figure 2). No obvious correlation existed between abundance, fingerprint quality, and sequence

FIGURE 2. Oligonucleotide fingerprinting of chickpea (*Cicer arietinum* L.). DNA was isolated from three randomly selected individual plants of each of four chickpea accessions (A: ILC 482; B: ILC 1272; C: ILC 1929: D: ILC 3279), digested with *Taq* I or *Dra* I and separated in 1.2% agarose gels (5 μg DNA per lane). The DNA in the gels was then denatured, neutralized, the gels dried, and consecutively hybridized to several microsatellite-complementary oligonucleotide probes. The results obtained with [GCGT]₄ and [GAAT]₄ are shown here. Note that [GCGT]₄ produces accession-specific patterns, while intra-accessional variability becomes apparent with [GAAT]₄. Positions of molecular weight markers are given in kilobases.

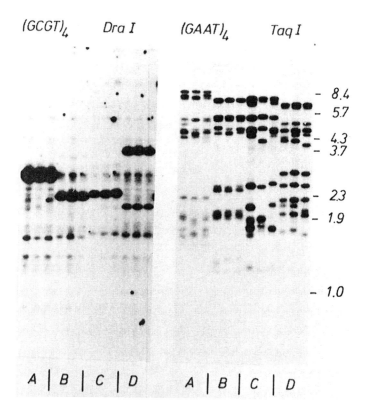

characteristics of a particular motif (Sharma et al., 1995). Depeiges et al. (1995) used oligonucleotide probes complementary to all possible microsatellites of the mono-, di-, tri- and tetranucleotide repeat types (49 in total) to screen *Eco*RI- and *Hind*III-digested *Arabidopsis* and yeast DNA by Southern analysis. Most probes did not reveal any signal or only a few bands, and only nine probes resulted in clear fingerprints with at least one of the two species, most notably [A]$_{18}$, [ATG]$_6$, [AAG]$_6$, [GTG]$_6$, [GGAT]$_4$ and [AAAC]$_4$. The scarcity of simple sequence repeats in *Arabidopsis thaliana* as revealed by oligonucleotide fingerprinting was already noted by Weising et al. (1995a) and Sharon et al. (1995). Genomes of little complexity such as those of yeast and *Arabidopsis* contain relatively fewer extended microsatellite stretches than larger genomes. Species with very large genomes, on the other hand, may harbour so many arrays that oligonucleotide fingerprinting results in a smear rather than in a clear-cut banding profile (as was observed in *Triticum aestivum* and other members of the grass family, *Pinus radiata* or *Allium cepa*; Weising et al., 1995a; Sharon et al., 1995).

The probes used in oligonucleotide fingerprinting most probably recognize microsatellite-like target sequences. However, the molecular basis of the observed polymorphisms is not yet perfectly clear. While sequenced microsatellites were generally found to be in a size range of about 20-100 bp (see below; STMS analysis), polymorphic RFLP fragments detected by microsatellite probes are often much larger (up to more than 10 kb), and their size may vary by several kilobases. Thus, variable numbers of tandemly repeated microsatellite units are probably not the only underlying cause for the observed polymorphisms. Since microsatellites are often embedded in other types of repeats (see Broun and Tanksley, 1993, 1996), minisatellites or interspersed repeats in the vicinity of the probe target sequences may play an important role in the generation of polymorphisms.

Oligonucleotide fingerprinting offers several advantages over other marker techniques, most notably its high sensitivity (down to an individual-specific level in some species), its excellent reproducibility, and the ability to rapidly screen the genome with few multilocus probes. However, there are also some obvious disadvantages, including a relatively complex laboratory protocol, the requirement of microgram amounts of DNA (as compared to nanogram amounts in PCR-based techniques), and the insufficient allelic information provided by multilocus banding patterns. The mitotic stability of oligonucleotide fingerprints made the technique highly useful for the unequivocal identification and discrimination of plant cultivars (Kaemmer et al., 1992) and micropropagated plants (Wallner et al., 1996). Substantial problems are, however, encountered when fingerprint data are applied to genetic mapping. One problem concerns mutation rate. In human minisatellites, mutation to new length alleles usually occurs at rates from 0.5-1% per gamete and generation (Jeffreys, Turner, and Debenham, 1991), but can be much higher at certain loci. Considerable proportions of nonparental (i.e., probably mutated) fingerprint bands were also observed in plants (Hüttel, 1996) and fungi (DeScenzo and Harring-ton, 1994), when microsatellite-complementary probes were used. A second problem for mapping is a certain tendency of oligonucleotide fingerprint bands to occur in clusters, unlike the much shorter PCR-detected microsatellites (see Bell and Ecker, 1994). Such clustering was observed in several organisms, including pea (Dirlewanger et al., 1994) and tomato (Arens et al., 1995; Broun and Tanksley, 1996).

In view of these problems, the unequivocal identification and discrimi-nation of genotypes (e.g., of cultivars and clones) is probably the most appropriate application area of oligonucleotide fingerprinting in plant breeding. Here, the inherent advantages of the technique such as high sensitivity and reproducibility are very important, while difficulties with allelic designation, high mutation rate and clustering of bands are of minor significance.

Microsatellite-Primed PCR with Unanchored Primers (MP-PCR; SPAR)

Microsatellite-primed PCR (*MP-PCR*) combines some elements of Alu-PCR (Sinnett et al., 1990), STMS and RAPD analysis (Williams et al., 1990). Microsatellite-complementary oligonucleotides are here *not* used as hybridization probes (as in oligonucleotide fingerprinting), but instead serve as single PCR primers. If inversely repeated microsatellites are present within an amplifiable distance from each other, the inter-repeat sequences are amplified (see Figure 1). The resulting PCR products are separated on agarose gels, and stained with ethidium bromide. MP-PCR was introduced by Meyer et al. (1993) who showed that strains and sero-types of the human fungal pathogen *Cryptococcus neoformans* could be differentiated by this approach. The technique was subsequently applied to other fungi (reviewed by Meyer and Mitchell, 1995) as well as to animals (Perring et al. 1993) and plants (Gupta et al., 1994; Weising et al., 1995b).

Using an MP-PCR variant called *SPAR* (single primer amplification reactions), Gupta et al. (1994) tested 23 primers complementary to di-, tri-, tetra- and pentanucleotide repeats for their ability to amplify genomic DNA across a panel of eukaryotes. The authors found that tetranucleotide repeat primers were most efficient in amplifying polymorphic patterns. GC- as well as AT-rich primers worked equally well. Primers representing a combination of two tetranucleotide repeats, or compound microsatel-lites, were also effective. Single base permutations were shown to yield different PCR fingerprints. Banding patterns of higher complexity were observed when the PCR was performed radioactively and products were separated on denaturing polyacrylamide gels. In particular, small-sized bands (< 300 bp) were well resolved under these conditions. Bands mapped as dominant markers in a segregating maize population. These results were confirmed by our own studies using a variety of oligonucleotides representing di-, tri-, and tetranucleotide repeats as PCR primers for the analysis of yeast, human and different plant DNAs (Weising et al., 1995b). Distinct and polymorphic banding patterns were obtained with most primers and in all investigated species including chickpea (exemplified in Figure 3). Between 1 and about 20 bands were obtained in each primer/template combination, usually in the range between 0.3 and 2 kb. However, dinucleotide repeats as well as AT-rich trinucleotide repeats often produced a smear, probably a consequence of the large copy number of this kind of microsatellites in plant DNA.

Meyer et al. (1993) stressed that MP-PCR combined some advantages of RAPD analysis (i.e., no need for sequence information), and of STMS analysis (i.e., use of high stringency annealing conditions, leading to more

FIGURE 3. Microsatellite-primed PCR analysis of intra- and interspecific variation in chickpea. Genomic DNA from individual plants was amplified using the unanchored trinucleotide repeats [CCG]$_5$, [CGA]$_5$ and [CAG]$_5$ as single primers. PCR products were separated on 1.5% agarose gels and stained with ethidium bromide. Lane 1: *C. arietinum* ILC 482; lanes 2-3: *C. arietinum* ILC 1272; lanes 4-5; *C. arietinum* ILC 3279; lane 6: *C. arietinum* ILC 1929; lane 7: *C. reticulatum* ILWC 123; lane 8: *C. reticulatum* ILWC 137; lane 9: *C. echinospermum* ILWC 179. Only low levels of variability are detected. Positions of molecular weight markers are given in kilobases.

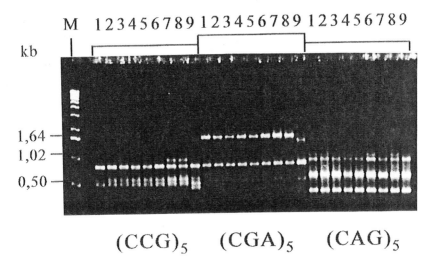

reproducible banding patterns). However, the validity of this statement was challenged by our own results (Weising et al., 1995b). In terms of reproducibility and sensitivity to reaction conditions, we found no significant advantage of MP-PCR as compared to RAPD analysis. Instead, the majority of MP-PCR bands were found to be generated by mismatch annealing of primers, indicating a higher similarity of MP-PCR to RAPD analysis than previously anticipated.

Microsatellite-Primed PCR with Anchored Primers (AMP-PCR; INTER-SSR-PCR)

In a more sophisticated variant of the MP-PCR technique, 5'- or 3'-anchored di- or trinucleotide repeats serve as single PCR primers, the amplification products are separated on polyacrylamide gels and banding patterns are revealed by autoradiography (Zietkiewicz, Rafalski, and Labuda, 1994; see Figure 1). In this approach called *Inter-SSR-PCR* or *AMP-PCR* (= anchored microsatellite-primed PCR), complex fingerprint-like patterns were obtained that revealed inter- as well as intraspecific polymorphisms from a wide variety of eukaryotic taxa (Zietkiewicz, Rafalski, and Labuda, 1994; Kostia et al., 1995; Wolff, Zietkiewicz, and Hofstra, 1995; Hüttel, 1996). In our own experiments on chickpea (Figure 4), we found banding pattern complexity as well as informativeness of AMP-PCR to be considerably higher as compared to MP-PCR. While MP-PCR usually resulted in species-specific banding, AMP-PCR also detected some intraspecific polymorphisms (Hüttel, 1996; Figures 3 and 4).

AMP-PCR has so far been applied to the identification of chrysanthemum (Wolff et al., 1995) and rapeseed cultivars (Charters et al., 1996), the assessment of genetic diversity among dent and popcorn inbred lines (Kantety et al., 1995), and the analysis of the genome orgins of finger millet, *Eleusine coracana* (Salimath et al., 1995). In the latter study, RFLPs, RAPDs and AMP-PCR were compared for their relative efficiency in detecting polymorphisms, and AMP-PCR proved to be most informative. In an extensive study on Douglas fir (*Pseudotsuga menziesii*) and sugi (*Cryptomeria japonica*), Tsumura, Ohba, and Strauss (1996) compared the efficiencies of 96 different microsatellite primers of various lengths and sequences, including dinucleotide repeats anchored at their 5' and 3' ends as well as unanchored tri-, tetra- and pentanucleotide repeats. More than 60% of the primers gave interpretable banding patterns after separation on agarose gels and staining with ethidium bromide. Anchored dinucleotide repeats based on GA and GT motifs produced the most useful banding profiles and were analyzed in more detail. Inheritance analysis

FIGURE 4. Anchored microsatellite-primed PCR analysis of intra- and interspecific variation in chickpea. Genomic DNA from individual plants was amplified using the oligonucleotides indicated below the figure as single primers. All primers consist of dinucleotide repeats carrying a degenerate "anchor" at their 5' ends. PCR products were separated on 1.5% agarose gels and stained with ethidium bromide. Lane 1-6: *C. arietinum* (lane 1: ILC 191; lane 2: ILC 200; lane 3: ILC 482; lane 4: ILC 1272; lane 5: ILC 1929; lane 6: ILC 3279); lane 7: *C. reticulatum* ILWC 36; lane 8: *C. echinospermum* ILWC 35. The level of intra- and interspecific variability is higher as compared to PCR with unanchored primers (see Figure 3), but lower as compared to oligonucleotide fingerprinting (Figure 2). Positions of molecular weight markers (in bp) are indicated by arrows. B = C, G or T; D = A, G or T; V = A, C or G.

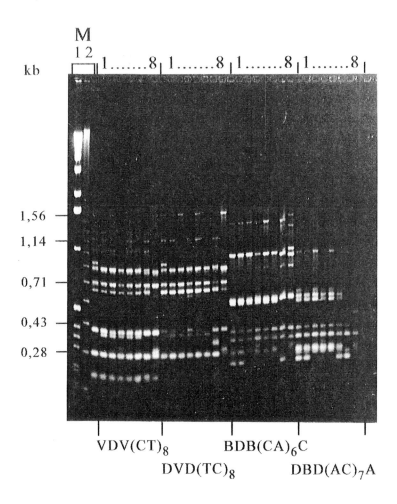

showed that all bands behaved as dominant markers (presence versus absence), and that 96% of all bands were inherited according to Mendelian expectations. RAPD markers, in contrast, usually show much larger extents of segregation distortion in conifers (e.g., more than 50% in *Picea abies*; Bucci and Menozzi, 1993).

Taken together, AMP-PCR has several advantages over the unanchored variants of MP-PCR described in the previous section. First, the way how primers are designed assures annealing of the primer only to the ends of a microsatellite, thus obviating internal priming and smear formation. Second, the anchor allows only a subset of the targeted inter-repeat regions to be amplified, thereby reducing the high number of PCR products expected from priming of dinucleotide inter-repeat regions to a set of about 10-50 easily resolvable bands. Pattern complexity can be tailored by applying different primer lengths and sequences. Third, the use of 5' anchors ensures that the targeted microsatellite is part of the product. Potential VNTR polymorphisms within the microsatellite will then contribute to the inter-repeat variation, which could considerably enhance the chance to observe a polymorphism.

Random Amplified Polymorphic Microsatellites (RAMP)

In a technique designated as *RAMP* (random amplified polymorphic microsatellites), Wu et al. (1994) used combinations of 5'-anchored microsatellite primers and arbitrary 10mer primers (i.e., RAPD primers; see Figure 1) to obtain screenable microsatellite polymorphisms from different *Arabidopsis thaliana* strains and ecotypes. The PCR program was designed to switch between high and low annealing temperatures, so that microsatellite loci were preferentially amplified. PCR products were separated on denaturing polyacrylamide gels. Since only the microsatellite-complementary primer was radiolabelled with ^{33}P-dATP, autoradiograms showed microsatellite-derived bands only. A considerable number of polymorphic bands were observed (10-30 polymorphic fragments per primer; 2-7 alleles in 11 different *Arabidopsis* ecotypes). The majority of these fragments (67 out of 104) behaved as codominant markers when tested in a segregating population (Wu et al., 1994).

The RAMP technique was also exploited for genetic mapping in barley. Becker and Heun (1995b) used 5'-anchored, labelled [GA]$_n$ primers in combination with 10-mer, 16-mer and 20-mer RAPD primers. The longer than usual RAPD primers were chosen to ensure comparable annealing temperatures for both primers, thus circumventing the necessity for the thermally asymmetric PCR profile used by Wu et al. (1994). To obtain additional polymorphisms, aliquots of the amplification products were digested with a restriction enzyme (*Mse*I), resulting in so-called dRAMPs. There were 0-11 polymorphisms per primer combination (mainly presence/absence polymorphisms). A total of 10 primer combinations resulted in 43 RAMPs and 17 dRAMPs that identified 40 new loci on a barley RFLP map. Mapping also demonstrated that some of the dRAMPs were actually derived from RAMPs. Only 7 loci defined by dRAMPs were unique, i.e., digestion of RAMP products was of no considerable advantage. More recently, RAMPs were also used to study the genetic relationships between barley cultivars (Sanchez de la Hoz et al., 1996). Silver-staining was applied in this study to visualize the electrophoresed PCR products. Staining certainly is technically more simple, but the omission of radioactive label led to banding profiles that included also RAPD fragments. Banding patterns were evaluated phenetically. Interestingly, dendrograms solely based on RAMP markers reflected the known pedigrees of cultivars much better than dendrograms based on RAPDs.

As compared to *Arabidopsis* and barley, only few intraspecific polymorphisms were detected in our own RAMP experiments on chickpea (Hüttel, 1996; Figure 5). Only few additional bands and polymorphisms could be obtained by inclusion of RAPD primers in an otherwise similar AMP-PCR reaction. Considering the relatively tricky experimental protocol and the poor yield of polymorphic bands, we no longer use the RAMP technique for generating molecular markers in chickpea.

Hybridization of Microsatellite Probes to RAPD or MP-PCR Fragments: The RAMPO Technique

Recently, a novel strategy was developed that combines arbitrarily or microsatellite-primed PCR with microsatellite hybridization (Cifarelli, Gallitelli, and Cellini, 1995; Richardson et al., 1995; Ender et al., 1996). In the first step of the procedure, genomic DNA is amplified with either a single arbitrary 10mer primer (as usually used in RAPD analysis), or a microsatellite-complementary 15- or 16mer primer (Gupta et al., 1994; Weising et al., 1995b). PCR products are then electrophoresed, photographed, blotted, and hybridized to a ^{32}P- or digoxigenin-labelled mono-, di-, tri- or tetranucleotide repeat probe such as $[A]_{16}$, $[GA]_8$, $[CAA]_5$ or $[GATA]_4$. Subsequent autoradiography reveals reproducible, probe-dependent fingerprints which are completely different from the ethidium bromide staining patterns, and which are usually polymorphic at an intraspecific level. This method was coined *RAMPO* (for *r*andom *a*mplified *m*icrosatellite *p*olymorphisms; Richardson et al., 1995), *RAHM* (for *r*andom *a*mplified *h*ybridization *m*icrosatellites; Cifarelli, Gallitelli, and Cellini, 1995) or *RAMS* (for *r*andomly *a*mplified *m*icrosatellites; Ender et al., 1996).

FIGURE 5. RAMP analysis of chickpea. Genomic DNA samples from individual plants of the chickpea (*Cicer arietinum*) accessions C-104 (1) and WR-315 (2) were amplified using the anchored microsatellite primer CAA[CT]$_6$ either singly or in combination with various RAPD primers (A = OPAQ-02; B = OPAQ-07; C = OPAQ-09; D = OPAQ-13; E = OPAQ-14; F = OPAQ-17; G = OPAQ-18; H = OPM-01; all from Operon Technologies, Alameda, USA). The anchored microsatellite primer was ^{33}P-labeled by T4 polynucleotide kinase. PCR products were separated on a denaturing 6% polyacrylamide gel and autoradiographed. Bands that are visible throughout *all* lanes are derived from the microsatellite primer only, bands specific for either A, B, C, D, E, F, G or H are derived from mixed priming. Only few polymorphisms between the two investigated chickpea accessions are detected. Positions of molecular weight markers given in bp.

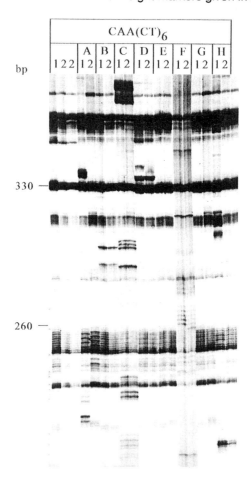

The occurrence of RAMPO bands may be explained as follows: Any RAPD or MP-PCR reaction probably creates many thousand different products of varying abundance. The majority of less abundant fragments will remain below the detection level of ethidium bromide staining. However, the ubiquitous presence of microsatellites in eukaryotic genomes provides a means of visualizing a subset of such minor amplification products by hybridization. The signal intensity of fragments harboring a certain microsatellite motif will depend both on the length of this motif and the abundance of the fragment. Hybridizing bands may be isolated, cloned, sequenced and either used as a probe on RAPD or RFLP gels (Cifarelli, Gallitelli, and Cellini, 1995), or flanking sequences exploited for the generation of STMS primers (Ender et al., 1996). Cloned RAMPO fragments generally contained microsatellites, which are often polymorphic within and between species. For example, one RAHM band detected with a $[CC]_{12}$ probe contained a short, interrupted $[C]_n$ motif which, when used as a probe, revealed a presence/absence type of polymorphism among different sugar beet cultivars (Cifarelli, Gallitelli, and Cellini, 1995).

An important advantage of this novel procedure is that the first step is PCR-based, and therefore applicable for studies where only nanogram amounts of DNA are available (as in small organisms, or endangered species, where nondestructive sampling is required). The RAMPO approach was extensively tested with different species of yams (Richardson et al., 1995; Ramser et al., submitted), olive, sugarbeet, and sunflower (Cifarelli, Gallitelli, and Cellini, 1995), but most likely works with all plant species. Since a similar strategy produced polymorphic fragments among different *Daphnia* species (RAMS; Ender et al., 1996), the technique is probably applicable to most if not all eukaryotes.

Locus-Specific Microsatellite Analysis (STMS Analysis)

All microsatellite-based marker techniques outlined so far share two main disadvantages: (1) With few exceptions (e.g., RAMPs; Wu et al., 1994), they generally yield *dominant* types of markers; homo- and heterozygous states are therefore hard or impossible to distinguish; (2) they generally produce *multilocus* markers; individual loci are therefore hard to identify. PCR amplification of individual microsatellites does not suffer from these disadvantages. Instead, locus-specific, codominant markers are obtained (see also Conclusions). Locus-specific microsatellite analysis involves screening a genomic library for microsatellite repeats, sequencing of "positive" clones, designing primers which flank the repeats, and amplifying genomic DNA with these primers (see Figure 1). The size of the resulting amplification products often varies by integral numbers of the basic repeat unit. To detect a polymorphism, PCR products are usually radiolabelled, separated on denaturing polyacrylamide gels, and detected by autoradiography. Fragments with size differences >4 bp can also be resolved on agarose gels, thus avoiding radioactivity (see below; Figures 6 and 7). Several acronyms have been used to describe this technique. In accordance with Beckmann and Soller (1990), we here refer to the technique as *STMS* (sequence-tagged microsatellite sites) analysis. Nowadays, STMS analysis is certainly the most popular way to exploit microsatellite polymorphisms for genetic studies, combining the high informativeness of microsatellite loci with the ease and speed of the PCR to obtain codominant markers.

The successful application of locus-specific amplification of tandem repeats was actually first demonstrated for minisatellites. In 1988, Alec Jeffreys and colleagues amplified several human minisatellite loci with the help of flanking primers (Jeffreys et al., 1988). One year later, four groups independently applied the same approach to simple tandem repeats, mainly of the $[CA]_n$ type (Litt and Luty, 1989; Smeets et al., 1989; Tautz, 1989; Weber and May, 1989). It was soon realized that these "microsatellites" (this term was created by Litt and Luty, 1989) are more suitable for PCR amplification than minisatellites. First, they are short (typically between 20 and 40, rarely up to 200 bp) and therefore easy to amplify (minisatellite arrays often turned out to be too long, i.e., 0.5 to 30 kb, for efficient amplification). Second, stretches of simple sequences are more abundant, and more evenly distributed throughout the genome as compared to minisatellites.

Weber (1990) was the first to investigate the informativeness of microsatellites of the $[CA]_n$ type. For the human genome, he showed that the level of polymorphism detected by PCR-amplified microsatellites depends on the number of the "perfect" (i.e., uninterrupted), tandemly repeated motifs. Below a certain threshold (i.e., 12 CA-repeats in his investigation), the microsatellites behaved mainly monomorphic. Above this threshold, however, the probability of polymorphism increased with microsatellite length. These findings were later supported by numerous other studies (also in plants), and a positive correlation of the repeat copy number with the number of different alleles was generally observed. Consequently, long, perfect arrays of microsatellites are preferred for the generation of markers, i.e., for the design and synthesis of flanking primers.

The primers used for the first generation of experiments were deduced from DNA data bases (e.g., Akkaya, Bhagwat, and Cregan, 1992). For the vast majority of species, however, data base entries are either limited or

FIGURE 6. STMS analysis of chickpea I: Initial test of locus-specific primer pairs. Genomic DNA samples of individual plants from four different accessions of *Cicer arietinum* (1-4), and one accession each of *C. reticulatum* (5) and *Pisum sativum* (6), were amplified with five different microsatellite-flanking primer pairs (CaSTS5 to CaSTS10). PCR products were separated in 2.0% agarose gels and stained with ethidium bromide. Lane 7 is a control lane (no template DNA). Amplification of chickpea DNA is observed with all primer pairs, CaSTS8 and CaSTS9 also produce a product with pea DNA. Some primer pairs reveal allelic size differences already on low resolution agarose gels. The presence of a single band within each lane is consistent with the highly inbred, homozygous state of chickpea accessions. Positions of molecular weight markers are given in bp.

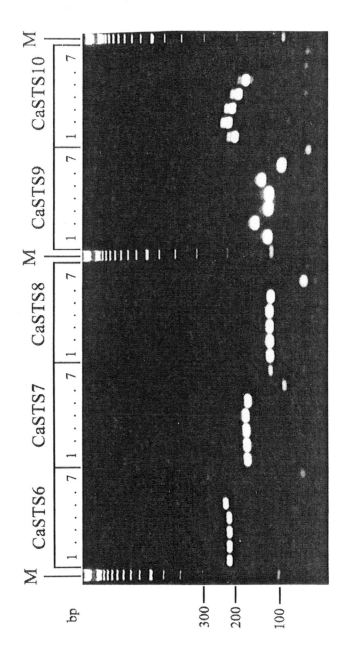

FIGURE 7. STMS analysis of chickpea II: Allelic variation in accessions originating from different countries. Genomic DNA samples of individual plants were amplified with the microsatellite-flanking primer pair CaSTS15. PCR products were separated in 2.0% agarose gels and stained with ethidium bromide. Only few samples are heterozygous (e.g., sample 3 from Iraq). Positions of molecular weight markers are given in bp.

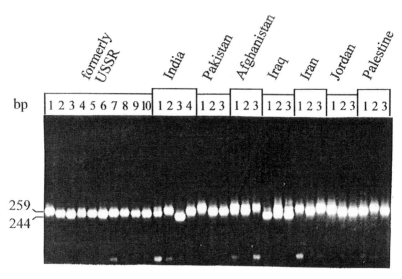

CaSTS15

even non-existing, and cloning is necessary to acquire sequence data which are needed for the design of flanking primers. Typically, size-selected genomic libraries (200 to 500 bp) are constructed by the following steps: (1) Isolation of genomic DNA; (2) digestion with one or more 4 base-specific restriction enzymes; (3) size-selection of restriction fragments by agarose gel electrophoresis, excision and purification of the desired size fraction; (4) ligation of the DNA into a suitable vector and transformation into a suitable *E. coli* strain; (5) screening for the presence of microsatellites by colony or plaque hybridization with a labelled probe; (6) isolation of positive clones and sequencing of the inserts; (7) design of suitable primers flanking the microsatellite repeat.

Establishing libraries with small, size-selected inserts is advantageous for microsatellite cloning for two reasons: (1) very long microsatellites are unstable in *E. coli,* and (2) positive clones can be sequenced without subcloning. A number of strategies were developed for the enrichment of microsatellites in genomic libraries. Such enrichment procedures are especially useful if libraries are screened with comparatively rare tri- and tetranucleotide repeat motifs. An efficient but somewhat cumbersome approach was presented by Ostrander et al. (1992) who created a small-insert phagemid library in an *E. coli* strain deficient in UTPase (*dut*) and uracil-N-glycosylase (*ung*) genes. As a consequence of the absence of the encoded enzymes, dUTP can efficiently compete with dTTP for the incorporation into DNA. Single-stranded phagemid DNA was isolated from such a library, second-strand synthesis was primed with $[CA]_n$ and $[TG]_n$ primers, and the products were used to transform a wild-type *E. coli* strain. Since there was a strong selection against single-stranded, uracil-containing DNA molecules, the resulting library primarily consisted of primer-extended, double-stranded products, resulting in an about 50-fold enrichment in CA-repeats.

Other enrichment strategies rely on hybridization selection of simple repeats prior to cloning (Karagyozov, Kalcheva, and Chapman, 1993; Armour et al., 1994; Kijas et al., 1994; Kandpal, Kandpal, and Weissman, 1994; Edwards et al., 1996). Hybridization selection usually involves the following steps: (1) Genomic DNA is fragmented, either by sonication, or by digestion with a restriction enzyme. (2) Genomic DNA fragments are ligated to adaptors that allow to perform a "whole genome PCR" at this or a later stage of the procedure. (3) Genomic DNA fragments are amplified, denatured and hybridized with single-stranded microsatellite sequences bound to a nylon membrane. (4) After washing off unbound DNA, hybridizing fragments enriched for microsatellites are eluted from the membrane by boiling or alkali treatment, reamplified using adaptor-complementary primers, and digested with a restriction enzyme to remove the adaptors. (5) DNA fragments are ligated into a suitable vector and transformed into a suitable *E. coli* strain. Random sequencing indicated the presence of a microsatellite in 50-70% of the clones obtained from these procedures (Armour et al., 1994; Edwards et al., 1996). A somewhat different hybridization selection strategy was described by Kijas et al. (1994) who replaced the nylon membrane with biotinylated, microsatellite-complementary oligonucleotides attached to streptavidin-coated magnetic particles. Microsatellite-containing DNA fragments are selectively bound to the magnetic beads, reamplified, restriction-digested and cloned. In this way, Kijas et al.

(1994) obtained a library from *Citrus* which was highly enriched for TAA-repeats.

PCR-amplified microsatellites are very powerful markers, because they are locus-specific, codominant, occur in large numbers and allow the unambiguous identification of alleles. However, although the advantages prevail, there are also some limitations. Most importantly, STMS analysis requires cloning and sequencing and is therefore very cost- and labour-intensive. Another main obstacle is the tediousness of the genotyping procedure. Standard protocols use radioisotopes and denaturing polyacrylamide gels to detect the amplified microsatellites. In many situations, however, allele sizes are sufficiently different to be resolved on high percentage agarose gels in combination with ethidium bromide staining (e.g., Bell and Ecker, 1994; Becker and Heun, 1995a; Hüttel, 1996; see Figures 6 and 7). High resolution without applying radioactivity is also provided by nondenaturing polyacrylamide gels in combination with either ethidium bromide (Scrimshaw, 1992) or silver staining (Klinkicht and Tautz, 1992; Neilan et al., 1994). A very promising alternative of STMS typing involves the use of fluorescent primers in combination with a semi-automated DNA sequencer (Ziegle et al., 1992; Schwengel et al., 1994). If an appropriate instrumentation is available, fluorescent PCR products can be detected by real-time laser scanning during gel electrophoresis. A major advantage of this technology is that different amplification reactions as well as a size marker (each labelled with a different fluorophore) can be combined into one lane during electrophoresis. Multiplex analysis of up to 24 different microsatellite loci per lane were already demonstrated (Schwengel et al., 1994). Although high throughputs are therefore possible, the main disadvantage of the technique is that expensive equipment is required.

Another disadvantage of the STMS technique is the commonly observed "stuttering" of bands. Instead of yielding one particular band, the enzymatic amplification of dinucleotide repeats commonly results in a cluster of "shadow bands" which are separated from each other by two basepair intervals. The additional bands are most probably the result of slippage events, which occur during replication by the *Taq* polymerase (Hauge and Litt, 1993). This artifact sometimes makes the interpretation of correct allele sizes difficult, especially if two alleles differ by two base pairs only. Ordering of alleles according to size may help assigning allelic states correctly (Saghai-Maroof et al., 1994). Slippage is less severe, and amplimers are more clearly resolved if microsatellites with tri- and tetrameric repeat units are amplified (Edwards et al., 1991; Kijas, Fowler, and Thomas, 1995).

Molecular Breeding with STMS Markers:
Genetic Mapping and Variety Protection

The first locus-specific microsatellite analysis in plants was published by Akkaya, Bhagwat, and Cregan (1992). The authors analyzed Gene Bank-derived AT- and ATT-repeats from soybean, and found between 6 and 8 alleles, depending on the locus. Since these initial studies, an ever increasing number of papers on PCR-amplified microsatellites in plants have appeared, and locus-specific microsatellite analysis is likely to be underway for all major crop plants. STMS analysis was used for many different applications, mainly in the areas of plant breeding and population genetics (for review, see Gupta et al., 1996; Powell, Machray, and Provan, 1996). We will focus here on the potential of STMS markers for genetic mapping and genotype identification.

The huge number of polymorphic loci combined with the even dispersal of microsatellites throughout the genome makes STMS analysis a particularly efficient tool for *genetic mapping*. For humans and some animal species, high resolution genetic maps were established that mainly, or even solely consist of thousands of PCR-amplified microsatellites (Jacob et al., 1995; Dib et al., 1996; Dietrich et al. 1996). The most comprehensive map of the mouse genome now contains 7,377 markers at an average spacing of 0.2 cM, which is equivalent to about 400 kb (Dietrich et al. 1996). In the near future, this extreme marker density will allow "chromosome landing," i.e., the direct isolation of genes from YAC libraries without any "walking" steps (Tanksley, Ganal, and Martin, 1995). In plants, genetic mapping with STMS markers is still lagging behind. Nevertheless, first attempts to assign microsatellite loci to linkage groups were reported from rice (Wu and Tanksley, 1993), *Arabidopsis* (Bell and Ecker, 1994), soybean (Morgante et al., 1994; Akkaya et al., 1995), barley (Becker and Heun, 1995a,b), maize (Taramino and Tingey, 1996), *Pinus radiata* (Devey et al., 1996) and tomato (Broun and Tanksley, 1996), and rapid map saturation may be expected at least in major crop plants such as maize, rice, soybean and tomato. Tight linkage of a microsatellite with a useful trait (the soybean mosaic virus resistance locus *Rsv1*) was already demonstrated for soybean (Yu et al., 1994; Yu, Saghai-Maroof, and Buss, 1996). Such "resistance markers" will provide invaluable tools for marker-assisted selection in breeding programs.

The potential of STMS markers for *genotype identification* shall be illustrated by two examples. In one study on soybean, most of the 96 different genotypes investigated could be discriminated by a set of 7 loci exhibiting 11-26 alleles per locus (Rongwen et al., 1995). It was hypothesized that 10 or 15 loci may be adequate to distinguish closely related genotypes, and produce unique DNA profiles suitable for variety identification and protection in soybean. In grapevine, a *cultivar identification system* was set up based on STMS loci analyzed with fluorescent primers and the GENESCAN software (Thomas, Cain, and Scott, 1994). This approach allowed the unequivocal identification of cultivars, paternity testing as well as the identification of duplicates using an electronic database. For variety identification, STMS markers are generally superior to RAPDs, MP-PCR and oligonucleotide fingerprinting, because alleles and genotypes can be assigned unambiguously, and data (i.e., primer sequences) can easily be distributed and reproduced among different laboratories.

CONCLUSIONS

Among the different microsatellite-based markers currently available, STMS markers come very close to the ideal molecular marker system. They combine most (but not all!) of the desired marker properties summarized in Table 1:

1. STMS markers are *highly polymorphic* and hence highly informative. In a recent paper on barley, an $[AT]_n$ microsatellite was shown to have 37 different alleles in a sample of about 200 accessions (Saghai-Maroof et al., 1994). We found 36 alleles of one particular locus in a limited set of kiwifruit accessions (Weising et al., 1996). This allelic variability is much higher than of any other marker currently available.
2. STMS markers are inherited in a *codominant* fashion. This allows one to discriminate between homo- and heterozygous states, and increases the efficiency of mapping and population genetic studies.
3. The use of sequencing gels for the separation of PCR-generated microsatellites allows the *unambiguous designation of alleles,* and thus a precise calculation of allele frequencies.
4. Microsatellites occur *ubiquitously* and *abundantly* in eukaryotic genomes.
5. Microsatellites are more or less *evenly dispersed* throughout the genome. This has been most comprehensively demonstrated in man and mouse, where 5,264 and 7,377 (mainly STMS) markers, respectively, have been placed on a map (Dib et al., 1996; Dietrich et al., 1996). Chromosomes were evenly covered by microsatellite markers in both species, though some local clustering occurred.
6. Since only a few microsatellites are transcribed, the majority of STMS markers will probably behave *selectively neutral.*

7. Provided that the adequate primers are available, typing assays via PCR are *fast and easy*, require only nanogram amounts of template DNA, and are amenable to automation.
8. The use of stringent PCR conditions guarantees *high reproducibility.*
9. Information about primer sequences can *easily be exchanged between laboratories.*

As outlined in a previous section, the main limitation of STMS analysis is the high input in terms of cost and labour that is required for the identification of informative microsatellite loci. This will certainly limit the large-scale use of STMS markers to a few important crops. However, the situation may improve considerably if primer binding sites and polymorphic microsatellite loci are conserved among different taxa. Thus STMS markers identified in one species can also be used in related taxa. In animals, studies in this direction have already yielded encouraging results. For example, primer binding sites were conserved among several species of whale (Schlötterer, Amos, and Tautz, 1991), artiodactyls (Moore et al., 1991; Pépin et al., 1995) and primates (Deka et al., 1994). In plants, the few studies performed until now indicate that microsatellite flanking sequences are at least conserved within genera [e.g. *Brassica* (Lagercrantz et al., 1993; Kresovich et al., 1995), *Citrus* (Kijas, Fowler, and Thomas, 1995), *Actinidia* (Weising et al., 1996) and chickpea (Hüttel, 1996, see Figure 6]. Additional studies on the conservation of microsatellites between taxa are clearly needed, and would also improve our knowledge about the evolution of this useful class of repetitive DNA.

REFERENCES

Akkaya, M.S., A.A. Bhagwat and P.B. Cregan. (1992). Length polymorphisms of simple sequence repeat DNA in soybean. *Genetics* 132: 1131-1139.

Akkaya, M.S., R.C. Shoemaker, J.E. Specht, A.A. Bhagwat and P.B. Cregan. (1995). Integration of simple sequence repeat DNA markers into a soybean linkage map. *Crop Science* 35: 1439-1445.

Ali, S., C.R. Müller and J.T. Epplen. (1986). DNA fingerprinting by oligonucleotide probes specific for simple repeats. *Human Genetics* 74: 239-243

Arens, P., P. Odinot, A.W. Van Heusden, P. Lindhout and B. Vosman. (1995). GATA- and GACA-repeats are not evenly distributed throughout the tomato genome. *Genome* 38: 84-90.

Armour, J.A.L., R. Neumann, S. Gobert and A.J. Jeffreys. (1994). Isolation of human simple repeat loci by hybridization selection. *Human and Molecular Genetics* 3: 599-605.

Becker, J. and M. Heun. (1995a). Barley microsatellites: allele variation and mapping. *Plant Molecular Biology* 27: 835-845.

Becker, J. and M. Heun. (1995b). Mapping of digested and undigested random amplified microsatellite polymorphisms in barley. *Genome* 38: 991-998.

Beckmann, J.S. and M. Soller. (1990). Toward a unified approach to genetic mapping of eukaryotes based on sequence tagged microsatellite sites. *Bio/ Technology* 8: 930-932.

Beckmann, J.S. and J.L. Weber. (1992). Survey of human and rat microsatellites. *Genomics* 12: 627-631.

Bell, C.J. and J.R. Ecker. (1994). Assignment of 30 microsatellite loci to the linkage map of *Arabidopsis. Genomics* 19: 137-144.

Broun, P. and S.D. Tanksley. (1993). Characterization of tomato clones with sequence similarity to human minisatellites 33.6 and 33.15. *Plant Molecular Biology* 23: 231-242.

Broun, P. and S.D. Tanksley. (1996). Characterization and genetic mapping of simple repeat sequences in the tomato genome. *Molecular and General Genetics* 250: 39-49.

Bucci, G. and P. Menozzi. (1993) Segregation analysis of random amplified polymorphic DNA (RAPD) markers in *Picea abies* Karst. *Molecular Ecology* 2: 227-232.

Charters, Y.M., A. Robertson, M.J. Wilkinson and G. Ramsay. (1996). PCR analysis of oilseed rape cultivars (*Brassica napus* L. ssp. *oleifera*) using 5'-anchored simple sequence repeat (SSR) primers. *Theoretical and Applied Genetics* 92: 442-447.

Cifarelli, R.A., M. Gallitelli and F. Cellini. (1995). Random amplified hybridization microsatellites (RAHM): isolation of a new class of microsatellite-containing DNA clones. *Nucleic Acids Research* 23: 3802-3803.

Condit, R. and S.P. Hubbell. (1991). Abundance and DNA sequence of two-base repeat regions in tropical tree genomes. *Genome* 34: 6-71.

Deka, R., M.D. Shriver, L.M. Yu, C.E. Aston, R. Chakraborty and R.E. Ferrell. (1994). Conservation of human chromosome 13 polymorphic microsatellite $(CA)_n$ repeats in chimpanzees. *Genomics* 22: 226-230.

Dennis, E.S., W.L. Gerlach and W.J. Peacock. (1980). Identical polypyrimidine-polypurine satellite DNAs in wheat and barley. *Heredity* 44: 349-366.

Depeiges, A., C. Goubely, A. Lenoir, S. Cocherel, G. Picard, M. Raynal, F. Grellet and M. Delseny. (1995). Identification of the most represented repeated motifs in *Arabidopsis thaliana* microsatellite loci. *Theoretical and Applied Genetics* 91: 160-168.

DeScenzo, R.A. and T.C. Harrington. (1994). Use of $(CAT)_5$ as a DNA fingerprinting probe for fungi. *Phytopathology* 84: 534-540.

Devey, M.E., J.C. Bell, D.N. Smith, D.B. Neale and G.F. Moran. (1996). A genetic linkage map for *Pinus radiata* based on RFLP, RAPD, and microsatellite markers. *Theoretical and Applied Genetics* 92: 673-679.

Dib, C., S. Fauré, C. Fizames, D. Samson, N. Drouot, A. Vignal, P. Millasseau, S. Marc, J. Hazan, E. Seboun, M. Lathrop, G. Gyapay, J. Morissette and J. Weissenbach. (1996). A comprehensive genetic map of the human genome on 5,264 microsatellites. *Nature* 380: 152-154.

Dietrich, W.F., J. Miller, R. Steen, M.A. Merchant, D. Damron-Boles, Z. Husain,

R. Dredge, M. Daly, K.A. Ingalis, T.J. O'Connor, C.A. Evans, M.M. DeAngelis, D.M. Levinson, L. Kruglyak, N. Goodman, N.G. Copeland, N.A. Jenkins, T.L. Hawkins, L. Stein, D.C. Page and E.S. Lander. (1996). A comprehensive genetic map of the mouse genome. *Nature* 380: 149-152.

Dirlewanger, E., P.G. Isaac, S. Ranade, M. Belajouza, R. Cousin and D. De Vienne. (1994). Restriction fragment length polymorphism analysis of loci associated with disease resistance genes and developmental traits in *Pisum sativum* L. *Theoretical and Applied Genetics* 88: 17-27.

Edwards, A., A. Civitello, H.A. Hammond and C.T. Caskey. (1991). DNA typing and genetic mapping with trimeric and tetrameric tandem repeats. *American Journal of Human Genetics* 49: 746-756.

Edwards, K.J., J.H.A. Barker, A. Daly, C. Jones and A. Karp. (1996). Microsatellite libraries enriched for several microsatellite sequences in plants. *BioTechniques* 20: 758-760.

Ender, A., K. Schwenk, T. Städler, B. Streit and B. Schierwater. (1996). RAPD identification of microsatellites in *Daphnia*. *Molecular Ecology* 5: 437-441.

Epplen, J.T. (1992). The methodology of multilocus DNA fingerprinting using radioactive or nonradioactive probes specific for simple repeat motifs. In *Advances in Electrophoresis*, Vol. 5, eds. A. Chrambach, M. Dunn and B.J. Radola. Weinheim, Germany: VCH, pp. 59-112.

Gupta, M., Y.S. Chyi, J. Romero-Severson and J.L. Owen. (1994). Amplification of DNA markers from evolutionary diverse genomes using single primers of simple-sequence repeats. *Theoretical and Applied Genetics* 89: 998-1006.

Gupta, P.K., H.S. Balyan, P.C. Sharma and B. Ramesh. (1996). Microsatellites in plants: A new class of molecular markers. *Current Science* 70: 45-54.

Hauge, X.Y. and M. Litt. (1993). A study of the origin of 'shadow bands' seen when typing dinucleotide repeat polymorphisms by the PCR. *Human Molecular Genetics* 2: 411-415.

Hüttel, B. (1996). Mikrosatelliten als molekulare Marker in der Kichererbse (*Cicer arietinum* L.). Ph.D. Thesis, University of Frankfurt, Germany.

Jacob, H.J., D.M. Brown, R.K. Bunker, M.J. Daly, V.J. Dzau, A. Goodman, G. Koike, V. Kren, T. Kurtz, A. Lernmark, G. Levan, Y. Mao, A. Pettersson, M. Pravenec, J.S. Simon, C. Szpirer, J. Szpirer, M.R. Trolliet, E.S. Winer and E.S. Lander. (1995). A genetic linkage map of the laboratory rat, *Rattus norvegicus*. *Nature Genetics* 9: 63-69.

Jeffreys, A.J., M. Turner and P. Debenham. (1991). The efficiency of multilocus DNA fingerprint probes for individualization and establishment of family relationships, determined from extensive casework. *American Journal of Human Genetics* 48: 824-840.

Jeffreys, A.J., V. Wilson, R. Neumann and J. Keyte. (1988). Amplification of human minisatellites by the polymerase chain reaction: towards DNA fingerprinting of single cells. *Nucleic Acids Research* 16: 10053-10071.

Kaemmer, D., R. Afza, K. Weising, G. Kahl and F.J. Novak. (1992). Oligonucleotide and amplification fingerprinting of wild species and cultivars of banana (*Musa* spp.). *Bio/Technology* 10: 1030-1035.

Kandpal, R.P., G. Kandpal and S.M. Weissman. (1994). Construction of libraries enriched for sequence repeats and jumping clones, and hybridization selection for region-specific markers. *Proceedings of National Academy of Sciences USA* 91: 88-92.

Kantety, R.V., X. Zeng, J.L. Bennetzen, B.E. Zehr. (1995). Assessment of genetic diversity in dent and popcorn (*Zea mays* L.) inbred lines using inter-simple sequence repeat (ISSR) amplification. *Molecular Breeding* 1: 365-373.

Karagyozov, L., I.D. Kalcheva and V.M. Chapman. (1993). Construction of random small-insert genomic libraries highly enriched for simple sequence repeats. *Nucleic Acids Research* 21: 3911-3912.

Kijas, J.M.H., J.C.S. Fowler, C.A. Garbett and M.R. Thomas. (1994). Enrichment of microsatellites from the *Citrus* genome using biotinylated oligonucleotide sequences bound to streptavidin-coated magnetic particles. *BioTechniques* 16: 658-662.

Kijas, J.M.H., J.C.S. Fowler and M.R. Thomas. (1995). An evaluation of sequence tagged microsatellite site markers for genetic analysis within *Citrus* and related species. *Genome* 38: 349-355.

Klinkicht, M. and D. Tautz. (1992). Detection of simple sequence length polymorphisms by silver staining. *Molecular Ecology* 1: 133-134.

Kostia, S., S.L. Varvio, P. Vakkari and P. Pulkkinen. (1995). Microsatellite sequences in a conifer, *Pinus sylvestris*. *Genome* 38: 1244-1248.

Kresovich, S., A.K. Szewc-McFadden, S.M. Bliek and J.R. McFerson. (1995). Abundance and characterization of simple-sequence repeats (SSRs) isolated from a size-fractionated genomic library of *Brassica napus* L. (rapeseed). *Theoretical and Applied Genetics* 91: 206-211.

Lagercrantz, U., H. Ellegren and L. Andersson. (1993). The abundance of various polymorphic microsatellite motifs differs between plants and vertebrates. *Nucleic Acids Research* 21: 1111-1115.

Litt, M. and J.A. Luty. (1989). A hypervariable microsatellite revealed by *in vitro* amplification of a dinucleotide repeat within the cardiac muscle actin gene. *American Journal of Human Genetics* 44: 397-401.

Meyer, W. and T.G. Michell. (1995). Polymerase chain reaction fingerprinting in fungi using single primers specific to minisatellites and simple repetitive DNA sequences: Strain variation in *Cryptococcus neoformans*. *Electrophoresis* 16: 1648-1656.

Meyer, W., T.G. Mitchell, E.Z. Freedman and R. Vilgalys. (1993). Hybridization probes for conventional DNA fingerprinting used as single primers in the polymerase chain reaction to distinguish strains of *Cryptococcus neoformans*. *Journal of Clinical Microbiology* 31: 2274-2280.

Moore, S.S., L.L. Sargeant, T.J. King, J.S. Mattick, M. Georges and D.J.S. Hetzel. (1991). The conservation of dinucleotide microsatellites among mammalian genomes allows the use of heterologous PCR primer pairs in closely related species. *Genomics* 10: 654-660.

Morgante, M. and A.M. Olivieri. (1993). PCR-amplified microsatellites as markers in plant genetics. *The Plant Journal* 3: 175-182.

Morgante, M., A. Rafalski, P. Biddle, S. Tingey and A.M. Olivieri. (1994).

Genetic mapping and variability of seven soybean simple sequence repeat loci. *Genome* 37: 763-769.

Neilan, B.A., D.A. Leigh, E. Rapley and B.L. McDonald. (1994). Microsatellite genome screening: rapid non-denaturing, non-isotopic dinucleotide repeat analysis. *BioTechniques* 17: 708-712.

Ostrander, E.A., P.M. Jong, J. Rine and G. Duyk. (1992). Construction of small-insert genomic DNA libraries highly enriched for microsatellite repeat sequences. *Proceedings of National Academy of Sciences USA* 89: 3419-3423.

Panaud, O., X. Chen and S.R. McCouch. (1995). Frequency of microsatellite sequences in rice (*Oryza sativa* L.). *Genome* 38: 1170-1176.

Pépin, L., Y. Amigues, A. Lépingle, J.L. Berthier, A. Bensaid and D. Vaiman. (1995). Sequence conservation of microsatellites between *Bos taurus* (cattle), *Capra hircus* (goat) and related species. Examples of use in parentage testing and phylogeny analysis. *Heredity* 74: 53-61.

Perring, T.M., A.D. Cooper, R.J. Rodriguez, C.A. Farrar and T.S. Bellows. (1993). Identification of a whitefly species by genomic and behavioral studies. *Science* 259: 74-77.

Powell, W., M. Morgante, R. McDevitt, G.G. Vendramin and J.A. Rafalski. (1995a). Polymorphic simple sequence repeat regions in chloroplast genomes: Applications to the population genetics of pines. *Proceedings of National Academy of Sciences USA* 92: 7759-7763.

Powell, W., G.C. Machray and J. Provan. (1996). Polymorphism revealed by simple sequence repeats. *Trends in Plant Science* 1: 215-222.

Powell, W., M. Morgante, C. Andre, J.W. McNicol, G.C. Machray, J.J. Doyle, S.V. Tingey and J.A. Rafalski. (1995b). Hypervariable microsatellites provide a general source of polymorphic DNA markers for the chloroplast genome. *Current Biology* 5: 1023-1029.

Richardson, T., S. Cato, J. Ramser, G. Kahl and K. Weising. (1995). Hybridization of microsatellites to RAPD: a new source of polymorphic markers. *Nucleic Acids Research* 23: 3798-3799.

Rongwen, J., M.S. Akkaya, A.A. Bhagwat, U. Lavi and P.B. Cregan. (1995). The use of microsatellite DNA markers for soybean genotype identification. *Theoretical and Applied Genetics* 90: 43-48.

Saghai-Maroof, M.A., R.M. Biyashev, G.P. Yang, Q. Zhang and R.W. Allard. (1994). Extraordinarily polymorphic microsatellite DNA in barley: species diversity, chromosomal locations, and population dynamics. *Proceedings of National Academy of Sciences USA* 91: 5466-5470.

Salimath, S.S., A.C. De Oliveira, I.D. Godwin and J.L. Bennetzen. (1995). Assessment of genome origins and genetic diversity in the genus *Eleusine* with DNA markers. *Genome* 38: 757-763.

Sanchez de la Hoz, M.P., J.A. Dávila, Y. Loarce and E. Ferrer. (1996). Simple sequence repeat primers used in polymerase chain reaction amplifications to study genetic diversity in barley. *Genome* 39: 112-117.

Schlötterer, C., B. Amos and D. Tautz. (1991). Conservation of polymorphic simple sequence loci in cetacean species. *Nature* 354: 63-65.

Schwengel, D.A., A.E. Jedlicka, E.J. Nanthakumar, J.L. Weber and R.C. Levitt. (1994). Comparison of fluorescence-based semi-automated genotyping of multiple microsatellite loci with autoradiographic techniques. *Genomics* 22: 46-54.

Scrimshaw, B.J. (1992). A simple nonradioactive procedure for visualization of $(dC\text{-}dA)_n$ dinucleotide repeat length polymorphisms. *BioTechniques* 13: 189.

Sharma, P.C., P. Winter, T. Bünger, B. Hüttel, F. Weigand, K. Weising and G. Kahl. (1995). Abundance and polymorphism of di-, tri- and tetra-nucleotide tandem repeats in chickpea (*Cicer arietinum* L.). *Theoretical and Applied Genetics* 90: 90-96.

Sharon, D., A. Adato, S. Mhameed, U. Lavi, J. Hillel, M. Gomolka, C. Epplen and J.T. Epplen. (1995). DNA fingerprints in plants using simple-sequence repeat and minisatellite probes. *HortScience* 30: 109-112.

Sinnett, D., J.M. Deragon, L.R. Simar and D. Labuda. (1990). Alumorphs–Human DNA polymorphisms detected by polymerase chain reaction using Alu-specific primers. *Genomics* 7: 331-334.

Smeets, H.J.M., H.G. Brunner, H.H. Ropers and B. Wieringa. (1989). Use of variable simple sequence motifs as genetic markers: application to study of myotonic dystrophy. *Human Genetics* 83: 245-251.

Tanksley, S.D., M.W. Ganal and G.B. Martin. (1995). Chromosome landing: a paradigm for map-based gene cloning in plants with large genomes. *Trends in Genetics* 11: 63-68.

Taramino, G. and S. Tingey. (1996). Simple sequence repeats for germplasm analysis and mapping in maize. *Genome* 39: 277-287.

Tautz, D. (1989). Hypervariability of simple sequences as a general source for polymorphic DNA markers. *Nucleic Acids Research* 17: 6463-6471.

Thomas, M.R., P. Cain and N.S. Scott. (1994). DNA typing of grapevines: a universal methodology and database for describing cultivars and evaluating genetic relatedness. *Plant Molecular Biology* 25: 939-949.

Tsumura, Y., K. Ohba and S.H. Strauss. (1996). Diversity and inheritance of inter-simple sequence repeat polymorphisms in Douglas-fir (*Pseudotsuga menziesii*) and sugi (*Cryptomeria japonica*). *Theoretical and Applied Genetics* 92: 40-45.

Wallner, E., K. Weising, R. Rompf, G. Kahl and B. Kopp. (1996). Oligonucleotide fingerprinting and RAPD analysis of *Achillea* species: characterization and long-term monitoring of micropropagated clones. *Plant Cell Reports* 15: 647-652.

Wang, Z., J.L. Weber, G. Zhong and S.D. Tanksley. (1994). Survey of plant short tandem DNA repeats. *Theoretical and Applied Genetics* 88: 1-6.

Weber, J.L. (1990). Informativeness of human $(dC\text{-}dA)_n \times (dG\text{-}dT)_n$ polymorphisms. *Genomics* 7: 524-530.

Weber, J.L. and P.E. May. (1989). Abundant class of human DNA polymorphisms which can be typed using the polymerase chain reaction. *American Journal of Human Genetics* 44: 388-396.

Weising, K., R.G. Atkinson and R.C. Gardner. (1995b). Genomic fingerprinting by microsatellite-primed PCR: a critical evaluation. *PCR Methods & Applications* 4: 249-255.

Weising, K., R.W.M. Fung, J. Keeling, R.G. Atkinson and R.C. Gardner. (1996). Cloning and characterization of microsatellite repeats from *Actinidia chinensis*. Molecular Breeding (in press).

Weising, K., D. Kaemmer, F. Weigand, J.T. Epplen and G. Kahl. (1992). Oligonucleotide fingerprinting reveals probe-dependent levels of informativeness in chickpea (*Cicer arietinum* L.). *Genome* 35: 436-442.

Weising, K. and G. Kahl. (1996) Hybridization-based microsatellite fingerprinting of plants and fungi. In: eds., G. Caetano-Anollés and P. Gresshoff. *DNA Markers: Protocols, Applications and Overviews*. New York: Wiley & Sons.

Weising, K., H. Nybom, K. Wolff and W. Meyer. (1995a). DNA Fingerprinting in Plants and Fungi. Boca Raton, Florida: CRC Press.

Weising, K., F. Weigand, A.J. Driesel, G. Kahl, H. Zischler and J.T. Epplen. (1989). Polymorphic simple GATA/GACA repeats in plant genomes. *Nucleic Acids Research* 17: 10128.

Williams, J.G.K., A.R. Kubelik, K.J. Livak, J.A. Rafalski and S.V. Tingey. (1990). DNA polymorphisms amplified by arbitrary primers are useful as genetic markers. *Nucleic Acids Research* 18: 6531-6535.

Winter, P. and G. Kahl. (1995). Molecular marker technologies for plant improvement. *World Journal of Microbiology and Biotechnology* 11: 438-448.

Wolff, K., E. Zietkiewicz and H. Hofstra. (1995). Identification of chrysanthemum cultivars and stability of fingerprint patterns. *Theoretical and Applied Genetics* 91: 439-447.

Wu, K., R. Jones, L. Danneberger and P.A. Scolnik. (1994). Detection of microsatellite polymorphisms without cloning. *Nucleic Acids Research* 22: 3257-3258.

Wu, K.S. and S.D. Tanksley. (1993). Abundance, polymorphism and genetic mapping of microsatellites in rice. *Molecular and General Genetics* 241: 225-235.

Yu, Y.G., M.A. Saghai-Maroof and G.R. Buss. (1996). Divergence and allelomorphic relationship of a soybean virus resistance gene based on tightly linked DNA microsatellite and RFLP markers. *Theoretical and Applied Genetics* 92: 64-69.

Yu, Y.G., A. Saghai-Maroof, G.R. Buss, P.J. Maughan and S.A. Tolin. (1994). RFLP and microsatellite mapping of a gene for soybean mosaic virus resistance. *Phytopathology* 84: 60-64.

Ziegle, J.S., Y. Su, K.P. Corcoran, L. Nie, P.E. Mayrand, L.B. Hoff, L.J. McBride, M.N. Kronick and S.R. Diehl. (1992). Application of automated DNA sizing technology for genotyping microsatellite loci. *Genomics* 14: 1026-1031.

Zietkiewicz, E., A. Rafalski and D. Labuda. (1994). Genome fingerprinting by simple sequence repeat (SSR)-anchored polymerase chain reaction amplification. *Genomics* 20: 176-183.

SUBMITTED: 09/23/96
ACCEPTED: 12/13/96

Weed Seed Bank Dynamics:
Implications to Weed Management

Douglas D. Buhler
Robert G. Hartzler
Frank Forcella

SUMMARY. Weeds continue to have major impacts on crop production in spite of efforts to eliminate them. Most weed species rely on seed for regeneration and persistence. The species composition and density of weed seed in the soil vary greatly and are closely linked to the cropping history of the land. Altering tillage practices changes patterns of soil disturbance and weed seed depth in the soil, which plays a role in weed species shifts. Crop rotation and weed control practices also impact the weed seed bank in the soil. Information on the weed seed bank should be a useful tool for integrated weed management. Decision aid models are being developed that use information on the composition of the weed seed bank to estimate weed populations, crop yield loss, and to recommend weed

---

Douglas D. Buhler, Research Agronomist, United States Department of Agriculture-Agricultural Research Service, National Soil Tilth Laboratory, 2150 Pammel Drive, Ames, IA 50011, USA. Robert G. Hartzler, Associate Professor, Department of Agronomy, Iowa State University, Ames, IA 50011, USA. Frank Forcella, Research Agronomist, United States Department of Agriculture-Agricultural Research Service, North Central Soil Conservation Laboratory, Morris, MN 56267, USA.

Address correspondence to: Douglas D. Buhler, United States Department of Agriculture-Agricultural Research Service, National Soil Tilth Laboratory, 2150 Pammel Drive, Ames, IA 50011, USA (E-mail: buhler@nstl.gov).

[Haworth co-indexing entry note]: "Weed Seed Bank Dynamics: Implications to Weed Management." Buhler, Douglas D., Robert G. Hartzler, and Frank Forcella. Co-published simultaneously in *Journal of Crop Production* (The Food Products Press, an imprint of The Haworth Press, Inc.) Vol. 1, No. 1 (#1), 1998, pp. 145-168; and: *Crop Sciences: Recent Advances* (ed: Amarjit S. Basra) The Food Products Press, an imprint of The Haworth Press, Inc., 1998, pp. 145-168. Single or multiple copies of this article are available for a fee from The Haworth Document Delivery Service [1-800-342-9678, 9:00 a.m. - 5:00 p.m. (EST). E-mail address: getinfo@haworth.com].

145

control tactics. Understanding weed seed bank dynamics can also be used to guide management practices. Improving and applying our understanding of weed seed and seed bank dynamics is essential to developing improved weed management systems. *[Article copies available for a fee from The Haworth Document Delivery Service: 1-800-342-9678. E-mail address: getinfo@haworth.com]*

KEYWORDS. Population dynamics, tillage systems, weed control, weed ecology, weed emergence

INTRODUCTION

Weeds have presented a challenge to crop production since the beginning of civilization. Weed control techniques, rather than management of populations, have been the focus of weed science for the past several decades. Weed management implies a shift away from strict reliance on control of existing weed problems and places greater emphasis on prevention of propagule production, reduction of weed emergence in a crop, and minimizing weed competition with the crop (Zimdahl, 1991). Weed management emphasizes integration of techniques to anticipate and manage problems rather than reacting to them after they are present. Weed management does not eliminate the need for control nor does it suggest that the best control techniques be abandoned. The goal is to maximize crop production where appropriate and optimize grower profit by integrating preventive techniques, scientific knowledge, and management skills. While additional knowledge is needed in all areas of weed management, the most important task of weed science is to increase knowledge of weed biology and ecology, creating a better understanding of "weediness." This knowledge will lead to the use of appropriate management techniques rather than prophylactic approaches that produce short-term results, but may create or worsen long-term problems.

Weed seed and seed bank dynamics play a major role in regulating weed communities of agricultural land. The purpose of this paper is to review processes that contribute to the formation of seed banks of annual weed species and the effects of selected management practices on those weed seed banks. We will also provide examples of how this information can be used to manage weed populations in agronomic crops.

WEED SEED BANK DYNAMICS

Most annual plant communities regenerate from seed stored in the soil seed bank. However, establishment of seedlings requires that seeds are in a proper physiological state for germination. This physiological state often occurs within a limited period in the life of the seed and must coincide with appropriate environmental conditions. Therefore, regeneration of plant communities from seed banks requires that seeds are in the right place at the right time. For most annual species, soil is the medium in which these processes occur (Thompson and Grime, 1979). Because some seeds persist in the soil for several years and new seed is added annually, the soil seed bank also serves as a genetic "memory" for the plant community (Cavers, 1995).

Crop management practices have major impacts on seed bank processes in annual weed species and regulate the development of weed communities. Regeneration strategies of commonly occurring weed species vary. For example, seeds of some weed species germinate soon after they are shed (Bazzaz, 1990). Seeds of these species typically have a short life in the soil and regeneration is highly dependent on annual seed production and dispersal. In other species, seed may remain in the soil for long periods with intermittent germination (Murdoch and Ellis, 1992). Some of these seeds are very long-lived. However, while reports of extreme longevity are impressive, these seeds usually represent a small proportion of the total seed bank (Wilson, 1988). In agronomic situations, we are most interested in the majority of seeds that germinate during the first few years after shed. Understanding the short-term dynamics of these seeds and their resultant weed populations will aid in predicting potential for crop yield loss and control costs.

The seed bank in the soil is the primary source of new infestations of annual weeds, which represent the majority of the weed pests in many agronomic crop production systems (Cavers, 1983). Therefore, weed seed bank characteristics influence both the weed populations that occur in a field and the success of weed management practices. Many processes are involved in the generation and regulation of the weed seed bank in the soil (Figure 1), such as the size, species composition, and distribution, which fluctuate depending on seed introductions and losses (Schweizer and Zimdahl, 1984; Burnside et al., 1986; Wilson, 1988). Management practices have major impacts on these processes and represent opportunities for regulating seed bank characteristics in crop production systems.

The species composition and density of weed seed in soil vary greatly and are closely linked to the cropping history of the land. Seed composition is influenced by past farming practices and varies from field to field (Robinson, 1949; Buhler et al., 1984; Fenner, 1985) and among areas within fields (Benoit, Kenkel, and Cavers, 1989; Benoit, Derksen, and Panneton, 1992; Mortensen, Johnson, and Young, 1993). Reports of seed

FIGURE 1. Weed Seed Cycle.

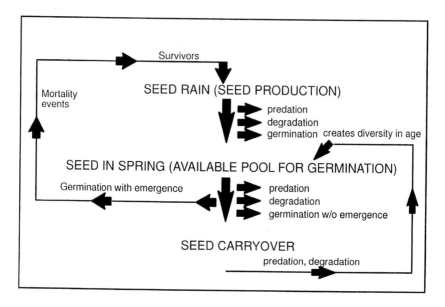

bank size in agricultural land range from near zero to as much as one million seeds/m^2 (Fenner, 1985). While each seed bank may be different, some similarities have emerged. Generally, seed banks are composed of many species, with a few dominant species comprising 70 to 90% of the total seed bank (Wilson, 1988). These species are the primary pests in agronomic systems due to resistance to control or their adaptation to the cropping system. A second group of species comprises 10 to 20% of the seed bank. These are generally species that are adapted to the geographic area, but not adapted to the production practices used on the land. The final group accounts for a very small percentage of the total seed and includes recalcitrant seed from previous seed banks, newly introduced species, and seed of the previous crop (Wilson, Kerr, and Nelson, 1985). This group undergoes constant change due to seed dispersal by the activities of humans and animals and by climatic factors such as wind and water.

Additions to the Seed Bank

New seeds may enter the seed bank through many sources, but the largest source is plants producing seed within the field (Cavers, 1983). A characteristic of many weed species is the potential for prolific seed production (Stevens, 1957). However, weeds present in agricultural fields usually produce fewer seeds due to competition from the crop, damage from herbicides, and other factors. Common cocklebur (*Xanthium strumarium* L.) growing without crop competition produced more than 7000 seeds/plant, whereas seed production of common cocklebur growing with soybean [*Glycine max* (L.) Merr.] was reduced to 1100 seeds (Senseman and Oliver, 1993). Velvetleaf (*Abutilon theophrasti* Medik.) seed production was reduced up to 82% by competition with soybean (Lindquist et al., 1995). Increased shading, which often occurs when weeds emerge later than the crop, also reduces seed production. For example, 76% shade reduced velvetleaf seed production up to 94% (Bello, Owen, and Hatterman-Valenti, 1995). Herbicide applications that do not kill plants may also reduce seed production. Sublethal doses of herbicide reduced seed production of several weed species by as much as 90% (Biniak and Aldrich, 1986; Salzman, Smith, and Talbert, 1988). Although seed production in most weed species can be reduced by management factors, seed production will likely be adequate, even with moderate weed infestations, to allow rapid increases in the seed bank (Schweizer and Zimdahl, 1984; Hartzler, 1996).

Seeds may also enter fields from external sources such as farm equipment, contaminated crop seed, animals, wind, or manure. The number of seeds introduced into the seed bank by these sources is smaller than those produced by weeds in the field; however, these sources are often important in establishing infestations of new species. Many weeds (e.g., Canada thistle [*Cirsium arvense* (L.) Scop.], horseweed [*Conyza canadensis* (L.) Cronq.], and dandelion (*Taraxacum officinale* Weber in Wiggers) have seeds adapted to wind dispersal. Dandelion and horseweed have become problems in no-tillage systems partially due to the wind transport of their seeds (Buhler, 1995).

Manure can be an important source of weed seed. While most of the seeds are killed when passing through the digestive tracts of animals, a small percentage typically survives (Harmon and Keim, 1934). A study of 20 New York dairy farms found that, on average, spreading manure introduced 350 weed seeds/m^2 (Mt. Pleasant and Schlather, 1994). This seemingly high number of seeds is relatively low compared with the number of seeds already present in the seed bank. If manure is spread on the fields where the feed was grown, the number of seeds returned to the field will be of little consequence. However, manure can be a source of new weed problems if feed is moved among farms and is contaminated with seed of species not currently present in the field (Eberlein et al., 1992).

Another mechanism of weed seed transport is farm machinery moving between fields. This mechanism has become increasingly important as machinery is moved greater distances due to increasing farm size. Movement of weed seed by combines and other harvest equipment is of particular concern (Currie and Peeper, 1988; McCanny and Cavers, 1988). Careful management can reduce the risk of spreading weed seed into non-infested fields. Preventive practices include working in infested fields last or thoroughly cleaning machinery after working in infested fields.

Seed Losses

Although seed of many weed species have the potential for long-term survival in the seed bank, most seeds have a relatively short life-span (Murdoch and Ellis, 1992). Factors accounting for the loss of weed seed in the soil include germination, decay, predation, and physical movement. The relative importance of these mechanisms varies with species and environmental conditions.

In weed management we are primarily interested in seeds that germinate and emerge. These seeds result in new plants that may reduce crop yields if not controlled. Sporadic germination in time and space (Forcella et al., 1992, 1996b) is a characteristic allowing weeds to survive despite our efforts to eradicate them. Dormancy is a primary mechanism regulating these variable germination patterns. Several types of seed dormancy exist (Nikolaeva, 1977) and most weed species possess one or more type. Several reviews of seed dormancy and related literature have been published recently (Egley and Duke, 1985; Lang et al., 1987; Taylorson, 1987; Bradbeer, 1988; Wilson, 1988; Dyer, 1995), so an extensive discussion will not be presented here. In a review on exploiting weed seed dormancy through agronomic practices (Dyer, 1995), Dyer concluded that management practices can influence dormancy. Many environmental and edaphic factors that affect seed germination are altered during tillage, planting, and harvesting. Even slight adjustments in planting date, cultivation timing, harvest method, or residue management may have significant effects on the dynamics of weed seed dormancy.

Although genetics regulates seed dormancy, dormancy in a population of genetically related seeds can vary due to environmental factors (Murdoch and Ellis, 1992). For example, seeds shed from the same plant may have different dormancy states depending upon environmental conditions at the time of seed development and seed position on the inflorescence (Dekker et al., 1996). To complicate matters further, nondormant seeds may be induced into secondary dormancy (Taylorson, 1987; Forcella et al., 1996b). Temperature and/or water potential has been shown to regulate secondary dormancy in the laboratory (Taylorson, 1982). For example, hydrated, non-dormant giant foxtail (*Setaria faberi* Herrm.) seeds were induced into secondary dormancy by exposure to 35°C. This may be relevant in the field because soil temperatures near the surface often reach 35°C early in the spring (Gupta, Larson, and Linden, 1983). The complexity of dormancy has limited our ability to predict weed emergence, but current research may be making progress in this area.

The percentage of seeds in the seed bank that germinate in a given year is influenced by the species and the environment encountered by the seeds. For common annual species in cultivated soil, approximately 1 to 40% of the seed bank will emerge in a given year (Roberts and Ricketts, 1979; Forcella et al., 1992, 1996b; Wilson and Lawson, 1992) with great variation both within and among weed species. The most commonly reported range of emergence under agronomic conditions is 3 to 6%. In field experiments conducted from 1991 through 1994 (Forcella et al., 1996b), information on weed emergence was collected for 22 site-years from Ohio to Colorado and Missouri to Minnesota. Average emergence percentages for some major species were: giant foxtail, 31%; velvetleaf, 28%; common ragweed (*Ambrosia artemisiifolia* L.), 15%; pigweed species (*Amaranthus* spp.), 3%; and common lambsquarters (*Chenopodium album* L.), 3%. Coefficients of variation for the species mean values ranged from 62 to 135%. Reasons for the high variation among site-years within species are not fully understood. However, for some species, induction of secondary dormancy by microclimatic variables was thought to play a major role. Understanding and predicting variation in emergence percentages will be key to developing accurate decision aids for weed management.

Seeds are an important food source for many insects, birds, and small mammals. In natural systems, more than 70% of seeds may be consumed by animals (Crawley, 1992). Seed predation is usually less in agricultural systems due to the intensive soil disturbance, seed burial by tillage, and lack of habitats for predators. However, studies have found significant weed seed loss from predation when seeds remained on the soil surface (Brust and House, 1988; Reader, 1991). As many as 69% of the weed seed was lost to predation in no-tillage soybean compared with 27% in conventional tillage (Brust and House, 1988). An undefined proportion of the seed decays in the soil after being infected by fungi or other microorganisms (Kremer, 1993). Research is being conducted to isolate microorganisms that are more efficient at infecting seeds in the soil. If found, these pathogens could be used to inoculate soils and reduce seed bank size.

Other mechanisms for seed loss exist, such as water moving through a field and movement with tillage and harvesting equipment. However, contributions to weed seed bank dynamics by these mechanisms appear to be minor.

MANAGEMENT IMPACTS ON THE WEED SEED BANK

Weed Management

The size of the seed bank can change dramatically in a short period (two to three years) because of seed losses and inputs to the seed bank. Weed seed densities can be greatly reduced by eliminating seed production for a few years; conversely, soil with low seed density can increase at a rapid rate if plants are allowed to produce seed.

A six-year study in Colorado evaluated changes in the weed seed bank in continuous-corn (*Zea mays* L.) (Schweizer and Zimdahl, 1984). The number of seeds in the seed bank dropped by approximately 70% after three years of annual atrazine [6-chloro-*N*-ethyl-*N'*-(1-methylethyl)-1,3,5-triazine-2,4-diamine] application plus interrow cultivation. Atrazine use was ceased in some plots after the first three years, and weeds were controlled with one or two cultivations. After three years of cultivation only, the seed bank was approximately 25 times greater than that where atrazine use and cultivation were continued. A similar study was conducted at five locations in Nebraska (Burnside et al., 1986). Broadleaf and grass seed density in the soil declined by 95% after a 5-year weed-free period. During the sixth year herbicide use was ceased and seed density increased to 90% of the original level at two of the five locations.

A study in Iowa evaluated the impact of a single year of velvetleaf seed production on future velvetleaf populations (Hartzler, 1996). Velvetleaf was planted in soybean at 0.2 and 0.4 plant/m^2 and allowed to produce seed. For the next four years, the field was maintained in a corn-soybean rotation. Velvetleaf plants were counted as they emerged and then removed to prevent additional seed production (Table 1). Each plant that produced seed in the first year resulted in more than 1000 new plants over the four-year experiment with emergence dropping by 80% between years 2 and 4 of the experiment. After four years, emergence from the seed bank accounted for an estimated 25% of the introduced seed.

These studies provide examples of the impact of weed management on the weed seed bank and illustrate the rapid decline in the seed bank when seed introductions are minimized or prevented. However, in most weed species a small number of seeds remain viable for long periods in the soil. When weed management practices are not entirely effective, these seeds can germinate, mature, and produce enough seed to replenish the seed bank.

TABLE 1. Effect of a Single Year's Velvetleaf Seed Rain on Emergence in Subsequent Years (Adapted from Hartzler, 1996).

Years after seed rain	Original velvetleaf population (plants/m^2)		
	0	0.2	0.4
	- - - - - - - - - - - - - plants/m^2 - - - - - - - - - - - - -		
1	6 (2)[a]	91 (32)	145 (41)
2	7 (2)	128 (51)	203 (78)
3	7 (11)	34 (11)	62 (20)
4	2 (3)	23 (8)	37 (19)

[a]Values in parentheses are the standard error of the mean.

Crop Rotation

Crop rotation has been cited extensively as an effective method of weed management (Liebman and Dyck, 1993 and references therein). Crop rotation is effective for weed management because selection pressure is diversified by changing patterns of disturbance. This diversification prevents the proliferation of weed species well suited to the practices associated with a single crop.

Few studies have characterized the effects of crop rotation on the weed seed bank. In ridge-tillage (Forcella and Lindstrom, 1988), soils harbored at least twice as many weed seeds under continuous corn than a corn/soybean rotation. Truncation of the ridges at the time of crop planting removed 35% of the weed seed from the ridges of continuous corn and 90% of the seed from ridges of the corn/soybean rotation. Ridging the soil just before canopy closure stimulated germination of weed seed. The resulting weed population produced up to 1000 seeds/m^2 in continuous corn and about 100 seeds/m^2 in the corn/soybean rotation.

Schreiber (1992) found that growing corn in a soybean/corn or soybean/wheat (*Triticum aestivum* L.)/corn rotation greatly reduced giant foxtail seed in the soil compared to corn grown continuously, regardless of herbicide use or tillage system. The effects of crop rotation and environmental conditions associated with years and locations were larger than tillage effects on weed species composition and abundance in two separate studies in Canada (Derksen et al., 1993; Thomas and Frick, 1993). Similarly, Ball (1992) reported that cropping sequence was the most dominant factor influencing species composition in weed seed banks.

The mechanisms by which crop rotation reduces the size of weed seed banks are likely related to the use of crop sequences that employ varying patterns of resource competition, allelopathic interference, soil disturbance, and mechanical damage. Proliferation of otherwise well adapted weed species is reduced by these processes, which provide a more diverse environment. Additional research is necessary to understand the mechanisms and exploit the responses of weed seed banks to crop rotation.

Tillage Systems

Tillage systems affect weed management, weed seed production, and patterns of soil disturbance. Therefore, changing tillage systems will change the distribution and density of weed seed in agricultural soils (Buhler, 1995). Weed species not previously observed have rapidly appeared in fields following elimination of preplant tillage (Wicks et al., 1994; Buhler, 1995). Species most rapidly and commonly observed are winter annual and biennial species. These weed species are unable to complete their life cycles in association with summer annual crops if the soil is disturbed before planting (Bazzaz, 1990). However, in no-tillage systems, seedlings that emerge in the fall are not destroyed by tillage and gain a competitive advantage over later emerging summer annual species.

Tillage is the primary cause of vertical seed movement in agricultural soils (Roberts, 1963; Cousens and Moss, 1990; Staricka et al., 1990). The effects of chisel and moldboard plowing on the vertical distribution of weed seed in the soil were evaluated using ceramic spheres (Staricka et al., 1990). Spheres, with similar size and density of weed seeds, were found to depths of 12 cm following chisel plowing and 32 cm following moldboard plowing. In chisel plowed plots, 48% of the spheres were found within 4 cm of the soil surface compared with 4% in moldboard plow plots.

Moldboard plow plots had fewer weed seeds in the upper 20 cm of soil than chisel plow or no-tillage after five years (Figure 2) (Yenish, Doll, and Buhler, 1992). Moldboard plowing resulted in the most uniform distribution of seed over soil depths. In the no-tillage system, more than 60% of all weed seeds were found in the upper 1 cm of soil and few seeds were found below 10 cm. The concentration of weed seeds in no-tillage decreased logrithmically with increasing depth. In the chisel plow system, more than 30% of the weed seeds were in the upper 1 cm of soil and seed concentration decreased linearly with depth. Pareja, Staniforth, and Pareja (1985) found 85% of all weed seeds in the upper 5 cm of soil in a reduced tillage system, but only 28% in the same zone in the moldboard plow system.

FIGURE 2. Distribution of Weed Seed in the Top 20 cm of Soil as Affected by Tillage Systems (Adapted from Yenish, Doll, and Buhler, 1992).

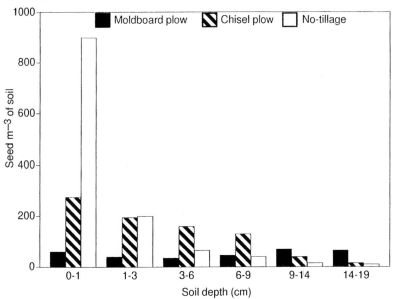

Seed and tracer distribution data (Pareja, Staniforth, and Pareja, 1985; Staricka et al., 1990; Yenish, Doll, and Buhler, 1992) substantiate differences in emergence depths of giant and green foxtail [*Setaria viridis* (L.) Beauv.] in different tillage systems in the field (Buhler and Mester, 1991). Mean seedling emergence depths were smallest in no-tillage, followed by chisel plow and then moldboard plow systems in two soil types over three years. At least 40% of the giant and green foxtail emergence was from the upper 1 cm of soil in no-tillage compared to about 25% in chisel plow and less than 15% in moldboard plow plots. As many as 25% of the foxtail plants that became established in the moldboard plow plots emerged from greater than 4 cm compared to about 10% in chisel plow and less than 5% in no-tillage.

Changes in seed depth in the soil and the corresponding differences in emergence depth may contribute to shifts among weed species under different tillage systems. In a greenhouse (Buhler, 1995), velvetleaf establishment from seed germinating on the soil surface was only 18% of seed planted 6 cm deep (Table 2). Giant foxtail seeds germinating on the soil surface had an establishment percentage similar to seeds planted 1 to 4 cm deep, but giant foxtail establishment was reduced 50% when seeds were planted 6 cm deep.

Buhler, Mester, and Kohler (1996) attempted to separate the effects of weed seed depth distribution in the soil and surface residue by establishing tilled and untilled plots with various levels of corn residue. Velvetleaf densities were greatest in tilled plots without residue, whereas redroot pigweed (*Amaranthus retroflexus* L.) densities were greatest when the residue was removed from untilled plots. The ambient corn residue reduced redroot pigweed densities up to 70% compared with plots with no residue. Giant foxtail densities were several times greater in plots that were not tilled when averaged over residue levels. It was concluded that vertical seed distribution in the seed bank plays a more important role in weed population shifts among tillage systems than surface residue.

The effect of tillage practices on the population dynamics of summer annual weed species is complex and involves several factors. However, seed depth in the soil may be the most important factor. Weed species that can germinate and become established when the seeds are at or near the soil surface have the greatest potential to proliferate under conservation

TABLE 2. Effect of Planting Depth on the Establishment and Growth of Velvetleaf and Giant Foxtail After 28 days in the Greenhouse (From Buhler, 1995).

Planting depth	Velvetleaf		Giant foxtail	
	Seedling establishment	Seedling height	Seedling establishment	Seedling height
cm	------- % of maximum observed -------			
0	18	45	86	100
1	73	82	100	76
2	73	100	86	83
4	91	82	81	89
6	100	70	50	24
LSD (0.05)	26	26	17	18

tillage systems. Deep burial of seed of small-seeded species by moldboard plowing reduces germination and emergence. Seeds of large-seeded species remain near the soil surface in conservation tillage systems, inhibiting their establishment and preventing them from contributing to reinfestation following subsequent tillage operations (Lueschen and Andersen, 1980; Warnes and Andersen, 1984).

WEED SEED, SEED BANKS, AND WEED MANAGEMENT

As discussed earlier in this paper, most seeds of annual weeds are not very persistent in agronomic production systems. In theory, managing weed seed banks should be easy: simply stop seed production and deplete the seed bank by managing the soil to provide the optimum environment for germination. In practice, managing seed banks is far more complex because of the difficulty in preventing seed production and introduction, the persistence of a small percentage of the seed bank, and the high seed production potential of many weed species.

Seed Bank Philosophy

Some weed scientists argue that allowing even a single weed to produce seed is detrimental to long-term farm profits (Norris, 1992). Most producers would agree that eliminating weed seed production is a worthy goal, possibly even an economically rational one. However, concerns with attempts to eliminate weed seed production include: (a) whether this goal can be attained by farmers over large areas of land given the dormancy and seed longevity characteristics of commonly occurring weed species and (b) if other problems such as increased costs for labor, equipment, and herbicide; weed resistance; increased soil erosion; water contamination with herbicides; and loss of habitat for wildlife and other beneficial organisms may be created in the process. It could be argued that we have been trying to eliminate seed banks for many years, and have failed. It may be more realistic to accept weed seed banks as an ever-present component of agricultural land and attempt to understand, interpret, and predict their behavior. Then, devise management systems that minimize the impacts of resultant weeds rather than seeking to eliminate the seed bank.

Control Efficacy

Knowledge of seed bank characteristics should be useful in predicting weed management efficacy. A study was conducted in two fields with large differences in weed seed bank size (Hartzler and Roth, 1993). In the first year of the study, plots were established with either 0 or 100% weed control. During the second year of the study, the influence of weed control the previous year on herbicide performance was evaluated. In the field with a large initial seedbank, metolachlor [2-chloro-N-(2-ethyl-6-methylphenyl)-N-(2-methoxy-1-methylethyl)acetamide] controlled 75% of the giant fox-tail in plots where heavy infestations (530 plants/m^2) had produced seed the previous year. In the same field, metolachlor provided 95% control in plots maintained weed free the previous year. In a field with a small seedbank, low grass populations (8 plants/m^2) producing seed the pre-vious year did not influence control. These results suggest that large changes in the seed bank impact weed control efficacy, but small changes may be of little consequence. This is supported by other research where large increases in weed densities reduced weed control with herbicides and mechanical practices (Winkle, Leavitt, and Burnside, 1981; Buhler, Gun-solus, and Ralston, 1992).

Development of Cultural and Control Practices

Periodicity of germination and emergence of different weed species is also an important aspect of weed management. The seed banks of most agricultural lands contain many weed species. Knowledge of when differ-ent weed species are likely to emerge is important in planning effective weed control programs (Ogg and Dawson, 1984). There are a typical periodicity and a period or periods of high emergence, which is character-istic for each species (Chepil, 1946; Roberts and Feast, 1970; Stoller and Wax, 1973; Grime et al., 1981; Hakansson, 1983; Ogg and Dawson, 1984; Wiese and Binning, 1987; Wilson et al., 1992; Popay et al., 1995). For example, Wilson et al. (1992) found a 20-day difference in the time of initial emergence for nine summer annual weed species grown at the same location.

The time of weed seedling emergence influences which species will be the most serious weeds with a given crop production practice or most susceptible to certain control measures. Stoller and Wax (1973) concluded that weed species that complete most of their emergence early are killed during soil preparation before planting corn or soybean. Ogg and Dawson (1984) observed distinct patterns of emergence for eight weed species and concluded that if a weed species has a restricted emergence pattern, the crop could be planted later and tillage used to destroy weed seedlings before planting. They also pointed out that knowledge of emergence pat-terns could be used for timing of cultivation and postemergence applica-tion of herbicides. Delaying soybean planting reduced weed populations and improved weed control with rotary hoeing and cultivation (Buhler and Gunsolus, 1996). Reductions in weed density due to delayed planting varied by species with a 25% reduction for pigweed species and nearly 80% for common lambsquarters.

Knowledge of the time of emergence of different weed species compared with the crop and the influence of tillage and other cultural practices on emergence can be useful in developing integrated weed management systems. Understanding the dynamics of weed emergence could be useful in determining the most effective timing for tillage and herbicide applications. If effective and selective control measures are not available, it would be unwise to plant a crop in a field infested with weeds having emergence patterns similar to the crop.

Decision Aids

Bioeconomic models that utilize seed bank and weed emergence data to recommend possible weed management strategies are a potential use of seed bank information (King et al., 1986; Swinton and King, 1994; Wiles et al., 1996). These models build on previous modeling efforts because they incorporate weed population dynamics into weed management while accommodating multiple weed species and a wide range of control tactics.

Field evaluation of bioeconomic models has shown their potential to reduce herbicide use while maintaining weed control and increasing economic returns. In irrigated corn in Colorado (Lybecker, Schweizer, and King, 1991), grain yields and gross incomes were similar for flexible, model-generated, and fixed weed management strategies. However, herbicide use was reduced and gross economic margins increased by use of the model. They concluded that a bioeconomic model employed to make weed management decisions in corn could maintain weed control, increase profits, and decrease herbicide use.

A bioeconomic decision aid for management of annual weeds in corn and soybean (Swinton and King, 1994), was tested in the field in western Minnesota (Forcella et al., 1996a). After applying model-recommended treatments to the same plots for four years, there were no increases in weed densities or decreases in corn or soybean yields. Model recommendations reduced weed control costs and resulted in an average annual herbicide active ingredient application of 1.1 kg/ha compared with 3.5 kg/ha with a standard treatment.

In a field evaluation of the same model (Swinton and King, 1994) in eastern Minnesota (Buhler et al., 1996a), model-generated treatments controlled weeds as well as a standard herbicide treatment (Table 3). Herbicide use decreased by 27% using seed bank data and 68% using postemer-

TABLE 3. Herbicide Load, Weed Control, Corn Yield, and Net Return to Weed Control Using a Bioeconomic Weed Management Model (Adapted from Buhler et al., 1996a).

Treatment[a]	Herbicide load	Weed control[b]	Corn yield	Net return
	kg/ha	%	kg/ha	$/ha
1991				
Seed bank model	3.9	96	11300	638
Seedling model	0.43	92	11100	603
Standard herbicide	5.1	98	11600	661
1992				
Seed bank model	4.0	93	7840	297
Seedling model	2.2	92	9380	410
Standard herbicide	5.1	93	9910	440
1993				
Seed bank model	3.3	94	10950	684
Seedling model	2.2	91	10600	621
Standard herbicide	5.1	97	12130	730

[a]Seed bank and seedling model treatments were generated by a bioeconomic model (Swinton and King, 1994) using soil seed and seedling densities, respectively.
[b]Pooled over species.

gence seedling data compared with a standard herbicide treatment. Net economic return to weed control was not increased by using model-generated control recommendations. In soybean (Buhler et al., 1996b), model-generated treatments controlled weeds as well as a standard herbicide treatment. Averaged over 3 years, quantity of herbicide active ingredients applied was decreased by 47% using seed bank information and 93% with postemergence seedling information compared with a standard soil-applied herbicide treatment. However, frequency of herbicide application was not reduced. Net economic return to weed control was increased 50% of the time using the model compared with a standard herbicide treatment.

The performance of decision aid models has been influenced by weed population characteristics and environmental conditions. Decision aids have been useful for choosing control tactics, but have not consistently reduced herbicide use, treatment frequency, or increased economic returns when weed densities were high. To maximize the potential benefits of a bioeconomic weed management model, it should be used in integrated weed management systems that maintain low to moderate weed densities. Further research on bioeconomic weed management models under a wider range of production, weed population, and economic conditions will define their utility for improving economic and environmental aspects of weed management.

Economic Thresholds

Economic thresholds have been criticized for not accounting for the impacts of seed production by sub-threshold populations of weeds on future weed management costs. Economic optimum thresholds (Cousens, 1987) incorporate the impact of seed production on future weed populations. Bauer and Mortensen (1992) calculated the economic optimum thresholds for velvetleaf and common sunflower (*Helianthus annuus* L.) in soybean to be 7.5-fold and 3.6-fold lower than the economic threshold, respectively. The larger ratio for velvetleaf reflected higher seed production and greater seed longevity compared with that of common sunflower.

Field evaluation of a threshold-based decision support model showed that weed density the following year may be increased by seed production from sub-threshold weed populations (Buhler et al., 1996a,b). Model-based treatments often resulted in weed densities greater than a standard herbicide the following year. In 1 of 3 years of research (Buhler et al., 1996a), model-generated treatments used in corn increased weed densities and reduced weed control and soybean yields the following year. Conversely, Forcella et al. (1996a) found that weed populations did not increase over 3 years with model-based treatments compared with standard herbicide practices.

Understanding the impacts of seed production from sub-threshold populations of weeds will be a key to the adoption of threshold-based weed control systems. It will be essential that we understand how much seed return can occur without increasing weed control costs or yield losses in succeeding years.

Light Response

Exposure to light breaks seed dormancy in many plant species, including several weeds. Photodormancy is primarily related to the presence and form of phytochrome in seeds (Salisbury and Ross, 1978). The light requirement is thought to be an evolutionary adaptation by small-seeded species that may not emerge if germination occurs more than a few cm below the soil surface (Salisbury and Ross, 1978; Pons, 1991). Because light can penetrate only a few mm in soil (Woolley and Stoller, 1978; Egley, 1986), dormancy may be induced in light-requiring seeds by shallow burial (Wesson and Wareing, 1969; Pons, 1991). Other factors such as soil temperature and moisture (Henson, 1970; Salisbury and Ross, 1978) and nutrient availability (Vincent and Roberts, 1977; Roberts and Benjamin, 1979; Egley, 1989) may interact with light to regulate photodormancy.

A major source of light for buried weed seed is the flash received during tillage. Increased weed seed germination after tillage in the light versus darkness was first documented in 1969 (Wesson and Wareing, 1969). In this study it was proposed that tillage during daylight increased weed populations. Tilling the soil during darkness or while excluding light has been shown to reduce weed populations in field research (Hartmann and Nezadel, 1990; Ascard, 1994; Buhler and Kohler, 1994; Scopel, Ballare, and Radosevich, 1994). Less emergence of several weed species was observed in plots tilled during darkness compared with those tilled during daylight (Buhler and Kohler, 1994). This suggests that eliminating light exposure of the soil during tillage may have practical application for management of light-sensitive weed species.

Additional research is needed to determine the light sensitivity of various weed species, the effect of light quality on seed germination, the effects of tillage depth and timing, types of tillage tools, and the effects of soil and environmental conditions. Because all weed species do not possess photodormancy, continuous use of tillage during darkness will select for weed species and biotypes with light-insensitive germination.

CONCLUSION AND RESEARCH NEEDS

Seed and seed bank dynamics regulate communities of many of our most important weed species. A better understanding of seed and seed bank dynamics is critical for the development of more efficient weed management systems. In the short-term, weed biology research will not eliminate the inputs currently used to manage weeds. However, the knowledge gained will likely provide the foundation for development of new strategies and more efficient techniques, resulting in more reliable weed management systems that are cost effective and pose less threat to the environment.

In the past, the primary concern of weed science has been control technology rather than understanding weedy species and their interaction with the agroecosystem. Weeds have been present since the beginnings of civilization and are not likely to disappear in the foreseeable future. All forms of disturbance result in survival and selection of the best adapted plant species. Any cropping system that exerts a continuous, strong selection pressure will cause a build-up of the best adapted weed species and biotypes. Development of improved cropping systems will require an approach that concentrates on the processes and patterns linking scientific disciplines to agricultural systems. Agricultural systems are composed of interacting production, environmental, biological, economic, and social components. These interactions require the study of not only the parts, but also the whole system. Long-term improvements in weed management and agricultural production systems will require a convergence of agriculture, ecology, economics, and sociology (Levins, 1986; Radosevich and Ghersa, 1992). Linkages among these disciplines will form the basis for successful, stable, and profitable cropping and weed management systems.

REFERENCES

Ascard, J. (1994). Soil cultivation in darkness reduced weed emergence. *Acta Horticulturae* 372:167-177.

Ball, D.A. (1992). Weed seedbank response to tillage, herbicides, and crop rotation sequence. *Weed Science* 40:654-659.

Bauer, T.A. and D.A. Mortensen. (1992). A comparison of economic and economic optimum thresholds for two annual weeds in soybean. *Weed Technology* 6:228-235.

Bazzaz, F.A. (1990). Plant-plant interactions in successional environments. In *Perspectives on Plant Competition*, eds. J.B. Grace and D. Tilman. San Diego, CA: Academic Press, pp. 239-263.

Bello, I.A., M.D.K. Owen, and H.M. Hatterman-Valenti. (1995). Effect of shade on velvetleaf (*Abutilon theophrasti*) growth, seed production, and dormancy. *Weed Technology* 9:452-455.

Benoit, D.L., D.A. Derksen, and B. Panneton. (1992). Innovative approaches to seedbank studies. *Weed Science* 40:660-669.

Benoit, D.L., N.C. Kenkel, and P.B. Cavers. (1989). Factors influencing the precision of soil seed bank estimates. *Canadian Journal of Botany* 67:2833-2840.

Biniak, B.M. and R.J. Aldrich. (1986). Reducing velvetleaf (*Abutilon theophrasti*) and giant foxtail (*Setaria faberi*) seed production with simulated-roller herbicide applications. *Weed Science* 34:256-259.

Bradbeer, J.W. (1988). *Seed Dormancy and Germination*. New York: Chapman and Hall.

Brust, G.E. and G.J. House. (1988). Weed seed destruction by arthropods and rodents in low-input soybean agroecosystems. *American Journal of Alternative Agriculture* 3:19-25.

Buhler, D.D. (1995). Influence of tillage systems on weed population dynamics and management in corn and soybean production in the central USA. *Crop Science* 35:1247-1257.

Buhler, D.D. and J.L. Gunsolus. (1996). Effect of date of preplant tillage and planting on weed populations and mechanical weed control in soybean (*Glycine max*). *Weed Science* 44:373-379.

Buhler, D.D., J.L. Gunsolus, and D.F. Ralston. (1992). Integrated weed management techniques to reduce herbicide inputs in soybean. *Agronomy Journal* 84:973-978.

Buhler, D.D., R.P. King, S.M. Swinton, J.L. Gunsolus, and F. Forcella. (1996a). Field evaluation of a bioeconomic model for weed management in corn (*Zea mays*). *Weed Science* (in press).

Buhler, D.D., R.P. King, S.M. Swinton, J.L. Gunsolus, and F. Forcella. (1996b). Field Evaluation of a bioeconomic model for weed management in soybean (*Glycine max*). *Weed Science* (in press).

Buhler, D.D. and K.A. Kohler. (1994). Tillage in the dark and emergence of annual weeds. *Proceedings of the North Central Weed Science Society* 49:142.

Buhler, D.D. and T.C. Mester. (1991). Effect of tillage systems on the emergence depth of giant foxtail (*Setaria faberi*) and green foxtail *(Setaria viridis)*. *Weed Science* 39:200-203.

Buhler, D.D., T.C. Mester, and K.A. Kohler. (1996). Effect of tillage and maize residue on the emergence of *Setaria faberi, Abutilon theophrasti, Amaranthus retroflexus, and Chenopodium album*. *Weed Research*. 36:153-165.

Buhler, D.D., R.E. Ramsel, O.C. Burnside, and G.A. Wicks. (1984). Survey of weeds in winter wheat in Nebraska–1980 and 1981. *Univ. of Neb. Agric. Res. Div. Pub. MP 49.*

Burnside, O.C., R.S. Moomaw, F.W. Roeth, G.A. Wicks, and R.G. Wilson. (1986). Weed seed demise in soil in weed-free corn (*Zea mays*) production across Nebraska. *Weed Science*. 34:248-251.

Cavers, P.B. (1995). Seed banks: memory in soil. *Canadian Journal of Soil Science* 75:11-13.

Cavers, P.B. (1983). Seed demography. *Canadian Journal of Botany* 61:3678-3590.

Cousens, R. (1987). Theory and reality of weed control thresholds. *Plant Protection Quarterly* 2:13-20.

Cousens, R. and S.R. Moss. (1990). A model of the effects of cultivation on the vertical distribution of weed seeds within the soil. *Weed Research* 30:61-70.

Crawley, M.J. (1992). Seed predators and plant population dynamics. In *Seeds: The Ecology of Regeneration in Plant Communities,* ed. M. Fenner. Wallingford, OX UK: CAB International, pp. 157-191.

Currie, R.S. and T.F. Peeper. (1988). Combine harvesting affects weed seed germination. *Weed Technology* 2:499-504.

Dekker, J., B. Dekker, H. Hilhorst, and C. Karssen. (1996). Weedy adaptation in *Setaria* spp.; IV. Changes in the germinative capacity of *S. faberii* (Poaceae) embryos with development from anthesis to after abscission. *American Journal of Botany* 83:979-991.

Derksen, D.A., G.P. Lafond, A.G. Thomas, H.A. Loeppky, and C.J. Swanton. (1993). Impact of agronomic practices on weed communities: tillage systems. *Weed Science* 41:409-417.

Dyer, W.E. (1995). Exploiting weed seed dormancy and germination requirements through agronomic practices. *Weed Science* 43:498-503.

Eberlein, C.V., K. Al-Khatib, M.J. Guttieri, and E.P. Fuerst. (1992). Distribution and characteristics of triazine-resistant Powell Amaranth (*Amaranthus powelli*) in Idaho. *Weed Science* 40:507-512.

Egley, G.H. (1989). Some effects of nitrate-treated soil upon the sensitivity of buried redroot pigweed (*Amaranthus retroflexus* L.) seeds to ethylene, temperature, light, and carbon dioxide. *Plant, Cell and Environment* 12:581-588.

Egley, G.H. (1986). Stimulation of weed seed germination in soil. *Reviews of Weed Science* 2:67-89.

Egley, G.H. and S.O. Duke. (1985). Physiology of weed seed dormancy and germination. In *Weed Physiology*, Vol. 1, ed. S.O. Duke. Boca Raton, FL: CRC Press, pp. 27-64.

Fenner, M. (1985). *Seed Ecology*. New York: Chapman Hall.

Forcella, F., R.P. King, S.M. Swinton, D.D. Buhler, and J.L. Gunsolus. (1996a). Multi-year validation of a decision aid for integrated weed management. *Weed Science*. 44:650-661.

Forcella, F. and M.J. Lindstrom. (1988). Movement and germination of weeds in ridge-till crop production systems. *Weed Science* 36:56-59.

Forcella, F., R.G. Wilson, J. Dekker, R.J. Kremer, J. Cardina, R.L. Anderson, D. Alm, K.A. Renner, R.G. Harvey, S. Clay, and D.D. Buhler. (1996b). Weed seedbank emergence across the corn belt, 1991-1994. *Weed Science* (in press).

Forcella, F., R.G. Wilson, K.A. Renner, J. Dekker, R.G. Harvey, D.A. Alm, D.D. Buhler, and J.A. Cardina. (1992). Weed seedbanks of the U.S. Cornbelt: magnitude, variation, emergence, and application. *Weed Science* 40:636-644.

Grime, J.P., G. Mason, A.V. Curtis, J. Rodman, S.R. Band, M.A.G. Mowforth, A.M. Neal, and S. Shaw. (1981). A comparative study of germination characteristics in a local flora. *Journal of Ecology* 69:1017-1059.

Gupta, S.C., W.E. Larson, and D.R. Linden. (1983). Effect of tillage and surface residues on soil temperature. I. Upper boundary temperature. *Soil Science Society of America Journal* 47:1212-1218.

Hakansson, S. (1983). Seasonal variation in the emergence of annual weeds–an introductory investigation in Sweden. *Weed Research* 23:313-324.

Harmon, G.W. and F.D. Keim. (1934). The percentage and viability of weed seeds recovered in the feces of farm animals and their longevity when buried in manure. *Journal of the American Society of Agronomy* 26:762-767.

Hartmann, K.M. and W. Nezadal. (1990). Photocontrol of weeds without herbicides. *Naturwissenschaften* 77:158-163.

Hartzler, R.G. (1996). Velvetleaf (*Abutilon theophrasti*) population dynamics following a single year's seed rain. *Weed Technology* 10:581-586.

Hartzler, R.G. and G.W. Roth. (1993). Effect of prior year's weed control on herbicide effectiveness in corn (*Zea mays*). *Weed Technology* 7:611-614.

Henson, I.E. (1970). The effects of light, potassium nitrate and temperature on the germination of *Chenopodium album* L. *Weed Research* 10:27-39.

Kremer, R.J. (1993). Management of weed seed banks with microorganisms. *Ecological Applications* 3:42-52.

King, R.P., D.W. Lybecker, E.E. Schweizer, and R.L. Zimdahl. (1986). Bioeconomic modeling to simulate weed control strategies for continuous corn (*Zea mays*). *Weed Science* 34:972-979.

Lang, A.G., J.D. Early, G.C. Martin, and R.L. Darnell. (1987). Endo-, para-, and ecodormancy: physiological terminology and classification for dormancy research. *Journal of Horticultural Science* 22:371-377.

Levins, R. (1986). Perspectives in integrated pest management: From an industrial to ecological model of pest management. In *Ecological Theory and Pest Management*. New York: John Wiley and Sons, pp. 1-18.

Liebman, M. and E. Dyck. (1993). Crop rotation and intercropping strategies for weed management. *Ecological Applications* 3:92-122.

Lindquist, J.L., B.D. Maxwell, D.D. Buhler, and J.L. Gunsolus. (1995). Velvetleaf (*Abutilon theophrasti*) recruitment, survival, seed production, and interference in soybean (*Glycine max*). *Weed Science* 43:226-232.

Lueschen, W.E. and R.N. Andersen. (1980). Longevity of velvetleaf (*Abutilon theophrasti*) seed in soil under agricultural practices. *Weed Science* 28: 341-346.

Lybecker, D.W., E.E. Schweizer, and R.P. King. (1991). Weed management decisions in corn based on bioeconomic modeling. *Weed Science* 39:124-129.

McCanny, S.J. and P.B. Cavers. (1988). Spread of proso millet (*Panicum miliaceum* L.) In Ontario, Canada. II. Dispersal by combines. *Weed Research* 28:67-72.

Mortensen, D.A., G.A. Johnson, and L.J. Young. (1993). Weed distribution in agricultural fields. In *Soil Specific Crop Management,* ed. P.C. Robert, R.H. Rust, and W.E. Larson. Madison, WI: American Society of Agronomy, pp. 113-123.

Mt. Pleasant, J. and K.J. Schlather. (1994). Incidence of weed seed in cow (*Bos* sp.) manure and its importance as a weed source for cropland. *Weed Technology* 8:304-310.

Murdoch, A.J. and R.H. Ellis. (1992). Longevity, viability and dormancy. In *Seeds: The Ecology of Regeneration in Plant Communities,* ed. M. Fenner. Wallingford, OX UK: CAB International, pp. 193-229.

Nikolaeva, M.G. (1977). Factors controlling the seed dormancy pattern. In *The Physiology and Biochemistry of Seed Dormancy and Germination,* ed. A.A. Khan. Amsterdam: North Holland Publishing, pp. 51-74.

Norris, R.F. (1992). Have ecological and biological studies improved weed control strategies? *Proceedings of the First International Weed Control Congress* 1:7-33.

Ogg, A.G., Jr. and J.H. Dawson. (1984). Time of emergence of eight weed species. *Weed Science* 32:327-335.

Pareja, M.R., D.W. Staniforth, and G.P. Pareja. (1985). Distribution of weed seed among soil structural units. *Weed Science* 33:182-189.

Pons, T.L. (1991). Induction of dark dormancy in seeds: its importance for the seed bank in the soil. *Functional Ecology* 5:669-675.

Popay, A.I., T.I. Cox, A. Ingle, and R. Kerr. (1995). Seasonal emergence of weeds in cultivated soil in New Zealand. *Weed Research* 35:429-436.

Radosevich, S.R. and C.M. Ghersa. (1992). Weeds, crops, and herbicides: A modern-day "neckriddle." *Weed Technology* 6:788-795.

Reader, R.J. (1991). Control of seedling emergence by ground cover: a potential mechanism involving seed predation. *Canadian Journal of Botany* 69: 2084-2087.

Roberts, E.H. and S.K. Benjamin. (1979). The interaction of light, nitrate and alternating temperature on the germination of *Chenopodium album, Capsella bursa-pastoris, and Poa annua* before and after chilling. *Seed Science and Technology* 7:379-392.

Roberts, H.A. (1963). Studies on the weeds of vegetable crops. III. Effect of different primary cultivations on the weed seeds in the soil. *Journal of Ecology* 51:83-95.

Roberts, H.A. and P.M. Feast. (1970). Seasonal distribution of emergence in some annual weeds. *Experimental Horticulture* 21:36-41.

Roberts, H.A. and M.E. Ricketts. (1979). Quantitative relationships between the weed flora after cultivation and the seed population in the soil. *Weed Research* 19:269-275.

Robinson, R.G. (1949). Annual weeds, their viable seed populations in the soil and their effects on yields of oats, wheat, and flax. *Agronomy Journal* 41:513-518.

Salisbury, F.B. and C.W. Ross. (1978). *Plant Physiology.* 2nd ed. Belmont, CA: Wadsworth Pub. Co., Inc.

Salzman, F.P., R.J. Smith, and R.E. Talbert. (1988). Suppression of red rice (*Oryza sativa*) seed production with fluazifop and quizalofop. *Weed Science* 36:800-803.

Schreiber, M.M. (1992). Influence of tillage, crop rotation, and weed management on giant foxtail (*Setaria faberi*) population dynamics and corn yield. *Weed Science* 40:645-653.

Schweizer, E.E. and R.L. Zimdahl. (1984). Weed seed decline in irrigated soil after six years of continuous corn (*Zea mays*) and herbicides. *Weed Science* 32:76-83.

Scopel, A.L., C.L. Ballare, and S.R. Radosevich. (1994). Photostimulation of seed germination during soil tillage. *New Phytologist* 126:145-152.

Senseman, S.A. and L.R. Oliver. (1993). Flowering patterns, seed production, and somatic polymorphism of three weed species. *Weed Science* 41:418-425.

Staricka, J.A., P.M. Burford, R.R. Allmaras, and W.W. Nelson. (1990). Tracing the vertical distribution of simulated shattered seeds as related to tillage. *Agronomy Journal* 82:1131-1134.

Stevens, O.A. (1957). Weights of seeds and numbers per plant. *Weeds* 5:46-55.

Stoller, E.W. and L.M. Wax. (1973). Periodicity of germination and emergence of some annual weeds. *Weed Science* 21:574-580.

Swinton, S.M. and R.P. King. (1994). A bioeconomic model for weed management in corn and soybean. *Agricultural Systems* 44:313-335.

Taylorson, R.B. (1987). Environmental and chemical manipulation of weed seed dormancy. *Reviews of Weed Science* 3:135-154.

Taylorson, R.B. (1982). Anesthetic effects on secondary dormancy and phytochrome responses in *Setaria faberi* seeds. *Plant Physiology* 70:882-886.

Thomas, A.G. and B.L. Frick. (1993). Influence of tillage systems on weed abundance in southwestern Ontario. *Weed Technology* 7:699-705.

Thompson, K. and J.P. Grime. (1979). Seasonal variation in the seed banks of herbaceous species in ten contrasting habitats. *Journal of Ecology* 67:893-921.

Vincent, E.M. and E.H. Roberts. (1977). The interaction of light, nitrate and alternating temperatures in promoting the germination of dormant seeds of common weed species. *Seed Science and Technology* 5:659-670.

Warnes, D.D. and R.N. Andersen. (1984). Decline of wild mustard (*Brassica kaber*) seeds in soil under various cultural and chemical practices. *Weed Science* 32:214-217.

Wesson, G. and P.F. Wareing. (1969). The induction of light sensitivity in weed seeds by burial. *Journal of Experimental Botany* 20:414-425.

Wicks, G.A., O.C. Burnside, and W.L. Felton. (1994). Weed control in conservation tillage systems. In *Managing Agricultural Residues,* ed. P.W. Unger. Boca Raton, FL: Lewis Publishers, pp. 211-244.

Wiese, A.M. and L.K. Binning. (1987). Calculating the threshold temperature of development for weeds. *Weed Science* 35:177-179.

Wiles, L.J., R.P. King, E.E. Schweizer, D.W. Lybecker, and S.M. Swinton. (1996). GWM: General weed management model. *Agricultural Systems* 50:355-376.

Wilson, B.J. and H.M. Lawson. (1992). Seedbank persistence and seedling emergence of seven weed species in autumn-sown crops following a single year's seeding. *Annals of Applied Biology* 120:105-116.

Wilson, R.G. (1988). Biology of weed seeds in the soil. In *Weed Management in Agroecosystems: Ecological Approaches,* eds. M.A. Altieri and M. Liebman. Boca Raton, FL: CRC Press, pp. 25-39.

Wilson, R.G., K.J. Jarvi, R.C. Seymour, J.F. Witkowski, S.D. Danielson, and R.F. Wright. (1992). Annual weed growth across Nebraska. *Univ. NE Agric. Res. Div., Res. Bull.* 314-F.

Wilson, R.G., E.D. Kerr, and L.A. Nelson. (1985). Potential for using weed seed content in the soil to predict future weed problems. *Weed Science* 33:171-175.

Winkle, M.E., J.R.C. Leavitt, and O.C. Burnside. (1981). Effects of weed density on herbicide absorption and bioactivity. *Weed Science* 29:405-409.

Woolley J.T. and E.W. Stoller. (1978). Light penetration and light-induced seed germination in soil. *Plant Physiology* 61:597-600.

Yenish, J.P., J.D. Doll, and D.D. Buhler. (1992). Effects of tillage on vertical distribution and viability of weed seed in soil. *Weed Science* 40:429-433.

Zimdahl, R.L. (1991). *Weed Science–A Plea for Thought.* U.S. Dep. Agric., Cooperative State Research Service, Symposium Preprint, Washington, DC.

SUBMITTED: 06/13/96
ACCEPTED: 11/26/96

Allelopathy and Its Implications
in Agroecosystems

R. K. Kohli
Daizy Batish
H. P. Singh

SUMMARY. Allelopathy includes both positive and negative effects of one plant on the other through environment, though most of the studies seem to focus on its deleterious impacts. It plays a key role in both natural and managed ecosystems. In agroecosystems, several weeds, crops, agroforestry trees and fruit trees have been shown to exert allelopathic influence on the crops, thus, affecting their germination and growth adversely. Some of the agricultural and horticultural crops affect their own seedlings grown in succession which is commonly known as replant problem/syndrome. Available studies indicate that allelochemicals act via bringing certain changes in physiological functions like respiration, photosynthesis and ion uptake. These, in turn, result in visible changes in seed germination, further growth reduction and overall performance of the target plants. The studies on interplant interactions assume significance in agroforestry programmes for selecting the types of crops complementary to the selective tree species. In the recent past, however, scientific attention has also been drawn to exploit the positive signif-

---

R. K. Kohli, Reader, Daizy Batish, Lecturer, Department of Botany, Panjab University, Chandigarh 160 014, India. H. P. Singh, Lecturer, Department of Botany, D.A.V. College, Chandigarh 160 011, India.

Address correspondence to: R. K. Kohli, Department of Botany, Panjab University, Chandigarh 160 014, India (E-mail: pulib@puchd.ren.nic.in).

[Haworth co-indexing entry note]: "Allelopathy and Its Implications in Agroecosystems." Kohli, R. K., Daizy Batish, and H. P. Singh. Co-published simultaneously in *Journal of Crop Production* (The Food Products Press, an imprint of The Haworth Press, Inc.) Vol. 1, No. 1 (#1), 1998, pp. 169-202; and: *Crop Sciences: Recent Advances* (ed: Amarjit S. Basra) The Food Products Press, an imprint of The Haworth Press, Inc., 1998, pp. 169-202. Single or multiple copies of this article are available for a fee from The Haworth Document Delivery Service [1-800-342-9678, 9:00 a.m. - 5:00 p.m. (EST). E-mail address: getinfo@haworth.com].

icant roles this phenomenon can play in enhancing crop productivity. In this context, we discuss the tremendous scope of allelopathy towards weed and pest management, apart from nitrogen conservation, and synthesis of novel agrochemicals based on natural product chemistry. The use of natural products of plant or microbial origin as pesticides/herbicides have gained much attention of the scientists as they offer many advantages over synthetic chemicals. The production of such chemicals should be enhanced by devising suitable protocols based on biotechnological procedures for their widespread utilization. *[Article copies available for a fee from The Haworth Document Delivery Service: 1-800-342-9678. E-mail address: getinfo@haworth.com]*

KEYWORDS. Allelopathy, agriculture, crop productivity, weed and pest management, nitrogen conservation, agrochemicals, biotechnological aspects

INTRODUCTION

Etymologically, the term allelopathy coined by Molisch (1937) using two Greek words 'allelo' and 'pathos' means reciprocal sufferings of two organisms. This included both detrimental and beneficial interactions between the plants mediated through chemicals released by the donor. However, in practice, allelopathy is generally used for detrimental plant-plant interactions. Rice (1984) defined allelopathy as any harmful or beneficial effects of the plants including microorganisms on another plant through the release of chemicals that escape into the environment. The negative or positive effects would, however, depend upon the concentration of chemicals reaching the receptor, or its inherent susceptibility (Putnam, 1988). It is a complex phenomenon involving a variety of interrelationships among plants. Virtually all plant parts such as leaves (Kumari and Kohli, 1987), roots (Horsley, 1977), pollen (Cruz-Ortega, Anaya, and Ramos, 1988), trichomes (Bansal, 1990), bark (Kohli, 1990) and seeds and fruits (Friedman, Rushkin, and Waller, 1982) have allelopathic potential.

Allelopathic interactions among individuals of same species termed as 'Intraspecific Toxicity' or 'Autotoxicity' or 'Autoallelopathy' have been reported in a number of species (Kumar, 1991). The advantage of autotoxicity is difficult to interpret, but it may encourage geographical distribution of the donor species, serve as an adaptation to induce dormancy and prevent decay of its seeds and propagules (Friedman and Waller, 1985). Contrarily, the term 'Teletoxicity' is used when suscept species are taxonomically different from donor or agent species (Kushal, 1987). Shafer and Garrison (1986) used another terminology of 'Heterotoxicity' and 'Autotoxicity' for effects of one plant on different or same species, respectively.

The chemicals through which allelopathic effect is imposed are known as allelochemicals or allelochemics (Whittaker, 1970). The term 'Semiochemicals' was also proposed for all the chemicals that bring interactions among plants (Law and Regnier, 1971). Grümmer (1955) had used different terminology for the allelochemicals depending upon nature of donor and receptor or suscept plant: Koline (donor and suscept both higher plants); Antibiotic (donor and suscept both microbes); Phytoncide (donor higher plant and suscept microbe) and Marasmin (donor microbe and suscept higher plant). However, because of their non-specific and non-descriptive nature, they are no longer used. Whitman (1988) divided allelochemicals into 4 categories namely allomones, kariomones, synomones and antimones, depending upon their effect on donor or receptor plant. Further, the simple terms like 'Saproinhibitins' and 'Phytoinhibitins' for the allelochemicals of microbe and higher plant origin were also suggested (Fuerst and Putnam, 1983). However, term 'Allelochemicals' is widely used for all chemicals involved in plant-plant interactions. Their release is facilitated by a variety of processes such as leaching from above ground plant parts, volatilization, root exudation, stem flow, microbial activity, ploughing of plant residues in soil and dry residue decomposition (Muller, 1969; Rice, 1984; Putnam and Tang, 1986; Daizy, 1990; Inderjit, 1996). Secondary metabolites such as terpenoids, phenolics, alkaloids, fatty acids, steroids and polyacetylenes may function as allelochemicals (Waller, 1987). The number and diversity of allelochemicals is increasing rapidly. Over 10,000 secondary metabolites have been recognized (Einhellig, 1989).

Allelopathy has been demonstrated in a number of plants employing various laboratory assays. However, only in a few cases has its operation under natural conditions been established (Newman, 1978; Willis, 1985; Williamson, 1990). Further, in a community, it is difficult to segregate allelopathy from other phenomena such as competition, as both of these may be operative simultaneously or sequentially. Muller (1969) used the term 'Interference' to include all allelopathy and competition based adverse effects of neighbouring plants. As a proof of allelopathy, Fuerst and Putnam (1983) prepared a protocol parallel to Koch's postulates (for proof of disease). Accordingly, the term implies the fulfillment of some conditions like: (a) laboratory bioassays of allelopathic plants must clearly describe symptoms or growth reductions of suscept plant; (b) the chemical/s responsible for inhibition should be identified and isolated; (c) same symptoms should appear if such chemical/s is/are supplied externally to a controlled system and (d) the chemical/s responsible for bringing about growth reductions must actually be detected in and around suscept plant.

Allelopathy plays a significant role under both natural and manipulated ecosystems (Rice, 1984). In the former it plays an active role in terrestrial as well as phytoplankton succession, vegetation patterning, imparting dormancy to the seeds and preventing the seed decay, while in the latter it results in a number of direct or indirect effects.

ALLELOPATHY IN AGRICULTURE

Allelopathic effect of plants in agriculture has been known for centuries. For the first time Theophrastus (300 B.C.) is reported to have observed destruction of weeds by chickpea (*Cicer arietinum*). Chemical exudates released from the crops caused soil sickness problem in agriculture and crop rotation was suggested to be the only solution to this problem (de Candolle, 1832). In the earlier reports, however, allelopathy was not scientifically demonstrated until Schreiner and Reed (1908) conducted some experiments and isolated a number of chemical compounds from plants and soil. After that a number of reports have indicated the allelopathic effect of weeds, crops, woody plants and fruit trees in agriculture (Rice, 1984, 1995). However, suitable manipulations of the phenomenon towards improvement of crop productivity and environmental protection through eco-friendly control of weeds, pests, crop diseases and conservation of nitrogen have gained prominent attention of scientists engaged in allelopathic research. Following text attempts to review the available reports in this direction.

Allelopathic Effects of Weeds on Agricultural Crops

One of the most serious problems of modern agriculture is crop losses caused by weeds. Worldwide 10% loss of agricultural production is caused by weeds alone (Altieri and Liebman, 1988). However, these studies are preliminary and are based upon impact of only dominant weeds, ignoring the total weed community that colonizes the crop fields. Allelopathic weeds can affect crops by a number of ways like delaying or preventing seed germination, reducing seedling growth, affecting crop symbionts and influencing crop pests. It is often difficult to segregate allelopathy from other mechanisms of interference operating simultaneously. However, with the development of new techniques, it is now possible to demonstrate allelopathy more convincingly. Some of the common reports on allelopathic effects of weeds in agriculture have been tabulated alphabetically and arranged with respect to their families and botanical names (Table 1).

Allelopathic Effects of Agricultural Crops

Several crops are known to exhibit strong allelopathic property (Rice, 1984, 1995). They either exert their influence simultaneously or on crops grown sequentially. Most of the allelopathic effects of the crops such as *Triticum aestivum, Oryza sativa, Zea mays, Brassica* spp. and *Secale cereale* are attributed to the decomposing residues left in the fields after harvesting. Several studies are available where the residues not only inhibit seedling growth of other plants but also of their own (Srivastava, Totey, and Prakash, 1986; Lodhi, Bilal, and Malik, 1987; Thorne et al., 1990). *Z. mays* residues reduce early growth of its own seedlings (Yakle and Cruse, 1984) which was attributed to the phytotoxicity caused by its aqueous extracts (Martin, McCoy, and Dick, 1990). This effect was, however, considerably reduced when the practice of rotation of *Z. mays* with *Glycine max* (soybean) was followed (Turco et al., 1990). Not only residues, but *Z. mays* pollen were also allelopathic (Jiménez et al., 1983). *Oryza sativa* (rice) also possesses allelopathic potential (Rice, 1984) exerted through its own residues (Chou and Lin, 1976). *Brassica* spp. are known to be allelopathic towards wheat in Australia (Mason-Sedun and Jessop, 1988). There are several reports on allelopathic crops (Table 2) which are now being managed in several ways for increasing crop production, controlling weeds and combating crop diseases (discussed later).

Allelopathic Effects of Trees in Agricultural Fields

Trees are fast becoming one of the major components of agroecosystems under the various agroforestry programmes, combining the best features of forestry and agriculture. To increase the overall productivity and profitability, this old and traditional practice has been revived during the last decade. Many of the agroforestry benefits have been tested fruitful while others remain to be proved (Sanchez, 1995). These benefits of agroforestry largely depend upon tree-crop combination, orientation and number of trees and their interference (competition or allelopathy). Competitive effects of trees in agroecosystems can be easily manipulated like opting for sequential agroforestry. Allelopathic effects are, however, difficult to be managed until specific studies are conducted in this regard. There are now several reports on negative impact of trees in agroecosystems raised as shelterbelts, wind breaks, boundary plantation, alleys or simply scattered in the fields. Some of these are tabulated in Table 3.

TABLE 1. List of Weeds Showing Allelopathic Influence on Crops

Weed Species (Donor)	Crop (Recipient)	Effects	Reference (s)
Dicotyledons **Amaranthaceae** *Amaranthus palmeri* S. Wats	*Allium cepa* L. *Daucus carota* L.	Plant residue reduces fresh weight and seedling growth.	Bradow and Connick Jr.,1987
A. retroflexus	*Zea mays* L. *Glycine max* (L.) Merr.	Water extracts inhibit hypocotyl growth of *G. max* and coleoptile growth in *Z. mays.*	Bhowmik and Doll, 1979
Alternanthera triandra (Lam.) Steud.	*Gylcine max* (L.) Merr. *Arachis hypogaea* L. *Phaseolus aureus* Roxb.	Growth of the test plants is reduced.	Tiwari et al., 1985
Celosia argentea L.	*Pennisetum typhoideum* Rich. *P. americanum* (L.) Leeke	Aqueous extracts of fresh leaves, stems and roots inhibit shoot and root growth.	Pandya, 1976
Digera arvensis Forsk.	*Pennisetum typhoides* Stapf. et. Hubb.	Inhibition of seed germination and radicle elongation.	Sarma, 1974
Asciepiadaceae *Asclepias syriaca* L.	*Sorghum vulgare* Pers.	Aqueous extracts inhibit the growth of *Sorghum.*	Rasmussen and Einhellig, 1975
Asteraceae *Ambrosia trifida* L.	*Raphanus sativus* L., Sorghum species.	Aqueous extracts inhibit seed germination and seedling growth.	Rasmussen and Einhellig, 1979
Antennaria microphylla Rydb.	*Lactuca sativa* L., *Euphorbia esula* L.	Inhibition of root elongation of test species.	Manners and Galitz, 1986
Bidens pilosa L.	*Lactuca sativa* L. *Phaseolus vulgaris* L. *Zea mays* L. *Sorghum bicolor* (L.) Moench.	Root exudates inhibit the seedling growth.	Stevens and Tang, 1985
Cirsium arvense (L.) Scop.	*C. arvense* (L.) Scop. *Trifolium subterraneum* L. *Lolium perenne* L. *Hordeum distichon* L.	Aqueous leachates from roots and foliage retard seed germination.	Bendall, 1975
Eupatorium odoratum L.	*Euphorbia heterophylla* L. *Vigna unguiculata* (L.) Walp.	Stem, leaf and root residues retard germination, leaf area and dry matter yield.	Fadayomi and Oyebade, 1984
E. edenophorum Spreng.	*Lantana camara* L.	Inhibition of seed germination.	Dhyani, 1978

Weed Species (Donor)	Crop (Recipient)	Effects	Reference (s)
Parthenium hysterophorus L.	*Triticum aestivum* L. *Phaseolus vulgaris* L. *Lycopersicum esculentum* (L.) Mill. *Vigna sinensis* L.	Chemics from weed and dry leaves mixed with soil reduce growth and nodulation in beans.	Kanchan and Jayachandra,1979a,b
P. hysterophorus L.	*Glycine max* (L.) Merr. *Tricitum aestivum* L. *Zea mays* L. *Lycopersicon esculentum* (L.) Mill.	Extracts and residues significantly reduce root and shoot dry weight.	Mersie and Singh, 1987b, 1988
P. hysterophorus L.	*Phaseolus aureus* Roxb. *P. mungo* L. *Cajanus cajan* L. *Lens esculentum* L.	Germination and growth was reduced.	Kohli and Batish, 1994
Xanthium strumarium L.	*Brassica campestris* L. *Lactuca sativa* L. *Pennisetum americanum* (L.) Leeke.	Aqueous extracts from different parts reduce germination, early growth and dry weight of test plants.	Inam, Hussain, and Farhat, 1987
Brassicaceae *Sisymbrium irio* L.	*Allium cepa* L.	Growth is affected.	Menges and Tamez, 1981
Cannabinaceae *Cannabis sativa* L.	*Brassica campestris* L. *Sorghum bicolor* L. *Trifolium foenum-graecum* *Vigna mungo*	Litter and rain leachates and volatiles from shoots retard germination, radicle growth, fresh and dry biomass and moisture content.	Inam, Hussain, and Farhat, 1989
Chenopodiaceae *Chenopodium album* L.	*Glycine max* (L.) Merr. *Zea mays* L.	Water extracts of residues inhibit root and coleoptile growth.	Bhowmik and Doll, 1979.
Cucurbitaceae *Citrullus colocynthis* Schard.	*Pennisetum typhoideum* Rich.	Aqueous extracts of fresh root, stem, leaf and fruit pulp cause delay in germination.	Bhandari and Sen, 1972
Ericaceae *Calluna vulgaris* (L.) Salisb.	*Trifolium pratense* L.	Aqueous extracts from leaves inhibit growth.	Ballester, Vieitez, and Vieitez, 1982
Daboecia cantabrica (Hudson) K. Koch.	*Trifolium pratense* L.	Aqueous extracts of flowers retard germination.	Ballester, Vieitez, and Vieitez, 1982
Erica australis L. *E. arborea* L.	*Trifolium pratense* L.	Aqueous extracts of fresh and dried flowers inhibit the root and hypocotyl growth.	Ballester, Vieitez, and Vieitez, 1979

TABLE 1 (continued)

Weed Species (Donor)	Crop (Recipient)	Effects	Reference (s)
E. vagans	*Trifolium pratense* L.	Aerial parts drastically affect the growth.	Ballester, Vieitez, and Vieitez, 1982
Eupborbiaceae *Euphorbia hirta* L.	*Arachis hypogaea* L., *Glycine max* (L.) Merr. *Phaseolus aureus* Roxb.	Secretions of the root affect the germination and early growth of test species.	Tiwari et al.,1985
Fumariaceae *Fumaria indica* Haussk.	*Triticum aestivum* L. var. Blue Silver.	Aqueous extracts from different parts affect seed germination and seedling growth.	Chaghtai, Sadiq, and Ibrar, 1986
Lamiaceae *Salvia syriaca* L.	*Cicer arietinum* L. *Hordeum vulgare* L. *Lens culinaris* Medic *Triticurm durum* Desf.	Aqueous extracts of shoot retard the seed germination and seedling development.	Abu-Irmaileh and Qasem, 1986.
Stachys parviflora Bth.	*Brassica campestris* L. *B. chinensis* L. *Lens culinaris* Medik *Phaseolus radiatus* L. *P. vulgaris* L. *Raphanus sativus* L. *Sorghum sudanense* (Piper) Stapf.	Aqueous extracts and litter of root, shoot and volatiles from entire plant inhibit germination and growth.	Hussain, Ahmed, and Akram, 1986
Malvaceae *Abutilon theophrastii* Medic.	*Glycine max* (L.) Merr. *Zea mays* L.	Residues are highly toxic and cause reduction in height, growth and fresh weight of shoots.	Bhowmik and Doll, 1979.
Oxalidaceae *Oxalis corniculata* L.	*Triticum aestivum* L.	Aqueous extracts of leaf and tuber reduce the seedling growth.	Leela, 1984
Papaveraceae *Argemone maxicana* L.	*Pennisetum typhoideum* Rich. *Triticum* spp *Brassica campestris* L. *Raphanus sativus* L.	Seedling growth is affected.	Sharma and Nathawat, 1987
Polygonaceae *Polygonum aviculare* L.	*Chenopodium album* L.	Seed germination and growth of hypocoryl and epicotyl are reduced.	Al Saadawi and Rice, 1982

Weed Species (Donor)	Crop (Recipient)	Effects	Reference (s)
P. aviculare L.	*Cynodon dactylon* Pers. and some nitrogen fixing bacteria.	Germination and seedling growth get reduced. Growth of nitrogen fixing bacteria is also affected.	Al Saadawi, Rice, and Karns, 1983
Solanaceae *Datura stramonium* L.	*Hordeum vulgare* L. *Triticum aestivum* L.	Alkaloid that leaches out of seeds retards seedling growth.	Lovett and Potts, 1987
Verbenaceae *Lantana camara* L.	*Abutilon theophrastii* Medic. *Glycine max* (L.) Merr. *Lepidium virginicum* L. *Zea mays* L.	Shoot residues affect shoot length in *Z. mays* and *A. theophrastii* while root length is reduced in all test plants.	Mersie and Singh, 1987a
Monocotyledons **Cyperaceae** *Cyperus esculentus* L.	*Glycine max* (L.) Merr. *Zea mays* L.	Plant residues and extracts reduce the dry weight.	Drost and Doll, 1980
C. rotundus	*Allium cepa* L. *Lycopersicum esculentum* (L.) Mill *Raphanus sativus* L.	Water extracts reduce the seedling survival of tested plants.	Meissner, Nel, and Smit, 1982
Liliaceae *Asphodelus tenuifolius* Cav.	*Triticum* spp.	Dry matter of plants gets reduced.	Porwal and Gupta, 1986
Poaceae *Agropyron repens* (L.) Beauv.	*Avena sativa* L. *Medicago sativa* L. *Zea mays* L. *Glycine max* (L.) Merr. *Lepidium sativum* L. *Phaseolus vulgaris* L.	Aqueous extracts of rhizome and shoot retard the seed germination and root growth.	Gabor and Veatch, 1981; Weston and Putnam, 1986
Avena fatua L.	*Triticum* spp.	Reduces shoot and ear length.	Porwal and Gupta, 1986
Cynodon dactylon (L.) Pers.	*Prunus persica* L.	Growth of newly planted trees is affected.	Waller, Skroach, and Monaco, 1985
Dicanthium annulatum Stapf.	*Brassica campestris* L. *Lactuca sativa* L. *Pennisetum typhoideum* Rich. *Setaria italica* Beauv.	Aqueous extracts from shoots and soil inhibit the germination and growth of test species.	Dirvi and Hussain, 1979
Eragrostis poaeoides Beauv.	*Brassica campestris* L. *Pennisetum arnericanum* (L.) Leeke *Setaria italica* Beauv.	Aqueous extracts from shoots and inflorescence inhibit germination and growth.	Hussain, Zaidi, and Chughtai, 1984

TABLE 1 (continued)

Weed Species (Donor)	Crop (Recipient)	Effects	Reference (s)
Festuca arundinacea Schreb.	Brassica nigra Koch. Lotus corniculatus L. Trifolium pratense L.	Aqueous leachates of leaves and roots inhibit the seedling growth.	Peters and Luu, 1985
F. arundinacea Schreb.	Lactuca sativa L.	Organic fractions inhibit seedling growth.	Buta, Spaulding, and Reed, 1987
Lolium perenne L.	Lactuca sativa L.	Aqueous leachates of seeds strongly inhibit germination and early seedling growth.	Buta, Spaulding, and Reed, 1987
Phalaris minor Retz.	Triticum sp.	Shoot length as well as dry weight get reduced.	Porwal and Gupta, 1986
Setaria faberii Hersm. S. glauca (L.) Beauv. S. viridis (L.) Beauv.	Glycine max (L.) Merr. Zea mays L.	Residues significantly reduce height, growth and fresh weight of shoots.	Bhowmik and Doll, 1979
Sorghum bicolor (L.) Moench.	Lepidium sativum L.	Aqueous extracts of herbage and roots inhibit seed germination.	Lehle and Putnam, 1982
S. halepense (L.) Pers.	Hordeum vulgare L.	Dacaying plants in soil inhibit root and shoot growth.	Horowitz and Friedman, 1971

Fruit Trees Exhibiting Allelopathy Against Crops

Very often the orchard owners face a problem known as Replant Syndrome which is caused due to allelopathic effects of trees on its own seedlings. This problem arises following the removal of old orchards such as apple, peach, plum, apricots and citrus (Proebsting and Gilmore, 1941; Börner, 1959). Fruit trees in agriculture may however, affect other crops too (Brown, Tang, and Nishimoto, 1983). It involves a complex chain of reactions involving various pests, nematodes, and plant pathogens (Sharma, Chitkara, and Daulta, 1994). Some of the reports on toxicity of fruit crops in agriculture are given in Table 4.

ALLELOPATHY FOR IMPROVED CROP PRODUCTION

Since its inception in 1937, the term allelopathy is viewed only in terms of negative interactions, but of late it is being realized that, if suitably

TABLE 2. List of Some Allelopathic Crops

Crop (Donor)	Crop (Recipient)	Effects	Reference (s)
Dicotyledons **Asteraceae**			
Helianthus annuus L.	*Glycine max* (L.) Merr. *Sorghum* spp.	Dried leaves when mixed with soil, inhibit emergence of seeds and further growth.	Irons and Bumside, 1982; Schon and Einhellig, 1982
H. annuus L.	*Triticum aestivum* L.	Wheat emergence gets reduced by 4 to 33 % in the presence of sunflower stubbles in field.	Purvis and Jones, 1990
Brassicaceae *Brassica campestris* L. subsp.Pekinensis	*Vigna radiata* (L.) Wilczek.	Aqueous extracts of residues inhibit seed germination and growth.	Kuo, Chou, and Park, 1981
Cruciferous Plants	*Brassica juncea* (L.) Czern. *B. nigra* (L.) Koch. *Lactuca sativa* L. *Triticum aestivum* L.	Volatile components reduce germination and affect root and coleoptile growth.	Oleszek, 1987
Raphanus sativus L.	*Lactuca sativa* L.	Root and top residue inhibit seed germination.	Beckmann and Noffke, 1978
Convolvulaceae *Ipomoea batatas* (L.) Lam.	*Cyperus esculentus* L. *Medicago sativa* L.	Aqueous and methanolic extracts retard germination and root growth and reduce dry matter when grown in soil of *Ipomoea*.	Harrison and Peterson, 1986
I. batatas (L.) Lam.	Sweet potato, proso millet weeds	Root periderm extracts inhibit seed germination.	Peterson and Harrison, 1991
Fabaceae *Glycine max* (L.) Merr.	*Brassica rapa* L. *Medicago sativus* L. *Raphanus sativus* L.	Inhibition of seed germination by aqueous extracts in all test plants.	Tsuzuki and Kawagoe, 1984
G. max (L.) Merr.	*Zea mays* L.	Seed germination and early growth get inhibited.	Martin, McCoy, and Dick, 1990
Lupinus alnus L.	*Amaranthus retroflexus* L. *Chenopodium album* L.	Root exudates reduce growth and increase catalase and peroxidase activity.	Dzyubenko and Petrenko, 1971
Medicago sativus L.	*Triticum* spp.	Water and alcoholic extracts reduce germination and growth.	Oleszek and Jurzysta, 1987
M. sativus L.	*Cucumis sativus* L.	Residues and roots inhibit seed germination and seedling growth.	Ells and McSay, 1991

TABLE 2 (continued)

Crop (Donor)	Crop (Recipient)	Effects	Reference (s)
Trifolium alexandrinum L. T. incarnatum	*Allium cepa* L. *Daucus carota* L. *Lycopersicon esculentum* Mill.	Volatiles from residues reduce germination and seedling growth.	Bradow and Connick Jr., 1990
T. pratense L.	*Triticum* spp.	Root extracts inhibit growth.	Oleszek and Jurzysta, 1987
T. repens L.	*Agrostis tenuis* Sibth. *Bromus catharticus* Vahl *Dactylis glomerata* L. *Festuca rubra* L. *Lolium perenne* L. *Medicago sativa* L. *Triticum hybridum* L. *Trifolium pratense* L. *T. repens* L.	Aqueous extracts of dried shoots inhibit seed germination and induce seedling abnormalities.	Macfarlane, Scott, and Jarvis, 1982
Rubiaceae *Coffea arabica* L.	*Festuca* spp. *Lactuca sativa* L. *Lolium multiflorum* Lam	Aqueous extracts of dried leaves and roots reduce germination and radicle growth.	Chou and Waller, 1980
Solanaceae *Nicotiana tabacum* L.	*Chamomilla recutita* L.	Aqueous extracts inhibit seed germination.	Bondev, Panaiot, and Mariana, 1983
Monocotyledons Poaceae Pennisetum clandestinum Hochst. ex. Chiov.	*Lactuca sativa* L. *Oryza sativa* L.	Fresh leaf extracts inhibit growth of test plants.	Chou et al. 1987
Secale cereale L.	*Echinochloa crus-galli* L. *Lactuca sativa* L. *Lepidium sativum* L. *Panicum miliaceum* L.	Residue of crop reduces the emergence of test plant.	Barnes and Putnam, 1986; Barnes et al., 1987
Sorghum bicolor L.	*Triticum aestivum* L.	Wheat emergence gets reduced by 10 to 31% in the presence of *Sorghum* stubbles in field.	Purvis and Jones, 1990
Triticum aestivum L.	*Gossypium hirsutum* L.	Reduces germination and dry matter in cotton.	Hicks et al., 1989

managed, this phenomenon may be exploited for enhancing the crop productivity. The number of reports indicating the improvement of crop production due to allelopathic interactions are increasing. This manipulation can be achieved by management of noxious agricultural weeds, control of crop diseases, management of pests and conservation of nitrogen in croplands.

TABLE 3. List of Trees Showing Negative Influence on Crops

Tree	Crop	Effects	Reference (s)
Anacardiaceae			
Mangifera indica L.	*Oryza sativa* L.	Grain yield is decreased by the trees growing in the middle of fields.	Tomar and Shrivastava, 1986
Caesalpiniaceae			
Cassia siamea Lam.	*Sorghum* spp.	Growth was reduced up to 30 m on the leeward side of 4-row shelterbelt.	Jensen, 1983
Casuarinaceae			
Dipterocarpus obtusifolius Teysm. ex Miq.	*Oryza sativa* L.	Trees planted on field bund reduce the yield of crop.	Vityakon, Sae-lee, and Surasak, 1993
Meliaceae			
Azadirachta indica L.	*Pennisetum typhoides* S. & H.	Grain and stover get reduced by 20% over a distance of 0.5 to 2.0 h by windbreak.	Long and Persaud, 1988
A. indica L.	*Triticum aestivum* L.	Reduces crop productivity on northern side up to 3 m.	Puri and Bangarwa, 1992
A. indica L.	*Pennisetum typhoides* S. & H	2-row windbreak reduces the growth and stover yield up to distance of 1.5 h.	Brenner, Beldt, and Jarvis, 1993
Mimosaceae			
Acacia arabica (Lamk.) Willd.	*Triticum aestivum* L.	Yield of wheat was reduced on all the sides of tree.	Sheikh and Haq, 1978
A. holosericea A. Cunn. ex G. Don	*Pennisetum glaucum* (L.) R. Br.	Yield is reduced up to 2 m distance from windbreak.	Lamers et al., 1993
Acacia nilotica (L.) Del.	*Triticum aestivum* L.	Wheat yield was reduced by scattered trees.	Sheikh, 1988; Puri and Bangarwa, 1992; Sharma, 1992
A. nilotica (L.) Del. *A. tortilis* (Forssk.) Hayne	*Pennisetum* spp.	Average grain yield was reduced.	Lamers et al., 1993; Sharma, Rathore, and Gupta, 1994
Albizia saman (Jacq.) F. Muell	*Oryza sativa* L.	Grain yield, biomass and number of tillers are reduced by the bund trees.	Sae-lee, Vityakon, and Prachaiyo, 1992
Inga edulis Mart.	*Oryza sativa* L.	Hedge row of *Inga* decreases the grain yield up to 1.5 m.	Salazar, Szott, and Palm, 1993

TABLE 3 (continued)

Tree	Crop	Effects	Reference (s)
Leucaena leucocephala (Lam.) de Wit	*Zea mays* L. *Ipomoea batatas* (L.) Lam.	Grain yield of maize, and tuber and vine yield of sweet potato was reduced near the tree hedges.	Karim, Savill, and Rhodes, 1991
L. leucocephala (Lam.) de Wit	*Oryza sativa* L.	Grain yield is reduced up to 1.5 m by the tree hedge.	Salazar, Szott, and Palm, 1993
Prosopis cineraria (L.) Druce	*Triticum aestivum* L.	Trees affect the yield up to 2 m and delay the grain maturity by a week.	Puri and Bangarwa, 1992
Myrtaceae *Eucalyptus tereticornis* Sm.	*Triticum aestivum* L. *Brassica* spp.	Windbreaks reduce the density and grain yield in adjoining 10 m wide strips.	Malik and Sharma, 1990
E. tereticornis Sm.	*Cicer arietinum* L. *Lens esculentum* L. *Brassica oleracea* L. *Trifolium alexandrinum* L. *Brassica campestris* L.	Economic yield of crops in 12 m strip southward of *Eucalyptus* shelterbelt gets reduced by half.	Kohii, Singh, and Verma, 1990; Singh and Kohli, 1992
E. citriodora Hook. *E. camaldulensis* Dehnh. *E. grandis* W. Hill. ex Maiden	*Vigna unguiculata* (L.) Walp. *Zea mays* L. *Sorghum bicolor* L.	Shoot height, number of leaves, fruits and grain yield per hectare is reduced.	Igboanugo, 1988
Papilionaceae *Dalbergia sissoo* Roxb. ex. D.C.	*Triticum* spp.	Grain yield was reduced up to 6 m from base of tree.	Sheikh and Haq, 1978; Puri and Bangarwa,1992.
Erythrina spp.	*Oryza sativa* L.	Yield was affected up to 1.5 m by tree hedge.	Salazar, Szott, and Palm, 1993
Rhamnaceae *Zizyphus rotundifolia* Lam.	*Pennisetum americanum* (L.) Leeke *Cyamopsis tetragonoloba* (L.)Taub.	Grain and straw yield of test species is reduced.	Sharma, Rathore, and Gupta, 1994
Salicaceae *Populus deltoides*	*Saccharum officinarum*	Yield is reduced up to 10 m from a row of poplar.	Sheikh and Haq, 1986
P. deltoides	*Triticum aestivum* L.	Grain yield is reduced under agrisilvicultural system.	Ralhan, Singh, and Dhanda, 1992; Singh, Dhanda, and Ralhan, 1993.

TABLE 4. List of Some Allelopathic Fruit Crops

Fruit Crop	Test Plant	Effect	Reference
Juglandaceae			
Juglans regia L.	*Brassica rapa* L. *Echinochloa frumentacea* (Roxb.) Link *Eleusine coracana* Gaertn. *Fagopyrum esculentum* Moench *Lepidium sativum* L. *Setaria italica* (L.) Beauv. *Raphanus sativus* L. *Glycine max* (L.) Merr.	Seed germination of test species is reduced by leaf, leaf litter and wood litter leachates.	Melkania, 1984
Persica vulgaris Mill. (=*Prunus persica* Batsch)	*P. vulgaris* Mill.	Phenolics from root residues cause soil sickness and inhibit growth of young peach trees.	Grakhov, 1990
Prunus cerasoides D. Don	*Eleusine coracana* Gaertn. *Hordeum vulgare* L. *Glycine max* (L.) Merr.	Seed germination, root and shoot growth, dry matter production and pigment content is decreased in soil from below the canopy and soil mulched with dry leaves.	Bhatt and Todaria, 1990
P. persica Batsch	*P. persica* Batsch	Bark of old peach root releases a cyanogenic glycoside–Amygdalin, which breaks down into cyanide causing harm to young peach trees.	Proebsting and Gilmore, 1941; Koch, 1955; Patrick, 1955; Patrick, Toussoun, and Koch, 1964
Pyrus malus L.	*P. malus* L.	Old apple root tissue releases Phlorizin in the soil which microbially breaks down into phloretin, *p*-hydro cinnamic acid, phloroglucinol & *p*-hydroxybenzoic acid. These compounds greatly harm the growth of apple itself.	Börner, 1959
Rubiaceae			
Coffea arabica L.	*Triticum aestivum* L.	Caffeine present in seed extracts inhibits the germination of wheat.	Evenari, 1949
C. arabica L.	*Lactuca sativa* L. *Lolium multiflorum* Lam. *Festuca* spp.	Phytotoxic phenolics and purine aikaloids present in leaves, stems, roots, fallen leaves, young seedlings reduce seed germination and radicle growth.	Chou and Waller, 1980

TABLE 4 (continued)

Fruit Crop	Test Plant	Effect	Reference
C. arabica L.	Amaranthus spinosus Linn.	Ethanolic extracts of leaves and seeds inhibit germination. Caffeine present in leaves inhibits mitotic division and growth of coffee seedlings.	Rizvi, Mukerji, and Mathur, 1980, 1981
Rutaceae Citrus aurantium L.	Chenopodium album L. Avena sativa L. Cynodon dactylon (L.) Pers. Amaranthus retroflexus L.	Seed germination andlor seedling growth of test species was reduced by aqueous extracts, decaying materials and volatile compounds of senescent and non-senescent leaves.	Al Saadawi and Al Rubeaa, 1985; Al Saadawi, Arif, and Al Rubeaa, 1985
C. aurantium L.	C. aurantium L.	Aqueous extracts of old citrus soil, roots, decaying senescent and non-senescent leaves, and residues of decaying roots reduce growth of roots and tops of sour orange seedlings and increase the root disease index.	Hassan, Al Saadawi, and El-Behadi,1989a, b

Management of Noxious Agricultural Weeds

In the United States alone, at least ten billion dollar crop loss is reported alone by weeds (Putnam and Weston, 1986). As a control measure, farmers have to rely upon synthetic herbicides which are not only costly but environmentally harmful, too. Nowadays the impetus is shifting towards the use of biologically renewable agents, especially the botanical herbicides. These are not only more systemic but are biodegradable, too. Weed suppression can be achieved through living crop plants or their residues (Putnam and DeFrank, 1983). The three approaches through which allelopathy could be manipulated for weed management are: (a) transfer of allelopathic principles into cultivars, (b) use of allelopathic rotational and companion crops, and (c) use of allelochemicals as herbicides/ pesticides.

Transfer of Allelopathic Principles into Cultivars

Wild varieties from which modern cultivars have been derived are known to possess allelopathic principles which impart them resistance towards weeds and pests. It is reported that wild accessions of *Avena* spp. (Fay and Duke, 1977), *Cucumis sativus* (Putnam and Duke, 1978), *Glycine max* (Massantini, Coporali, and Zellini, 1977), *Helianthus annuus* (Leather, 1987) and *Brassica* spp. (Sarmah, Narwal, and Yadava, 1992) are capable of suppressing weed growth. In other words, the genes responsible for the synthesis of allelochemicals must be active in such plants. During course of cultivation and selection of high yielding varieties, such genes were either rendered weak or eliminated resulting in loss of allelopathic attributes. These, however, can be incorporated in the present day cultivars either through genetic recombination techniques or conventional plant breeding methods.

Use of Allelopathic Rotational and Companion Crops

The practice of crop rotation which minimizes the danger to crops by pests and diseases has declined with the advent of inorganic fertilizers and synthetic herbicides. The overdependence of modern agriculture on these chemicals is harmful to environment and public health. In annual or perennial cropping system, the companion and rotational crops capable of suppressing weeds through allelopathic potential can provide a non-herbicide mechanism for their control (Liebman and Dyck, 1993). Not only these crops but their residues, too, possess weed suppressing potential. The use of rye (*Secale cereale*) as a companion crop has been found to decrease the incidence of a number of weeds (Putnam, DeFrank, and Barnes, 1983). Likewise, *Sorghum bicolor* (Putnam and DeFrank, 1983), *Helianthus annuus* (Leather, 1987), *Hordeum vulgare* (Putnam, DeFrank, and Barnes, 1983) and Cruciferous plants (Oleszek, 1987) have also been found to decrease the weed incidence if planted as companion crop or in rotation. Further research is, however, needed to screen the more companion crops with weed suppressing ability.

Allelochemicals as Herbicides/Pesticides

Allelochemicals from higher plants can be successfully exploited for weed or pest control. This method is safe and effective since natural products are biodegradable and unlike most of the synthetic herbicides do not accumulate in soil as persistent pollutants. Besides, they have many other advantages like greater shelf-life, wide range of storage conditions, greater ease of application, broader environmental range and less space requirement.

Scopolin and scopoletin phytotoxins from *Celtis laevigata* are reported to suppress *Amaranthus palmeri* (Lodhi and Rice, 1971). Caffeine–a known allelochemical from *Coffea arabica,* completely inhibits the germination of spiny amaranth (*A. spinosus*) without affecting black gram (Rizvi, Mukerji, and Mathur, 1981). α-terthienyl (or α-T) produced by members of family asteraceae has been found to be effective against common milkweed *Asclepias syriaca* (Campbell et al., 1982) and acts as a contact herbicide against several other broad leaf weeds (Lambert et al., 1991). Artemisinin–a sesquiterpene lactone of *Artemisia annua,* inhibits the root and shoot growth of *Amaranthus retroflexus, Ipomoea lacunosa* and *Portulaca oleracea* at 33 mM (Duke et al., 1987) and has the same inhibitory level as that of glyphosate (Chen, Polatnick, and Leather, 1991). Cnicin which imparts allelopathic property to *Centaurea maculosa* inhibits germination of *Agropyron cristatum, A. spictatum* and *A. glaucum* (Kelsey and Locken, 1987). Perez (1990) observed that hydroxamic acid namely 2,4-Dihydroxy-7-methoxy-1,4-benzoxazin-3-one (DIMBOA) and its decomposition product namely 6-methoxy-benzoxazolin-2-one (MBOA) inhibits root growth of wild oat (*Avena fatua*)–a noxious agricultural weed. Ailanthone–a quassinoid compound extracted from root bark of *Ailanthus altissima,* has been reported to possess post-emergent herbicidal property similar to glyphosate and paraquat (Heisey, 1996).

Control of Plant Diseases

Allelopathy can be used for biological control of plant diseases. Several studies conducted during the last decade have successfully demonstrated the control of plant diseases using either allelopathic rotational crops or pure/crude allelochemicals (Rice, 1995).

Crop rotation besides suppressing the weeds also reduces the incidence of diseases caused by soil-borne pathogens to sub-optimal or harmful levels. This practice has been recommended in China for the control of cotton wilt disease caused by *Fusarium* sp., where the planting of peppermint (*Mentha haplocalyx* var. *Piperscens*) in rotation with cotton markedly decreases the incidence of wilt because of its volatile oils (Li, 1988).

The other way of controlling plant diseases is by using allelochemicals either in crude or purified form. The significant reports in this regard are the use of neem (Ghewande, 1989), eucalyptus (Singh and Dwivedi, 1990), tobacco (Menetrez et al., 1990), ginger (Endo, Kanno, and Oshima, 1990), tagetes (Kishore and Dwivedi, 1991) and *Salvia* species (Qureshi, Ahmed, and Kapadia, 1989). These may be in the form of plant aqueous extracts, organic fractions, volatile and non volatile fractions, oils or organic compounds. The plant diseases can also be controlled by the antagonistic effects of antibiotics produced by other microorganisms (Rice, 1995). With the advancement of genetic engineering and DNA recombinant technology, more and more research is being conducted in this regard.

Pest Management

Pest management is another challenge to sustainable agriculture production. To reduce the excessive use of pesticides, other alternatives such as increased pest resistance in plants or environmentally safe chemicals should be developed. Pyrethrum from *Chrysanthemum cinerariefolium*, rotenoids in *Derris, Lonchocarpus* and *Tephrosia* and nicotinoids from *Nicotiana* sp. are some of the classical examples of the insecticides which have stimulated further research in this direction. Hydroxamic acid (DIMBOA) present in maize imparts it resistance to European corn borer (Klun, Tipton, and Brindley, 1967). Over 2000 plant species belonging to different families possess toxic principles against insects (Crosby, 1971). Several compounds of a diverse nature including azadirachtin found in neem (*Azadirachta indica*) impart it pesticidal properties against over 300 pests and are thus receiving global attention. A number of microbial toxins have also been successfully screened for pest management (Devakumar and Parmar, 1993). Combined efforts of scientists from different fields are needed to exploit the vast array of allelochemicals or semiochemicals of microbial or higher plant origin.

Conservation of Nitrogen in Croplands

Conserving nitrogen in the soil is very important, and vital for sustaining crop productivity and management of natural resources. The inhibition of nitrification which converts ammonium (less leachable form) to nitrate (more leachable form) helps in conserving nitrogen as well as energy in the arable lands. Of late, the importance of its inhibition is realized so much that farmers have even started adding synthetic chemicals to achieve it. Nitrapyrin is one such product which is commercially available for this purpose. Some studies indicate that allelochemicals, particularly tannins, phenolic acids and flavonoids, from living plants or their residues or exudates inhibit nitrification (Rice and Pancholy, 1973, 1974). Mulches of barley, wheat and oat when added to the pots having *Picea mariana* seedlings increased the ammonium content by reducing the number of ammonium oxidisers, thus, lowering the loss of nitrogen in soil. This effect was attributed to the presence of phenolic acids in the mulches (Jobidon, Thibault, and Fortin, 1989). Likewise, improvement of degraded soils through stimulation of nitrogen conservation has been reported by leaf litter of *Elaeagnus angustifolia* (Llinares et al., 1993). The phenomenon of allelopathy can, therefore, be successfully exploited to conserve soil nitrogen. However, more needs to be done in this regard.

ALLELOPATHY–
A FAST EMERGING BIOTECHNOLOGICAL TOOL

Allelopathy can be successfully exploited as a biotechnological tool for weed and pest management by transferring allelopathic genes in crops, enhancing the production of natural toxins or their products as herbicides/pesticides through tissue or cell culture, DNA recombinant technology and microbial fermentation, and synthesizing and commercializing chemicals based on natural product chemistry.

As already discussed, the allelopathic genes from the wild accessions of modern day cultivars can be transferred to them by gene recombinant technology and traditional breeding practices. This may enable cultivars to resist the effect of weeds, pathogens, nematodes, and insects, etc. There is hardly any breakthrough in this regard but some transgenic tobacco and tomato plants have been reported to possess greater pest resistance (Barton, Whiteley, and Yang, 1987; Fischhoff et al., 1987). Besides, the production of natural plant toxins responsible for weed or pest suppression can be enhanced through microbial fermentation technology.

The synthesis of herbicides and pesticides based on natural product (microbial as well as higher plant) chemistry has opened a new era in weed and pest management. The progress in this regard gained momentum with the successful marketing and commercialization of Bialaphos as a natural herbicide under the trade name of Herbiaceae in Japan (Hatzios, 1987). Bialaphos is a fermentation product from *Streptomyces hygroscopicus* and *S. viridochromogenes* and has remarkable biodegradability in soil (Jobidon, 1991). It acts as a broad spectrum herbicide against a number of grasses and broad-leaf weeds. Rhizobitoxine–an allelochemical from *Bradyrhizobium japonicum* found in soybean nodules, shows selective toxicity against weeds which is comparable to amitrole–a synthetic herbicide (Owens, 1973). Synthesis of Cinmethyline by Shell in USA with similar chemistry as that of cineole–a higher plant product has further stimulated the interest of scientists to develop more chemical analogues using biorational designs. Some of these have been synthesized (Table 5) while others are in various stages of development. Manufacturers in Germany, USA, Switzerland and Japan have commercialized these products. Compounds likely to be commercialized in future are artemisinin–an already known antimalarial drug (Chen, Polatnick, and Leather, 1991) and ailanthone (Heisey, 1996).

Allelochemicals may also be used directly as herbicides, e.g., parthenin, ailanthone, cineole, etc. Parthenin–a sesquiterpene lactone of noxious weed *Parthenium hysterophorus* has been used as an aquatic weed suppressant (Pandey, 1996). Efforts are being made to use it as weedicide

TABLE 5. List of Some Commercially Synthesized Herbicides

Natural Compound	Origin	Herbicide (Trade Name)	Manufacturer
Anisomycin	Saproinhibitin	Methoxyphenone	Nihon, Japan
Benzoxazinones	Phytoinhibitin	Banzanin	BASF, Germany
Bialaphos	Saproinhibitin	Herbiaceae	Japan
Cineole	Phytoinhibitin	Cinmethyline	Shell, USA
Fusaric acid	Saproinhibitin	Picloram	DOW, USA
Iprexil	Saproinhibitin	Benzodox	Gulf-USA
Moniliformin	Saproinhibitin	3,4-dibutoxy moniliformin	Ciba-Geigy, Switzerland
Phosphinothricin	Saproinhibitin	Glufosinate	Hoechst, Germany
Quinolinic acid	Phytoinhibitin	Quinclorac	BASF, Germany

against terrestrial weeds. It has already been reported to change respiratory ability of higher plants (Kohli, Daizy, and Verma, 1993). However, for success of natural compounds as herbicides, it is necessary to understand their toxicology, cost of production, efficacy and selectivity in the field and successful licensing and patenting (Duke and Lydon, 1987).

Another biotechnological aspect of allelopathy is to use allelochemicals as growth regulators. This aspect has found less applicability, yet a number of sesquiterpene lactones, some of which are known allelochemicals, are known to possess this property (Fischer, Weidenhamer, and Bradow, 1989; Chen and Leather, 1990). Their activity is comparable to known plant growth regulators which are otherwise expensive (Batish et al., 1996). Therefore, plant products or allelochemicals may provide a cheap source of growth regulators with less complicated extraction techniques.

MECHANISM OF ACTION OF ALLELOCHEMICALS

Most of the allelopathic research pertains to the visible effects on germination and growth of the target or test species. Little attention has been paid to the cellular events following physiological changes in the plant system. Nevertheless, some studies are available in this regard which, by and large, involve phenolic compounds, particularly benzoic and cinnamic acids. In case of terpenoid lactones such as parthenin and artemisinin some studies are available showing the changes in respiratory metabolism (Duke et al., 1987; Kohli, Daizy, and Verma, 1993). Some of the physiological events influenced by allelochemicals leading to the visible effects on target species are cell division (Muller, 1965), biosynthesis of plant constituents (Van Sumere et al., 1972), changes in hormonal levels (Lee, Starratt, and Jevnikar, 1982), enzyme activities (Daizy, 1990; Devi and Prasad, 1992), water relations (Barkosky and Einhellig, 1993), photosynthesis (Daizy and Kohli, 1991; Einhellig and Rasmussen, 1993), respiration (Hejl, Einhellig, and Rasmussen, 1993; Kohli, Daizy, and Verma, 1993), and ion uptake (Booker, Blum, and Fiscus, 1992).

The studies on understanding the mode of action of allelochemicals are scanty and form one of the foremost and ongoing areas of research. The lack of information in this may be attributed to the following reasons:

- Involvement of more than one chemical in bringing about growth transformation through synergistic or additive effects.
- Transformation of allelochemicals in soil environment through microbial action leading to the formation of unidentified compounds, the mode of action of which is difficult to interpret.
- Retention of allelochemicals by soil particles or humus complex for some time.
- Diverse nature of allelochemicals makes the subject more complex.

CONCLUSIONS

The science of allelopathy involves complexities of biochemical interactions among plants. Its demonstration under field conditions is not as simple as it seems to be. However, if properly understood, this phenomenon may be successfully exploited for enhancing crop productivity. Some of the areas where prospects of allelopathic research are more demanding include the following:

i. In order to restrict the use of the synthetic and damaging agrochemicals towards sustainable agriculture, allelochemicals of both microbial and higher plant origin may provide excellent alternatives for integrated crop protection programs. They are comparatively cheap and environmentally safe. Screening of such chemicals with suitable experimental designs is therefore the most challenging area of research.

ii. Investigations should be made to screen allelopathic crops with selective toxicity towards weeds and such crops should be used in crop rotation. Farmers should be encouraged to follow this practice as the excessive use of synthetic chemicals has deleteriously affected the soil health. Some success has already been made but this deserves more attention.

iii. The use of recombinant DNA techniques through which DNA of crop plants can be successfully modified to obtain the desired results such as resistance towards pests and capability to suppress weeds; to enhance the production of natural compounds which could be used as weedicides, pesticides or growth regulators.

REFERENCES

Abu-Irmaileh, B.E. and J.R. Qasem. (1986). Aqueous extract effects of *Salvia syriaca* L. in various lines of four crops. *Dirasat* 13: 147-170.

Al Saadawi, I.S. and A.J. Al Rubeaa. (1985). Allelopathic effects of *Citrus aurantium* L. I. Vegetational patterning. *Journal of Chemical Ecology* 11: 1515-1525.

Al Saadawi, I.S. and E.L. Rice. (1982). Allelopathic effects of *Polygonum aviculare* L. I. Vegetational patterning. *Journal of Chemical Ecology* 8: 993-1009.

Al Saadawi, I.S., E.L. Rice and T.K.B. Karns. (1983). Allelopathic effects of *Polygonum aviculare* L. III. Isolation, characterization and biological activities of phytotoxins other than phenols. *Journal of Chemical Ecology* 9: 761-774.

Al Saadawi, I.S., M.B. Arif and A.J. Al Rubeaa. (1985). Allelopathic effects of *Citrus aurantium* L. II. Isolation, characterization, and biological activities of phytotoxins. *Journal of Chemical Ecology* 11: 1527-1534.

Altieri, A.M. and M. Liebman. (1988). *Weed Management in Agroecosystems: Ecological Approaches*. Florida: CRC Press.

Ballester, A., A.M. Vieitez and E. Vieitez. (1979). The allelopathic potential of *Erica australis* L. and *E. arborea* L. *Botanical Gazette* 40: 433-436.

Ballester, A., A.M. Vieitez and E. Vieitez. (1982). Allelopathic potential of *Erica vagans, Calluna vulgaris* and *Daboecia cantabrica*. *Journal of Chemical Ecology* 8: 851-858.

Bansal, G.L. (1990). Allelopathic potential of Linseed on buttercup (*Ranunculus arvensis* L.). In *Plant Science Research in India,* eds. M.L. Trivedi, B.S. Gill and S.S. Saini. New Delhi: Today and Tomorrow Printers and Publishers, pp. 801-805.

Barkosky, R.R. and F.A. Einhellig (1993). Effect of salicylic acid on plant-water relationships. *Journal of Chemical Ecology* 19: 237-247.

Barnes, J.P. and A.R. Putnam. (1986). Evidence for allelopathy by residues and aqueous extracts of rye (*Secale cereale*). *Weed Science* 34: 384-390.

Barnes, J.P., A.R. Putnam, B.A. Burke and A.J. Aasen. (1987). Isolation and characterization of allelochemicals in rye herbage. *Phytochemistry* 26: 1385-1390.

Barton, K.A., H.R. Whiteley and N.S. Yang. (1987). *Bacillus thuringiensis* delta endo-toxin expressed in *Nicotiana tabacum* provides resistance to lepidopteran insects. *Plant Physiology* 85: 1103-1109.

Batish, D.R., R.K. Kohli, D.B. Saxena and H.P. Singh. (1996). Growth regulatory response of parthenin and its derivatives. *Plant Growth Regulation* (Sent after revision).

Beckmann, E.O. and K. Noffke. (1978). The effect on lettuce of the addition of radish plant residues to various substances. *Gartenbauwissenchaft* 43: 70-78.

Bendall, G.M. (1975). The allelopathic activity of California thistle (*Cirsium arvense* (L.) Scop.) in Tasmania. *Weed Research* 15: 77-81.

Bhandari, M.C. and D.N. Sen. (1972). Growth regulation specificity exhibited by substances present in the fruit pulp of *Citrullus lanatus* (Thunb.) Mansf. *Zeitschrift fuer Naturforschung* 27: 72-75.

Bhatt, B.P. and N.P. Todaria. (1990). Studies on the allelopathic effects of some agroforestry tree crops of Garhwal Himalaya. *Agroforestry Systems* 12: 251-255.

Bhowmik, P.C. and J.D. Doll. (1979). Evaluation of allelopathic effects of selected weed species on corn and soybeans. *Proceedings of the North Central Weed Control Conference* 34: 43-45.

Bondev, I.A., P.P. Panaiot and I.L. Mariana. (1983). Allelopathic relations between germination of tobacco (*Nicotiana tobacum*) and chamomilla (*Chamomilla recutita*). *Ekologiya* (Sofia) 0: 14-25.

Booker, F.L., U. Blum and E.L. Fiscus. (1992). Short-term effects of ferulic acid on ion uptake and water relations in cucumber seedlings. *Journal of Experimental Botany* 93: 649-655.

Börner, H. (1959). The apple replant problem. 1. The excretion of phlorizin from apple root residues. *Contributions of Boyce Thompson Institute* 20: 39-59.

Bradow, J.M. and W.J. Connick, Jr. (1987). Allelochemics from palmer amaranth, *Amaranthus palmeri* S. Wats. *Journal of Chemical Ecology* 13: 185-202.

Bradow, J.M. and W.J. Connick, Jr. (1990). Volatile seed germination inhibitors from plant residues. *Journal of Chemical Ecology* 16: 645-666.

Brenner, A.J., R.J.V.D. Beldt and P.G. Jarvis. (1993). Tree-crop interface competition in a semi-arid sahelian windbreak. In *Proceedings of 4th International Symposium on Windbreaks and Agroforestry*, held in July 26-30, 1993 at Hedeselskabet, Viborg, Denmark. pp. 15-23.

Brown, R.L., C.S. Tang and R.K. Nishimoto. (1983). Growth inhibition from guava root exudates. *HortScience* 18: 316-318.

Buta, J.G., D.W. Spaulding and A.N. Reed. (1987). Differential growth responses of fractionated turfgrass seed leachates. *HortScience* 22: 1317-1319.

Campbell, G., J.D.H. Lambert, T. Arnason and G.H.N. Towers. (1982). Allelopathic properties of α-terthienyl and phenylheptatriyne, naturally occurring compounds from species of Asteraceae. *Journal of Chemical Ecology* 8: 961-972.

Chaghtai, S.M., A. Sadiq and M. Ibrar. (1986). Phytotoxicity of *Fumaria indica* on wheat (*Triticum aestivum* cv. Blue Silver). *Pakistan Journal of Botany* 18: 59-64.

Chen, P.K. and G. Leather. (1990). Plant growth regulatory activities of artemisinin and its related compounds. *Journal of Chemical Ecology* 16: 1867-1876.

Chen, P.K., M. Polatnick and G. Leather. (1991) Comparative study of artemisinin, 2,4-D, and glyphosate. *Journal of Agricultural and Food Chemistry* 39: 991-994.

Chou, C.H, S. Hwang, C. Peng, Y. Wang, F. Hsu and N. Chung. (1987). The selective allelopathic interaction of a pasture forest intercropping in Taiwan. *Plant and Soil* 98: 31-41.

Chou, C.H. and G.R. Waller. (1980). Possible allelopathic constituents of *Coffea arabica*. *Journal of Chemical Ecology* 6: 643-654.

Chou, C.H. and H.J. Lin. (1976). Autointoxication mechanisms of *Oryza sativa*. I. Phytotoxic effects of decomposing rice residues in soil. *Journal of Chemical Ecology* 2: 353-367.

Crosby, D.G. (1971). In *Naturally Occurring Insecticides,* eds. M. Jacobson and D.G. Crosby. New York: Marcel Dekker, pp. 177-242

Cruz-Ortega, R., A.L. Anaya and L. Ramos. (1988). Effects of allelopathic compounds of corn pollen on respiration and cell division of watermelon. *Journal of Chemical Ecology* 14: 71-86.

Daizy Rani and R.K. Kohli. (1991). Fresh matter is not an appropriate relation unit for chlorophyll content: experience from experiments on effects of herbicide and allelopathic substance. *Photosynthetica* 25: 655-658.

Daizy Rani. (1990). *Phytotoxic Properties of Parthenium hysterophorus* L. Ph.D. Thesis, Panjab University, Chandigarh, India.

de Candolle, M.A.P. (1832). *Physiologie Vegetale.* Tome III. Paris: Bechet Jeune, Lib. Fac. Med., pp. 1474-75.

Devakumar, C. and B.S. Parmar. (1993). Pesticides of higher plant and microbial origin. In *Botanical and Pesticides,* eds. B.S. Parmar and C. Devakumar. SPS Publication No. 4. New Delhi: Society of Pesticide Science, India and Westvill Publishing House, pp. 1-73.

Devi, S.R. and M.N.V. Prasad. (1992). Effect of ferulic acid on growth and hydrolytic enzyme activities of germinating maize seeds. *Journal of Chemical Ecology* 18: 1981-1990.

Dhyani, S.K. (1978). Allelopathic potential of *Eupatorium adenophorum* on seed germination of *Lantana camara* var. aculeata. *Indian Journal of Forestry* 1: 113.

Dirvi, G.A. and F. Hussain. (1979). Allelopathic effects of *Dichanthium annulatum* on some cultivated plants. *Pakistan Journal of Scientific and Industrial Research* 22: 194-197.

Drost, D.C. and J.D. Doll. (1980). The allelopathic effect of yellow nutsedge

(*Cyperus esculentus*) on corn (*Zea mays*) and soybeans (*Glycine max*). *Weed Science* 28: 229-233.

Duke, S.O. and J. Lydon. (1987). Herbicides from natural compounds. *Weed Technology* 1: 122-28.

Duke, S.O., K.C. Vaughn, E.M. Groom and H.N. Elsholy. (1987). Artemisinin, a constituent of annual wormwood (*Artemisia annua*) is a selective phytotoxin. *Weed Science* 35: 499-505.

Dzyubenko, N.N. and N.I. Petrenko. (1971). On biochemical interactions of cultivated plants and weeds. In *Physiological-Biochemical Basis of Plant Interaction in Phytocenoses,* ed. A.M. Grodzinsky. Naukova, Dumka: Kiev. Vol. 2, pp. 60-66.

Einhellig, F.A. (1989). Interactive effects of allelochemicals and environmental stress. In *Phytochemical Ecology: Allelochemicals, Mycotoxins and Insect Pheromones and Allomones,* eds. C.H. Chou and G.R. Waller. Taipei, China: Academia Sinica Monograph Series No. 9, pp. 101-118.

Einhellig, F.A. and J.A. Rasmussen (1993). Effect of root exudate sorgoleone on photosynthesis. *Journal of Chemical Ecology* 19: 369-375.

Ells, J.E. and A.E. McSay. (1991). Allelopathic effects of alfalfa plant residues on emergence and growth of cucumber seedlings. *HortScience* 26: 368-370.

Endo, K., E. Kanno and Y. Oshima. (1990). Structures of antifungal diarylheptenones, gingerenones A, B, C and isogingerenone B, isolated from the rhizomes of *Zingiber officinalis*. *Phytochemistry* 29: 797-799.

Evenari, M. (1949). Germination inhibitors. *Botanical Review* 15: 153-194.

Fadayomi, O. and E.O. Oyebade. (1984). An investigation of allelopathy in Siam Weed (*Eupatorium odoratum*). *Geobios* 11: 145-150.

Fay, P.K. and W.B. Duke. (1977). An assessment of allelopathic potential in *Avena* germplasm. *Weed Science* 25: 224-228.

Fischer, N.H., J.D. Weidenhamer and J.M. Bradow. (1989). Inhibition and promotion of germination by several sesquiterpenes. *Journal of Chemical Ecology* 15: 1785-1793.

Fischhoff, D.A., K.S. Bowdish, F.J. Perlak, P.G. Marene, S.M. McCoormick, J.G. Niedermeyer, D.A. Duff, K. Kusano-Kratzmer, E.J. Mayer, D.E. Rochester, S.G. Ragors and R.T. Fraley, R.T. (1987). Insect tolerant transgenic tomato plants. *Biotechnology* 5: 807-813.

Friedman, J. and G.R. Waller. (1985). Allelopathy and autotoxicity. *Trends in Biochemical Sciences* 10: 47-50.

Friedman, J., E. Rushkin and G.R. Waller. (1982). Highly potent germination inhibitors in aqueous eluate of fruits of bishop's weed (*Ammi majus* L.) and avoidance of autoinhibiton. *Journal of Chemical Ecology* 8: 55-65.

Fuerst, E.P. and A.R. Putnam. (1983). Separating the competitive and allelopathic components of interference: Theoretical principles. *Journal of Chemical Ecology* 9: 937-944.

Gabor, W.E. and C. Veatch. (1981). Isolation of a phytotoxin from quackgrass (*Agropyron repens*) rhizomes. *Weed Science* 29: 155-159.

Ghewande, M.P. (1989). Management of foliar diseases of groundnut (*Arachis hypogaea*) using plant extracts. *Indian Journal of Agricultural Sciences* 59: 133-134.

Grakhov, V.P. (1990). Allelopathic function of phenolic compounds in *Persica vulgaris* Mill. *Ukrainskii-Botanichnii-Zhurnal* 47: 98-100.

Grümmer, G. (1955). *Die gegenseitige Beeinflussung hoherer Pflanzen–Allelopathie*. Jena: Fischer.

Harrison, H.F., Jr. and J.K. Peterson. (1986). Allelopathic effects of sweet potatoes (*Ipomoea batatas*) on yellow nutsedge (*Cyperus esculentus*) and alfalfa (*Medicago sativa*). *Weed Science* 34: 623-627.

Hassan, M.S., I.S. Al Saadawi and A.M. El-Behadi. (1989a). Citrus replant problem in Iraq. II. Possible role of allelopathy. *Plant and Soil* 116: 157-160.

Hassan, M.S., I.S. Al Saadawi and A.M. El-Behadi. (1989b). Citrus replant problem in Iraq. III. Interactive effect of soil fungi and allelopathy. *Plant and Soil* 116: 161-166.

Hatzios, K.K. (1987). Biotechnology application in weed management: Now and in the future. *Advances in Agronomy* 41: 325-375.

Heisey, R.M. (1996). Identification of an allelopathic compound from *Ailanthus altissima* (Simaroubaceae) and characterization of its herbicidal activity. *American Journal of Botany* 83: 192-200.

Hejl, A.M., F.A. Einhellig and J.A. Rasmussen. (1993). Effect of juglone on growth, photosynthesis and respiration. *Journal of Chemical Ecology* 19: 559-568.

Hicks, S.K., C.W. Wendt, J.R. Gannaway and R.B. Baker. (1989). Allelopathic effects of wheat straw on cotton germination, emergence and yield. *Crop Science* 29: 1057-1061.

Horowitz, M. and T. Friedman. (1971). Biological activity of subterranean residues of *Cynodon dactylon* L., *Sorghum halepense* L., and *Cyperus rotundus* L. *Weed Research* 11: 88-93.

Horsley, S.B. (1977). Allelopathic interference among plants. II. Physiological modes of action. In *Proceedings Fourth North American Forest Biology Workshop,* eds. H.E. Wilcox and A.F. Hamer. Syracuse, New York: State University of New York, pp. 93-136.

Hussain, F., M.I. Zaidi and S.R. Chughtai. (1984). Allelopathy effects of Pakistani weeds: *Eragrostis poaeoides*. *Pakistan Journal of Scientific and Industrial Research* 27: 159-164.

Hussain, F., N. Ahmed and M. Akram. (1986). Allelopathy as expressed by *Stachys parviflora*. *Pakistan Journal of Scientific and Industrial Research* 29: 458-460.

Igboanugo, A.B.I. (1988). Effect of some eucalypts on yields of *Vigna unguiculata* L. Walp., *Zea mays* L. and *Sorghum bicolar* L. *Agriculture Ecosystem and Environment* 24: 453-458.

Inam, B., F. Hussain and B. Farhat. (1987). Allelopathic effects of Pakistani weeds: *Xanthium strumarium* L. *Pakistan Journal of Scientific and Industrial Research* 30: 530-533.

Inam, B., F. Hussain and B. Farhat. (1989). *Cannabis sativa* L. is allelopathic. *Pakistan Journal of Scientific and Industrial Research* 32: 617-620.

Inderjit. (1996). Plant phenolics in allelopathy. *Botanical Review* 62: 186-202.

Irons, S.M. and O.C. Burnside. (1982). Competitive and allelopathic effects of sunflower (*Helianthus annuus*). *Weed Science* 30: 372-377.

Jensen, A.M. (1983). *Shelterbelt Effects in Tropical and Temperate Zones*. Manuscript reports, IDRC-MR80-e Ottawa, Canada: International Development Research Centre, pp. 61.

Jiménez, J.J., K. Schultz, A.L. Anaya, J. Hernández and O. Espejo. (1983). Allelopathy potential of corn pollen. *Journal of Chemical Ecology* 9: 1011-1025.

Jobidon, R, J.R. Thibault and J.A. Fortin. (1989). Phytotoxic effect of barley, oat, and wheat-straw mulches in eastern Quebec forest plantations. 2. Effects on nitrification and black spruce (*Picea mariana*) seedling growth. *Forest Ecology and Management* 29: 295-310.

Jobidon, R. (1991). Potential use of bialaphos, a microbially produced phytotoxin, to control red raspberry in forest plantations and its effect on black spruce. *Canadian Journal of Forest Research* 21: 489-497.

Kanchan, S.D. and Jayachandra. (1979a). Allelopathic effects of *Parthenium hysterophorus* L. I. Exudation of inhibitors through roots. *Plant and Soil* 53: 27-35.

Kanchan, S.D. and Jayachandra. (1979b). Allelopathic effects of *Parthenium hysterophorus* L. III. Inhibitory effects of weed residues. *Plant and Soil* 53: 37-47.

Karim, A.B., P.S. Savill and E.R. Rhodes. (1991). The effect of young *Leucaena leucocephala* (Lam.) de Wit hedges on the growth and yield of maize, sweet potato and cowpea in an agroforestry system in Sierra Leone. *Agroforestry Systems* 16: 203-211.

Kelsey, R.G. and I.J. Locken. (1987). Phytotoxic properties of cnicin, a sesquiterpene lactone from *Centaurea maculosa* (spotted knapweed). *Journal of Chemical Ecology* 13: 19-33.

Kishore, N. and R.S. Dwivedi. (1991). Fungitoxicity of the essential oil of *Tagetes erecta* L. against *Pythium aphanidermatum* Fitz. the damping-off pathogen. *Flavour and Fragrance Journal* 6: 291-294.

Klun, J.A., C.L. Tipton and T.A. Brindley. (1967). 2,4-dihydroxy-7-methoxy-1,4-benzoaxazin-3-one (DIMBOA), an active agent in the resistance of maize to the European corn borer. *Journal of Economic Entomology* 60: 1529-1533.

Koch, L.W. (1955). The Peach replant problem in Ontario. I. Symptomatology and distribution. *Canadian Journal of Botany* 33: 450-460.

Kohli, R.K. (1990). *Allelopathic Potential of Eucalyptus*. India: MAB-DoEn Project Report.

Kohli, R.K. and D.R. Batish. (1994). Exhibition of allelopathy by *Parthenium hysterophorus* L. in agroecosystems. *Tropical Ecology* 35: 295-307.

Kohli, R.K., D. Singh and R.C. Verma. (1990). Influence of eucalypt shelterbelt on winter season agroecosystems. *Agriculture Ecosystem and Environment* 33: 23-31.

Kohli, R.K., Daizy Rani and R.C. Verma. (1993). A mathematical model to predict the tissue response to parthenin–an allelochemical. *Biologia Plantarum* 35: 567-576.

Kumar, R. (1991). Studies on autotoxicity in *Lantana camara* L. Ph.D. Thesis, Panjab University, Chandigarh, India.

Kumari, A. and R.K. Kohli. (1987). Autotoxicity of ragweed parthenium (*Parthenium hysterophorus*). *Weed Science* 35: 629-632.

Kuo, C.G., M.H. Chou and H.G. Park. (1981). Effect of chinese cabbage (*Brassica campestris* subsp. Pekinensis) residue on mung bean (*Vigna radiata*). *Plant and Soil* 61: 473-478.

Kushal Bala (1987). Physiological and biochemical aspects of teletoxicity and eradication of *Lantana camara* L. Ph.D. Thesis, Panjab University, Chandigarh, India.

Lambert, J.D.H., G. Campbell, J.T. Arnason and W. Majak. (1991). Herbicidal properties of α-terthienyl, a naturally occurring phytotoxin. *Canadian Journal of Plant Science* 71: 215-218.

Lamers, J.P.A., K. Michels, B.E. Allison and R.J. Vandenbeldt. (1993). Agronomic and socio-ecnomic aspects of windbreaks in Southwest-Niger. In *Proceedings of the 4th International Symposium on Windbreaks and Agroforestry*, held at Hedeselskabet, Viborg, Denmark, pp. 28-30.

Law, J.H. and F.E. Regnier. (1971). Pheromones. *Annual Review of Biochemistry* 40: 533-548.

Leather, G.R. (1987). Weed control using allelopathic sunflowers and herbicide. *Plant and Soil* 98: 17-23.

Lee, T.T., A.N. Starratt and J.J. Jevnikar (1982). Regulation of enzymic oxidation of Indole-3-acetic acid by phenols: structure-activity relationship. *Phytochemistry* 21: 517-523.

Leela, D. (1984). Allelopathy in *Oxalis corniculata*. *Pesticides* 18: 16-19.

Lehle, F.R. and A.R. Putnam. (1982). Quantification of allelopathic potential of Sorghum residues by novel indexing of Richards' function fitted to cumulative cress seed germination curves. *Journal of Chemical Ecology* 69: 1212-1216.

Li, Y. (1988). A Preliminary study on use of the medical herb peppermint in control of cotton fusarium wilt. *Scientia Agricultae Sinica* 21: 65-69.

Liebman, M. and E. Dyck. (1993). Crop rotation and intercropping strategies for weed management. *Ecological Applications* 3: 92-122.

Llinares, F., D. Muñoz-Mingarro, J.M. Pozuelo, B. Ramos and F. Bermedez de Castro. (1993). Microbial inhibition and nitrification potential in soils incubated with *Elaeagnus angustifolia* L. leaf litter. *Geomicrobiology Journal* 11: 149-156.

Lodhi, M.A.K. and E.L. Rice. (1971). Allelopathic effects of *Celtis laevigata*. *Bulletin of the Torrey Botanical Club* 98: 83-89.

Lodhi, M.A.K., R. Bilal and K.A. Malik. (1987). Allelopathy in agroecosystems: wheat phytotoxicity and its possible roles in crop rotation. *Journal of Chemical Ecology* 13: 1881-1891.

Long, S.P. and N. Persaud. (1988). Influence of neem (*Azadirachta indica*) windbreaks on millet yield, microclimate and water use in Niger, West Africa. In

Dryland Agriculture–A Global Perspective, eds. P.W. Unger, T.V. Sneed, W.R. Jordon and R. Jensen. Texas: Texas Agricultural Experimental Station, pp. 313-314.

Lovett, J.V. and W.C. Potts. (1987). Primary effects of allelochemicals of *Datura stramonium* L. *Plant and Soil* 98: 137-144.

Macfarlane, M.J., D. Scott, and P. Jarvis. (1982). Allelopathic effects of white clover. I. Germination and chemical bioassay. *New Zealand Journal of Agricultural Research* 25: 503-510.

Malik, R.S. and S.K. Sharma. (1990). Moisture extraction and crop yield as a function of distance from a row of *Eucalyptus tereticornis. Agroforestry Systems* 12: 187-195.

Manners, G.D. and D.S. Galitz. (1986). Allelopathy of small everlasting (*Antennaria microphylla*). I. Identification of constituents phytotoxic to leafy spurge (*Euphorbia esula*). *Weed Science* 34: 8-12.

Martin, V.L., E.L. McCoy and W.A. Dick. (1990). Allelopathy of crop residues influences corn seed germination and early growth. *Agronomy Journal* 82: 555-560.

Mason-Sudun, W. and R.S. Jessop. (1988). Differential phytotoxicity among species and cultivars of the genus *Brassica* to wheat. *Plant and Soil* 107: 69-80.

Massantini, F., F. Coporali and G. Zellini. (1977). Evidence for allelopathic control of weeds in lines of Soybean. *Proceedings EWRS Symposium on Methods of Weed Control and Their Integration* 1: 23-28.

Meissner, R., P.C. Nel and N.S.H. Smit. (1982). The residual effects of *Cyperus rotundus* on certain crop plants. *Agroplantae* 14: 47-53.

Melkania, N.P. (1984). Influence of leaf leachates of certain woody species on agricultural crops. *Indian Journal of Ecology* 11: 82-86.

Menetrez, M.L., H.W. Spurr, Jr., D.A. Danehower and D.R. Lawson. (1990). Influence of tobacco leaf surface chemicals on *Peronospora tabacina* Adam sporangia. *Journal of Chemical Ecology* 16: 1565-1576.

Menges, R.M. and S. Tamez. (1981). Response of onion (*Allium cepa*) to annual weeds and postemergence herbicides. *Weed Science* 29: 74-79.

Mersie, W. and M. Singh. (1987a). Allelopathic effect of *Lantana* on some agronomic crops and weeds. *Plant and Soil* 98: 25-30.

Mersie, W. and M. Singh. (1987b). Allelopathic effect of parthenium (*Parthenium hysterophorus* L.) extract and residue on some agronomic crops and weeds. *Journal of Chemical Ecology* 13: 1739-1747.

Mersie, W. and M. Singh. (1988). Effect of phenolic acids and ragweed parthenium (*Parthenium hysterophorus*) extracts on tomato (*Lycopersicon esculentum*) growth and nutrient and chlorophyll content. *Weed Science* 36: 278-281.

Molisch, H. (1937). *Der Einflus einer Pflanze auf die andere-Allelopathie.* Jena: Gustave Fishcher Verlag.

Muller, C.H. (1965). Inhibitory terpenes volatilized from *Salvia* shrubs. *Bulletin of the Torrey Botanical Club* 92: 38-45.

Muller, C.H. (1969). Allelopathy as a factor in ecological process. *Vegetatio* 18: 348-357.

Newman, E.I. (1978). Allelopathy: Adaptation or accident. In *Biochemical Aspects of Plant and Animal Coevolution,* ed. J.B. Harborne. New York: Academic Press, pp. 328-342.

Oleszek, W. (1987). Allelopathic effects of volatiles from some Cruciferae species on lettuce, barnyard grass and wheat growth. *Plant and Soil* 102: 271-273.

Oleszek, W. and M. Jurzysta. (1987). The allelopathic potential of alfalfa root medicagenic acid glycosides and their fate in soil environments. *Plant and Soil* 98: 67-80.

Owens, L.D. (1973). Herbicidal potential of rhizobitoxine. *Weed Science* 21: 63-66.

Pandey, D.K. (1996). Phytotoxicity of sesquiterpene lactone parthenin on aquatic weeds. *Journal of Chemical Ecology* 22: 151-160

Pandya, S.M. (1976). Effect of *Celosia argentea* Linn. extracts on dry weight of bajra seedlings. *Geobios* 3: 137-138.

Patrick, Z.A. (1955). The peach replant problem in Ontario. II. Toxic substances from microbial decomposition products of peach root residues. *Canadian Journal of Botany* 33: 461-486.

Patrick, Z.A., T.A. Toussoun and L.W. Koch. (1964). Effect of crop residue decomposition products on plant roots. *Annual Review of Phytopathology* 2: 267-292.

Perez, F.J. (1990). Allelopathic effects of hydroxamic acids from cereals on *Avena sativa* and *A. fatua. Phytochemistry* 29: 773-776.

Peters, E.J. and K.T. Luu. (1985). Allelopathy in tall fescue. In *The Chemistry of Allelopathy,* ed. A.C. Thompson. Washington DC: American Chemical Society, ACS Symposium Series No. 268, pp. 273-283.

Peterson, J.K. and H.F. Harrison, Jr. (1991). Differential inhibition of seed germination by sweet potato (*Ipomoea batatas*) root periderm extracts. *Weed Science* 39: 119-123.

Porwal, M.K. and O.P. Gupta. (1986). Allelopathic influence of winter weeds on germination and growth of wheat. *Journal of Tropical Agriculture* 4: 276-279.

Proebsting, E.L. and A.E. Gilmore. (1941). The relation of peach root toxicity to the re-establishing of peach orchards. *Proceedings of American Society for Horticultural Science* 38: 21-26.

Puri, S. and K.S. Bangarwa. (1992). Effect of trees on the yield of irrigated wheat crop in semi-arid regions. *Agroforestry Systems* 20: 229-241.

Purvis, C.E. and G.P.D. Jones. (1990). Differential response of wheat to retain crop stubbles. II. Other factors influencing allelopathic potential, intraspecific variation, soil type and stubble quantity. *Australian Journal of Agricultural Research* 41: 243-252.

Putnam, A.R. (1988). Allelopathy: Problems and opportunities in weed management. In *Weed Management in Agroecosystems: Ecological Approaches,* eds. M.A. Altieri and M. Liebman. Florida: CRC Press, pp. 77-88.

Putnam, A.R. and C.S. Tang. (1986). Allelopathy: State of the Science. In *The Science of Allelopathy,* eds. A.R. Putnam and C.S. Tang. New York: John Wiley and Sons, pp. 1-19.

Putnam, A.R. and J. DeFrank. (1983). Use of phytotoxic plant residues for selective weed control. *Crop Protection* 2: 173-181.

Putnam, A.R. and L.A. Weston. (1986). Adverse impacts of allelopathy in agricultural systems. In *The Science of Allelopathy,* eds. A.R. Putnam and C.S. Tang. New York: John Wiley and Sons, pp. 43-56.

Putnam, A.R. and W.B. Duke. (1978). Allelopathy in agroecosystems. *Annual Review of Phytopathology* 16: 431-451.

Putnam, A.R., J. DeFrank and J.P. Barnes. (1983). Exploitation of allelopathy for weed control in annual and perennial cropping systems. *Journal of Chemical Ecology* 9: 1001-1010.

Qureshi, I.H., S. Ahmed and Z. Kapadia. (1989). Antimicrobial activity of *Salvia splendens. Pakistan Journal of Scientific and Industrial Research* 32: 597-599.

Ralhan, P.K., A. Singh and R.S. Dhanda. (1992). Performance of wheat as intercrop under poplar (*Populus deltoides* Bartr.) plantations in Punjab (India). *Agroforestry Systems* 19: 217-222.

Rasmussen, J.A. and F.A. Einhellig. (1975). Non-competitive effects of common milkweed (*Asclepias syriaca* L.) on germination and growth of sorghum. *American Midland Naturalist* 94: 478-483.

Rasmussen, J.A. and F.A. Einhellig. (1979). Allelochemic effects of leaf extracts of *Ambrosia trifida* (Compositae). *Southwestern Naturalist* 24: 637-644.

Rice, E.L. (1984). *Allelopathy.* New York: Academic Press.

Rice, E.L. (1995). *Biological Control of Weeds and Plant diseases–Advances in Applied Allelopathy.* USA: University of Oklahoma Press.

Rice, E.L. and S.K. Pancholy. (1973). Inhibition of nitrification by climax ecosystems. II. Additional evidence and possible role of tannins. *American Journal of Botany* 60: 691-702.

Rice, E.L. and S.K. Pancholy. (1974). Inhibition of nitrification by climax ecosystems. III. Inhibitors other than tannins. *American Journal of Botany* 61: 1095-1103.

Rizvi, S.J.H., D. Mukerji and S.N. Mathur. (1980). A new report on a possible source of natural herbicide. *Indian Journal of Experimental Biology* 18: 777-778.

Rizvi, S.J.H., D. Mukerji and S.N. Mathur. (1981). Selective phytotoxicity of 1,3,7-trimethylxanthine between *Phaseolus mungo* and some weeds. *Agricultural and Biological Chemistry* 45: 1255-1256.

Sae-lee, S., P. Vityakon and B. Prachaiyo. (1992). Effects of trees on paddy bund on soil fertility and rice growth in Northeast Thailand. *Agroforestry Systems* 18: 213-223.

Salazar, A., L.T. Szott and C.A. Palm. (1993). Crop-tree interactions in alley croping systems on alluvial soils of the Upper Amazon Basin. *Agroforestry Systems* 22: 67-82.

Sanchez, P.A. (1995). Science in Agroforestry. *Agroforestry Systems* 30: 5-55.

Sarma, K.K.V. (1974). Allelopathic potential of *Digera arvensis* Forsk. on *Pennisetum typhoides* Stapf. et. Hubb. *Geobios* 1: 137.

Sarmah, M.K., S.S. Narwal and J.S. Yadava. (1992). Smothering effect of *Brassica* species on weeds. In *Proceedings First National Symposium on Allelopa-*

thy in Agroecosystems, eds. P. Tauro and S.S. Narwal. Hisar: Indian Society of Allelopathy, pp. 51-55

Schon, M.K. and F.A. Einhellig. (1982). The allelopathic effects of cultivated Sunflower on grain Sorghum. *Botanical Gazette* 143: 505-510.

Schreiner, O. and H.S. Reed. (1908). The toxic action of certain organic plant constituents. *Botanical Gazette* 45: 73-102.

Shafer, W.E. and S.A. Garrison. (1986). Allelopathic effects of soil incorporated *Asparagus* roots on lettuce, tomato and asparagus seedling emergence. *Hort-Science* 21: 82-84.

Sharma, B.M., S.S. Rathore and J.P. Gupta. (1994). Compatibility studies on *Acacia tortilis* and *Zizyphus rotundifolia* with field crops under arid conditions. *Indian Forester* 120: 423-429.

Sharma, K.K. (1992). Wheat cultivation in association with *Acacia nilotica* (L.) Willd. ex Del. field bund plantation—a case study. *Agroforestry Systems* 17: 43-51.

Sharma, S., S.D. Chitkara and B.S. Daulta. (1994). Allelopathy in fruit crops. In *Allelopathy in Agriculture and Forestry*, eds. S.S. Narwal and P. Tauro. Jodhpur, India: Scientific Publishers, pp. 145-152.

Sharma, V. and G.S. Nathawat. (1987). Allelopathic effect of *Argemone mexicana* L. on species of *Triticum, Brassica, Raphanus* and *Pennisetum*. *Current Science* 56: 427-428.

Sheikh, M.I. (1988). Planting and establishment of windbreaks in arid areas. *Agriculture Ecosystem and Environment* 22/23: 405-423.

Sheikh, M.I. and R. Haq. (1978). Effect of shade of *Acacia arabica* and *Dalbergia sissoo* on the yield of wheat. *Pakistan Journal of Forestry* 29: 183-185.

Sheikh, M.I. and R. Haq. (1986). *Effect of Size, Placement and Composition of Windbreaks for Optimum Production of Annual Crops and Woods*. Final Technical Report. Peshawar, Pakistan: PFI, pp. 125.

Singh, A., R.S. Dhanda and P.K. Ralhan. (1993). Performance of wheat varieties under poplar (*Populus deltoides* Bartr.) plantations in Punjab (India). *Agroforestry Systems* 22: 83-86.

Singh, D. and R.K. Kohli. (1992). Impact of *Eucalyptus tereticornis* Sm. shelterbelts on crops. *Agroforestry Systems* 20: 253-266.

Singh, R.K. and R.S. Dwivedi. (1990). Fungicidal properties of neem and blue gum against *Sclerotium rolfsii* Sacc., a foot-rot pathogen of barley. *Acta Botanica Indica* 18: 260-262.

Srivastava, P.C., N.G. Totey and O. Prakash. (1986). Effect of straw extracts on water absortion and germination of wheat (*Triticum aestivum* L var. RR-21) seeds. *Plant and Soil* 91: 143-145.

Stevens, G.A., Jr. and C.S. Tang. (1985). Inhibition of seedling growth of crop species by recirculating root exudates of *Bidens pilosa*. *Journal of Chemical Ecology* 11: 166-167.

Theophrastus. (ca 300 B.C.). *Enquiry into Plants and Minor Works on Odours and Weather Signs*. 2 Vol., Translated to English by A. Hort. London: W. Neinemaw.

Thorne, R.L.Z., G.R. Waller, J.K. McPherson, E.G. Krenzer, Jr. and C.C. Young.

(1990). Autotoxic effects of old and new wheat straw in conventional-tillage and no-tillage wheat soil. *Botanical Bulletin Academia Sinica* 31: 35-49.

Tiwari, S.P., S.S. Bhandauri, M.S. Voddoria and D. Dayal. (1985). Allelopathic effects of weeds on soybean, groundnut and green gram. *Current Science* 54: 434-435.

Tomar, G.S. and S.K. Shrivastava. (1986). Preliminary studies of rice cultivation in association with trees. In *Agroforestry Systems: A new challenge,* eds. P.K. Khosla, S. Puri and D.K. Khurana. New Delhi: Indian Society of Tree Scientists, pp. 207-212.

Tsuzuki, E. and H. Kawagoe. (1984). Studies of allelopathy among higher plants. 4. On allelopathy in leguminous crops. *Bulletin Faculty of Agriculture, Miyazaki University* 31: 189-196.

Turco, R.F., M. Bischoff, D.P. Breakwell and D.R. Griffith. (1990). Contribution of soil-borne bacteria to the rotation effect of corn. *Plant and Soil* 122: 115-120.

Van Sumere, C.F., J. Cottenie, J. Degreef and J. Kint. (1972). Biochemical studies in relation to the possible germination regulatory role of naturally occurring coumarin and phenolics. *Recent Advances in Phytochemistry* 4: 165-221.

Vityakon, P., S. Sae-lee and S. Surasak. (1993). Effects of tree leaf litter and shading on growth and yield of paddy rice in Northeast Thailand. *Kasetsart Journal (Natural Science)* 27: 219-222.

Waller, G.R. (1987). *Allelochemicals: Role in Agriculture and Forestry.* ACS Symposium Series No. 330. Washington, DC: American Chemical Society.

Waller, S.C., W.A. Skroach and T.J. Monaco. (1985). Common bermudagrass (*Cynodon dactylon*) interference in newly planted peach (*Prunus persica* cv. Norman) trees. *Weed Science* 33: 50-56.

Weston, L.A. and A.R. Putnam. (1986). Inhibition of legume seedling growth by residues and extracts of quackgrass (*Agropyron repens*). *Weed Science* 34: 366-372.

Whitman, D.W. (1988). Allelochemical interactions between plants, herbivores and their predators. In *Novel Aspects of Insect-Plant Interactions,* eds. P. Barbosa and D.K. Letourneau. New York: John Wiley and Sons, pp. 11-64.

Whittaker, R.H. (1970). The biochemical ecology of higher plants. In *Chemical Ecology,* eds. E. Sondheimer and J.B. Simeone. New York: Academic Press, pp. 43-70.

Williamson, G.B. (1990). Allelopathy, Koch's postulates and the neck riddle. In *Perspectives on Plant Competition,* eds. J.B. Grace and D. Tilman. New York: Academic Press, pp. 143-162.

Willis, R.J. (1985). The historical basis of the concept of allelopathy. *Journal of Historical Biology* 18: 71-109.

Yakle, G.A. and R.M. Cruse. (1984). Effects of fresh and decomposing corn plant residue extracts on corn seedling development. *Soil Science Society of America Journal* 48: 1143-1146.

SUBMITTED: 01/03/97
ACCEPTED: 01/18/97

Hormonal Mechanisms
of Dormancy Induction
in Developing Seeds

M. Th. Le Page-Degivry

SUMMARY. Primary dormancy develops in seeds during their maturation on the mother plant. In recent years, the use of hormone mutants and the manipulation of endogenous hormone levels by their biosynthetic inhibitors have led to a new approach of the hormonal regulation of the onset of dormancy. There is good evidence to show that ABA synthesis at axis level is an absolute requirement for the induction and maintenance of dormancy. This continued *de novo* synthesized ABA could positively control the expression of specific ABA-responsive genes. Among the set of late embryo abundant proteins, some appear to be good candidates as "dormancy proteins." This review aims to correlate the preliminary results obtained by molecular biology with the recent advances in understanding the hormonal control of dormancy induction in developing seeds. *[Article copies available for a fee from The Haworth Document Delivery Service: 1-800-342-9678. E-mail address: getinfo@haworth.com]*

KEYWORDS. Abscisic acid, dormancy, gene expression, induction, dormancy proteins, gibberellin, hormone sensitivity

M. Th. Le Page-Degivry, Professor, Laboratoire de Physiologie Végétale, Faculté des Sciences, Université de Nice-Sophia Antipolis, 0-6108 Nice, Cedex 2, France.

[Haworth co-indexing entry note]: "Hormonal Mechanisms of Dormancy Induction in Developing Seeds." Le Page-Degivry, M. Th. Co-published simultaneously in *Journal of Crop Production* (The Food Products Press, an imprint of The Haworth Press, Inc.) Vol. 1, No. 1 (#1), 1998, pp. 203-222; and: *Crop Sciences: Recent Advances* (ed: Amarjit S. Basra) The Food Products Press, an imprint of The Haworth Press, Inc., 1998, pp. 203-222. Single or multiple copies of this article are available for a fee from The Haworth Document Delivery Service [1-800-342-9678, 9:00 a.m. - 5:00 p.m. (EST). E-mail address: getinfo@haworth.com].

INTRODUCTION

Seed dormancy implies that a given seed fails to germinate under conducive conditions and that some restrictive requirement has been imposed on germination. The dormant stage is transient in a seed's life; seeds enter and emerge from dormancy. For several decades, the study of the internal factors that regulate seed dormancy was mainly approached by a comparison of the biochemical and physiological behavior of dormant and dormancy-released mature seeds. One of the most studied research fields was the involvement of plant hormones in this regulation. The hormone balance theory (Amen, 1968) suggested that dormancy depended on the interaction between naturally occurring growth-inhibiting and growth-promoting substances. However, the frequent lack of correlation between hormone levels and the physiological state, the delay between the physiological responses to environmental factors and the variations of hormone levels, and the lack of specificity of hormone action led to the value of this theory being questioned (Walton, 1980; Black, 1983). In recent years, the use of hormone mutants (in which the biosynthesis of hormones or the sensitivity to hormones are impaired) and the manipulation of endogenous hormone levels by their biosynthetic inhibitors have led to interesting new insights into the problem. A large part of this review deals with the hormonal regulation of the onset of dormancy on the mother plant.

Control of dormancy by the regulation of gene expression was proposed by Tuan and Bonner (1964). More recently, the involvement of plant hormones in the regulation of gene expression has received much support (Skriver and Mundy, 1990). In the light of recent results, mainly obtained in cereals, the relationship between hormones, gene expression and dormancy will be discussed.

ONSET OF PRIMARY DORMANCY DURING SEED DEVELOPMENT ON THE MOTHER PLANT

Timing of the Onset of Dormancy

Primary dormancy develops in seeds still connected to the mother plant, before the end of seed maturation (Bewley and Black, 1982). Concerning the seed-coat imposed dormancy of *Arabidopsis thaliana,* Karssen et al. (1983) showed that the dormancy was initiated between 12 to 15 days after pollination. For *Helianthus annuus* embryos (Le Page-Degivry, Barthe and Garello, 1990), the development of dormancy took place progressively within the third week after pollination, i.e., at the mid stage of seed development.

In *Pyrus malus* (Thévenot and Durand, 1973), a typical example of deep embryo dormancy broken by a long chilling treatment was demonstrated. Very young embryos isolated when their fresh weight did not exceed 5 mg, were unable to germinate; however, a cold treatment (60-80 days at 5°C) could induce their germination. Thus as soon as they could be isolated, those young embryos were already dormant and a transition between non-dormant and dormant state could not be studied.

Influence of Environmental Factors

Many studies have highlighted the influence of environmental factors on acquisition of seed dormancy on the mother plant. For example, the germinability of seeds of different genotypes of *Arabidopsis thaliana* (Karssen, Hilhorst, and Koornneef, 1990) was markedly different when studied during their development either in winter or in summer. High temperature, low humidity, and long photoperiods during grain development in barley reduced the dormancy level of the mature grain (Strand, 1989; Schuwrink, Van Duijn, and Heidekamp, 1993).

How the temperature influences seed dormancy has been studied in some cases. In *Rosa,* De Vries and Dubois (1987) showed that plants raised under relatively high temperatures (22 to 26°C) had a favorable influence on achene germination as compared to lower temperatures (10-14-18°C). The endocarp thickness was determined by temperature and played a major role in determining achene germinability (Gudin et al., 1990). In wheat, the level of dormancy in mature grain was also markedly influenced by the temperature during seed development (Black, Butler and Hughes, 1987). High temperatures increase embryo responsiveness to abscisic acid (ABA) manifested in decreased germinability (Walker-Simmons, 1989; Basra et al., 1993).

INDUCTION OF DORMANCY BY ABA

Changes in ABA content during the development of fruits, seeds or embryos, reveal generally a similar pattern: a fairly steep rise during development followed by a sharp drop as the seeds mature (reviewed in King, 1982; Black, 1983, 1991). This behavior was observed in dormant seeds as well as in seeds which are non-dormant at maturity.

Inverse Correlation Between ABA Level and Germinability Throughout the Development of Non-Dormant Seeds

One of the suggested roles of ABA is the inhibition of precocious germination. Developmental variation in ABA content and its correlation to germinabilty of excised embryos has been found in many species; *Pisum sativum* (Eeuwens and Schwabe, 1975), *Glycine max* (Quebedeaux, Sweetser, and Rowell, 1976), *Phaseolus vulgaris* (Van Onckelen et al., 1980). However, it is often difficult to obtain precocious germination of isolated embryos before the stage corresponding to the maximum ABA content. A study with *Phaseolus vulgaris* embryos (Prévost and Le Page-Degivry, 1985) showed that, as remarked by Walbot, Clutter, and Sussex (1972), a nutritive supply was necessary to support the precocious germination of the youngest embryos. Furthermore, while germination was possible on a nutritive medium for embryos isolated at all stages, the time required before germination was subject to large variations: in such circumstances, the total germination percentage achieved in a given time gave an incomplete picture of the germination capacity of the embryos. An inverse correlation between ABA content and the duration of the lag phase preceding the germination could be demonstrated throughout the embryo development.

Moreover, this lag phase could be shortened and immature embryos induced to germinate precociously by treatments which depleted the endogenous pool of ABA, e.g., by washing the embryos or allowing the embryos to dry slowly (Ackerson, 1984; Prévost and Le Page-Degivry, 1985). The extent of germination corresponded to the length of treatments which in turn affected the level of endogenous ABA.

Involvement of ABA in the Onset of Dormancy

Such a correlation between ABA levels and germinability was not found in developing dormant embryos. For example, in wild-type seeds of *Arabidopsis thaliana,* the ABA level reached its maximum half-way through development and the seeds developed dormancy 4-5 days later (Karssen et al., 1983). In *Helianthus annuus,* endogenous ABA levels, which increased sharply in the first half of the developmental period, fell precisely when embryo dormancy became established (Le Page-Degivry, Barthe, and Garello, 1990).

However, ABA is absolutely required for the development of seed dormancy. Mutant lines of *Arabidopsis thaliana,* characterized by symptoms of withering and the absence of seed dormancy, showed much lower levels of endogenous ABA in developing seeds than the wild-type. At maturity they contained the same low ABA levels as wild-type. This observation that differences in dormancy are not a result of actual ABA levels in mature seeds but the result of differences in ABA levels during seed development led Karssen et al. (1983) to introduce the concept of dormancy induction.

The same relationship was demonstrated in sunflower (Le Page-Degivry, Barthe, and Garello, 1990), using fluridone, a pyridone inhibitor of ABA biosynthesis. When fluridone was applied early on the achenes before the increase of ABA level, it prevented the development of embryo dormancy but could not do so when applied later after the rise in ABA level. Dormancy thus appears to be dependent on ABA synthesis but not concomitant with its accumulation.

Changes in Sensitivity to ABA

As there was no correlation between ABA levels and germinability during dormancy induction, the hypothesis of a change in embryo sensitivity to ABA was proposed. Besides mutants with reduced ABA levels, mutants with a reduced sensitivity to ABA were isolated in a variety of species. In *Arabidopsis,* three genetically distinct "ABA-insensitive" loci (*abi 1, abi 2, abi 3*) were identified (Koorneef, Reuling, and Karssen, 1984). Mutations at all three loci resulted in reduced seed dormancy although all these mutants had a normal or even slightly higher endogenous ABA content than wild-type.

Changes in ABA sensitivity also occurred in developing wheat seeds, measured by the capability of ABA to block embryonic germination (Walker-Simmons, 1987; Morris et al., 1989). While the embryos were very sensitive to ABA during the earlier stages of development, those from the non-dormant cultivar lost this sensitivity as the grain entered the desiccation stage. Embryos from dormant cultivar continued to show sensitivity to ABA even upon grain desiccation. A 10 to 100-fold greater concentration of ABA was required to block germination of embryos from non-dormant compared to dormant seeds, even though initial ABA levels were similar in the two types of seeds.

Necessity of a Second Factor in Addition to ABA to Induce Dormancy

The nature of the changes in sensitivity to ABA is a matter of debate. The altered sensitivity may be due to a reduced availability of a receptor or a reduced binding-capacity of the receptor to ABA or changes in some proteins involved in the transduction cascade.

In maize, viviparous kernels could be induced in wild-type by fluridone treatment which inhibited ABA synthesis in both cob and embryo tissue, as also in viviparous mutants (class II and III) having reduced levels of ABA. In these two types of ABA-deficient kernels, it was only possible to induce dormancy by treating them with exogenous ABA during a rather narrow time window near stage 4 of embryogenesis (Hole, Smith, and Cobb, 1989), suggesting that there is a developmental period beyond which the embryos are insensitive to ABA.

Efforts to induce embryo dormancy in another class of maize mutants (*vp* 1) which had normal levels of ABA and were relatively insensitive to exogenous ABA have been completely unsuccessful. The *vp* 1 gene was cloned and had been shown to encode a *vp* protein acting as a potential transcription activator (McCarty et al., 1991). Smith et al. (1989) suggested that the induction of dormancy required both ABA and the *vp* protein. ABA could not act as a dormancy inducing factor, prior to the time that the *vp* protein was synthesized or in tissues in which *vp* was repressed.

Recently the *Arabidopsis abi* 3 gene was isolated (Giraudat et al., 1992). It appeared that *abi* 3 protein participated in an abscisic acid perception/transduction cascade and since it contained a putative nuclear targeting sequence, it was possibly involved in ABA-regulated gene expression. At this stage, it remains an open question whether the *abi* 3 protein in *Arabidopsis* is the exact functional counterpart of the maize *vp* 1 protein. Studies with several alleles of *abi* 3 (Parcy et al., 1994) indicated that *abi* 3 participates in a complex network regulating the various programs of gene expression during seed development and probably interacts with various ABA-dependent and ABA-independent pathways.

Further evidence of factors involved in dormancy other than ABA is pointed out in *Arabidopsis* by the mutation *fusca* 3 (Keith et al., 1994) which prevents the establishment of seed dormancy without altering perception or response to ABA.

Dual Source of ABA

Since ABA accumulated during maturation in non-dormant as well as in dormant seeds, another pathway of research explored the respective role of ABA according to its two possible origins: maternal or embryonic. Indeed, reciprocal crosses between wild-type and *aba* mutants showed that ABA in *Arabidopsis thaliana* had both a maternal and an embryonic origin (Karssen et al., 1983). The genotype of the mother plant regulated a sharp rise in ABA content half-way through seed development (maternal ABA). The genotype of the embryo and endosperm was responsible for a second ABA fraction (embryonic ABA) which reached much lower levels but persisted for some time after the maxima in maternal ABA. In wild-type, the small amount of embryonic ABA might be hidden under the large amount of maternal ABA. The onset of dormancy was well correlated with the presence of the embryonic ABA fraction and not with the maternal ABA (Karssen et al., 1983).

In developing tomato seeds, Groot and Karssen (1992) clearly demonstrated the dual source of ABA. But if the ABA fraction of embryo is decisive for the induction of dormancy, defoliation experiments proved that maternal ABA is also involved in the dormancy control. In *Zea mays* (Hole, Smith, and Cobb, 1989), the maternal component which includes production of ABA by the cob tissue is also capable of dormancy induction.

The involvement of head tissue in the onset of embryo dormancy was also demonstrated in *Helianthus annuus* (Barthe and Le Page-Degivry, 1993). Whereas young isolated embryos (10 DAP) were able to germinate to a high percentage, keeping the achenes inside a block of head tissue cultured *in vitro* (Wolswinkel, 1987) allowed the onset of embryo dormancy. On the contrary, when fluridone was introduced into the culture medium of these blocks, i.e., when ABA synthesis was inhibited both in head tissue and in embryos, vivipary could be observed (Le Page-Degivry et al., 1993).

Requirement of in situ *ABA Synthesis*

The role of maternal ABA could be considered by studying the mode of action of exogenous ABA. In *aba* mutants of *Arabidopsis thaliana,* ABA originating from exogenous spray application to the plants, or applications to the roots, did not influence dormancy induction. Karssen et al. (1983) concluded that for dormancy induction, ABA needs to be synthesized close to its site of action or to be localized in a specific subcellar compartment not accessible to ABA transported from outside the embryo.

In sunflower (Le Page-Degivry and Garello, 1992), addition of ABA to the culture medium of young isolated embryos led to an inhibition of their germination. In non-dormant mature or developing seeds of *Brassica napus* (Schöpfer and Plachy, 1984: Finkelstein et al., 1987), it was suggested that the primary action of ABA involved the control of water uptake. However, its inhibition required the continued presence of the hormone in the culture medium. This temporary inhibition is distinct from dormancy which, during development *in situ,* is induced by ABA in less than one week (Le Page-Degivry and Garello, 1992). A preculture in the presence of ABA (5 or 50 µM) for 5 or 10 days was unable to induce dormancy of young isolated embryos; it became effective immediately prior to the natural induction of dormancy. This observation could obviously be explained by a change in sensitivity to ABA. However, when fluridone was applied on these ABA-sensitive embryos, it allowed germination to occur and exogenous ABA applied simultaneously with fluridone led again to the inhibition of the germination but could not reinduce dormancy.

Radioimmunological analysis showed that after first day of culture, the level of ABA increased in dormant embryo axes cultured in control conditions while an application of fluridone totally inhibited this increase; therefore in both *Helianthus* and *Zea mays* embryos (Gage, Fong, and Zeevaart, 1989), ABA is synthesized via the indirect pathway requiring a xanthophyll precursor. If this ABA synthesis was inhibited by fluridone, exogenous ABA could not reinduce dormancy even at a developmental stage where the second dormancy inducing factor was present along with ABA. It is clear that the synthesis of the inhibitor is required in order to induce and to maintain dormancy. *De novo* synthesis of ABA was also demonstrated in embryos isolated either from dormant grains of *Hordeum distichum* (Wang, Heimovaara-Dijkstra, and Van Duijn, 1995) or from dormant seeds of *Fagus sylvatica* (Le Page-Degivry et al., 1996a).

Eventual Role of ABA Conjugates

In mature dormant embryo of *Pyrus malus*, Le Page-Degivry and Bulard (1979) showed that during the embryo culture, free ABA was released from conjugates stored in cotyledons and was transported to the root, where its accumulation caused growth inhibition.

In order to study the eventual involvement of ABA conjugates in the induction of embryo dormancy of *Helianthus*, $(\pm)[2-^{14}C]ABA$, prepared from $(\pm)-[12-^{14}C]ABA$ by high performance liquid chromatography on a chiral column, was fed to isolated immature embryos (Barthe et al., 1993). The major metabolite which accumulated with incubation time was characterized as dihydrophaseic acid glucopyranoside, however, ABAGE was detected only in very small amounts. In the same way, radioimmunological analysis showed that the endogenous level of ABAGE remained low throughout maturation (Barthe and Le Page-Degivry, 1993). These results did not support the hypothesis that, when free ABA decreased, it accumulated under a bound form which could constitute a potential source of ABA during imbibition.

However, unlike *Helianthus*, some seeds showed levels of bound ABA as high as those of free ABA in cotyledons and embryonic axes throughout maturation. The levels of free and bound ABA paralleled one another in peach seeds (Bonamy and Dennis, 1977) and in grape seeds (Broquedis, 1983). ABA conjugation is usually considered as an irreversible process on account of the stability of ABA esters observed in leaves following recovery from stress. Such behaviour strongly contrasts with the properties of similar conjugates of other plant hormones in germinating seeds

(Sembdner, Atzorn, and Schneider, 1994): upon imbibition of maize kernels, both auxin and gibberellin conjugates might act as reversible storage
forms. During *Fagus* embryo culture, ABA conjugates formed during seed
maturation could be hydrolyzed within the first days of imbibition (Le
Page-Degivry et al., 1996b). The study of the physiological involvement
of these conjugates in the induction and maintenance of dormancy is
currently being investigated in our laboratory.

ROLE OF GIBBERELLINS

Gibberellins (GAs) also accumulate in developing seed which is also a
good material for GA metabolism studies (reviewed in Phinney, 1984;
Takahashi, Phinney, and McMillan, 1991). The active metabolism made
the changes in GA levels difficult to analyze in detail. Alternatively, the
role of GAs has been studied by the use of mutants.

No Simultaneous Action of ABA and GAs at Any Stage of Seed Life

In *Arabidopsis thaliana* (Karssen et al., 1983), GA deficient mutants
were isolated by selection for seeds that did not germinate without application of GA (Koornneef and Van der Veen, 1980). Application of 10 μM
GA strongly stimulated precocious germination of these *ga* mutants during the first half of development but not later, as in the wild-type seeds.
Karssen and Lacka (1986) concluded that in an *Aba/Aba* background, it is
irrelevant for the induction of dormancy whether or not the seeds are able
to synthesize GA_3. Therefore, they proposed that ABA and GA_3 never
acted simultaneously at any stage of seed life, ABA inducing dormancy
during seed development and GAs acting during germination (Karssen,
Hilhorst, and Koornneef, 1990). The GA requirement of the germination
process would depend on the action of ABA during dormancy induction.
A higher GA requirement would not be due to higher ABA levels in ripe
seeds but would be caused by the much stronger physiological blockade of
the germination process in the ABA-containing genotypes. Thus high
endogenous ABA caused deep dormancy which required high GA levels
whereas low endogenous ABA resulted in lightly dormant seeds which
required low GA levels to germinate.

Changes in GA Sensitivity During Dormancy Induction

Ontogenic changes in GA sensitivity were observed in excised developing sunflower embryos associated with the different stages of the develop-

ment of embryo dormancy (Le Page-Degivry et al., 1996c). At the precise moment when dormancy became established, the sensitivity to GAs was very low and gradually increased during further maturation. These changes, which occurred during the development of the same embryos, cannot be explained by different ABA amounts during the induction of dormancy as suggested for the different genotypes of *Arabidopsis thaliana* (Karssen and Lacka, 1986). However, they could be correlated with the capability of embryos to synthesize ABA: the more the ABA synthesis decreased, the embryos need less GAs to germinate. A similar correlation was demonstrated during release of dormancy by dry storage (Bianco, Garello, and Le Page-Degivry, 1994). Moreover, manipulation of gibberellin levels by spraying gibberellin biosynthesis inhibitors on leaves of the mother plant transiently increased sensitivity to ABA in developing seeds of *Brassica napus* (Juricic, Orlando, and Le Page-Degivry, 1995).

In *Arabidopsis,* non-dormant mutants which do not need GA for germination have been selected on the basis of their resistance to GA biosynthesis inhibitors resulting in the isolation of mutants hypersensitive to GAs (Jacobsen and Olszewski, 1993). Mutants with reduced seed dormancy (Leon-Kloosterziel et al., 1996) also showed a reduced sensitivity to the gibberellin biosynthesis inhibitor tetcyclasis, leading the authors to assume that mutated genes control a step in the induction of dormancy which is most likely induced by ABA and is expressed as an increase of the gibberellin requirement for germination.

DEVELOPMENTALLY REGULATED GENES AND SEED DORMANCY

Repression of Genes Expressed During Germination

Tuan and Bonner (1964) proposed that dormancy was associated with repression of the genome and that the genome was unmasked when dormancy was broken. Results concerning some classes of RNA were in agreement with this hypothesis. Jarvis, Frankland, and Cherry (1968) suggested that the major change in gene expression during breakage of dormancy in hazel (*Corylus avellana*) seed by GA_3 was a specific increase in the rate of mRNA synthesis. On the other hand, dry storage of seed or ABA treatment which both induced dormancy (Jarvis and Shannon, 1981), inhibited synthesis of poly(A) RNA. In *Acer platanoides* (Slater and Bryant, 1987), stratification stimulated RNA polymerase I activity, responsible for rRNA synthesis in embryo axes, and the resulting increase in rRNA content was associated with breakage of seed dormancy.

ABA has been shown to repress several genes that are usually expressed during germination, including α-amylase (Garcia-Maya, Chapman and Black, 1990) and aleurain, a thiol protease expressed in the aleurone layer of germinating barley seeds (Rogers, Dena, and Heck, 1985). However, these identified genes for reserve-mobilizing enzymes are associated with secondary events of germination rather than with the initiation of germination. Liu et al. (1992) characterized a gene that was expressed in the endosperm of developing barley seeds and in cultured embryos, the expression of which was reduced to undetectable levels by treatment with ABA.

Most of the results deal with soluble proteins. However, in *Echinochloa crus galli,* the transition of the seeds from a dormant to a non-dormant state, by high temperature and alcohol treatment, was associated with changes in the composition and synthesis of membrane-bound proteins (Di Nola, Mischke, and Taylorson, 1990). The synthesis of a 23kD protein was strongly increased upon release from dormancy. These results suggest that the plasma membrane constitutes the first site in the seed cells at which the stimulus from external factors affecting seed dormancy is detected.

Positive Control Resulting in Continued Expression of Specific ABA-Responsive Genes

More recently, the possibility that dormancy proteins are specifically expressed in dormant seeds has received some interesting support. In seeds, besides rigorous developmental expression of storage protein genes, another class of plant genes exhibits a more flexible expression pattern in response to environmental and developmental cues. Examples of such genes include those that encode late embryo-abundant proteins (LEA) whose mRNAs accumulate to high levels late in embryogenesis and persist in the dry seed until their degradation during the germination process; they can also be expressed in plantlets in response to environmental cues including water stress or exposure to exogenous ABA (Galau, Hughes, and Dure, 1986). The function of these genes remains unknown, but they appeared to be good candidates as dormancy proteins since dormancy is induced by ABA during the embryogenesis and is expressed during the subsequent imbibition.

In barley, Hong, Barg, and Ho (1992) studied a mRNA species, HVA 1, (LEA) both during seed development and upon imbibition. The expression of HVA 1 mRNA appeared to be under developmental and organ-specific regulation during seed development. The level of this mRNA increased at about 25 DAP and persisted to seed maturity in the embryo and aleurone layers of the mature seed. This expression coincided with the time when ABA levels were high during seed development. The levels of HVA 1 mRNA declined rapidly during the imbibition of non-dormant seeds, whereas it remained unchanged in the embryos of dormant seeds for at least 5 days after imbibition.

Very few analyses have been conducted during dormancy induction and dormancy proteins were mainly studied by comparing gene expression in dry or imbibed dormant and non-dormant mature seeds. Embryos isolated from mature dry dormant and non-dormant grains have relatively high levels of transcripts. However, differences between *in vitro* translation products of mRNA extracted from dry dormant and non-dormant embryos could not be observed in wheat (Morris et al., 1991) nor in *Avena fatua* (Dyer, 1993), indicating that differential gene expression does not occur during after-ripening.

Upon imbibition, the expression pattern of several mRNAs was temporally modulated. The comparison of embryonic mRNA and soluble protein populations of dormant and non-dormant caryopses of *Avena fatua* (Dyer, 1993), suggested that differential gene expression during the first hours of imbibition might be responsible for maintaining dormancy. Two dormancy-associated mRNAs were differentially overexpressed in dormant embryos after 3 h of imbibition. Similar dormancy-associated changes in two soluble proteins were observed during imbibition.

In wheat (Walker-Simmons, Crane, and Yao, 1989; Ried and Walker-Simmons, 1990; Kawakami, Kawabata, and Noda, 1992), new protein synthesis reached a maximum rate upon the 12th hour of imbibition. cDNA clones identified for mRNA transcripts, abundant in dormant but rare in non-dormant axes at 12 h, were classified into five major gene families (Morris et al., 1991). These genes included three types of LEA proteins: Em (Morris et al., 1991), dehydrin and group 3 LEA (Ried and Walker-Simmons, 1993). For all these clones, gene expression was maintained and even increased in the dormant grain axes (Morris et al., 1991) while transcription levels declined in the non-dormant axes following imbibition. As in axes, prolonged maintenance of transcript levels for the selected clones was positively correlated with imbibed dormant grain. Embryonic transcript levels were maintained in the dormant imbibed whole grains for 2 days while levels declined steadily in the germinating non-dormant seeds.

With wild oat (Li and Foley, 1994), over twenty *in vivo* labelled and *in vitro* translated polypeptides were determined to be more abundant in dormant than in after-ripened embryos. GA$_3$ treatment of dormant caryopses which breaks dormancy could lower the transcript levels in dormant embryos; when the germination of after-ripened caryopses was inhibited by high temperatures, the decline in abundance of the transcripts was arrested (Li and Foley, 1995). DNA sequence analyses indicated that some of these cDNA clones encode LEA proteins. Moreover, one clone showed over 90% homology in amino acid sequence with respect to aldose reductase gene cloned by Bartels et al. (1991) and expressed during the desiccation phase in barley developing embryos.

Likewise, a set of specific gene transcripts was also maintained at elevated levels in hydrated dormant seeds of *Bromus secalinas* (Goldmark et al., 1992) while transcript levels declined in non-dormant seeds. One of them, which is embryo-specific, is also expressed in other weed grasses (Goldmark et al., 1992), and in crop plants. The deduced protein has no known homology with genes previously reported.

The localization of dormancy-associated proteins was attempted by some authors. Ried and Walker-Simmons (1990) observed that embryonic axes themselves differed in their capability for synthesis of heat-stable proteins according to whether they were isolated from dormant or nondormant grain. On the other hand, Bevington and Hance (1993) showed that in *Acer saccharum,* stratification altered the expression of several proteins in the cotyledons. Some of them, whose synthesis was repressed when dormancy was broken, could be associated with cotyledon-imposed growth inhibition of the axes.

Moreover, mRNA and polypeptides specifically associated with dormant state were shown to be ABA-inducible. In wheat (Morris et al., 1991), axes isolated from dormant mature grains compared to non-dormant ones, are far more responsive to ABA as defined by enhancement of transcript expression in response to lower ABA levels. A prolonged and active synthesis of ABA-responsive proteins was detected in imbibed dormant grain axes which was not observed in germinating embryos. An examination of the structural requirement of the ABA recognition response in wheat (Walker-Simmons et al., 1992) showed that all the ABA analogs that inhibited germination induced as ABA-responsive gene belonging to the group 3 LEAs. Pretreatment of *Avena sativa* embryos by ABA for at least 14 h was necessary to induce subsequent inhibition of germination and to effect protein synthesis (Corbineau, Poljakoff-Mayber, and Côme, 1991). Synthesis of these ABA-inducible proteins disappeared when embryos were transferred to water. The continued presence of ABA was therefore necessary to maintain their synthesis.

CONCLUSIONS

Molecular studies suggest that dormancy is under positive control resulting in the continued expression of specific genes, at least some of which are ABA-responsive. This suggestion can be discussed in the light of some characteristics of the hormonal control described in the first part of this review.

We demonstrated that dormant sunflower embryos might be distinguished form non-dormant ones by their capacity to synthesize ABA in their axes during the first 24 h of imbibition (Le Page-Degivry and Garello, 1992); this *in situ* synthesis could justify the continuous presence of ABA necessary to maintain the synthesis of dormancy-associated proteins.

It, therefore, appears that where ABA is acting to induce or maintain dormancy, it is also causing the formation of a spectrum of proteins, including some of those which provoke the initiation of dormancy. The function of these ABA-responsive gene products has not been established: whether they actually restrict germination or are only associated with an ABA-regulated process, cannot be decided until it is possible to block their production or to introduce them into non-dormant axes.

Furthermore, in dormant barley grains (Hong, Barg, and Ho, 1992), addition of GA_3 breaks seed dormancy and causes the disappearance of HVA 1 mRNA, showing that an interaction between ABA and GA_3 at the level of gene expression may be responsible for the physiological behavior of the grain.

Moreover, these ABA-responsive proteins are not limited to the developing seeds since they are detected upon imbibition. The distinction between ABA effects that are developmentally regulated and those that are induced by environmental stress or exogenous ABA is under investigation at the molecular level. In maize, ABA alone is sufficient to modulate accumulation of group 3 LEAs (Thomann et al., 1992) whereas other proteins, storage protein or group 1 LEA (McCarty et al., 1991), required ABA and *vp*1 to act in concert as suggested for dormancy induction in this species (Smith et al., 1989).

This review shows that a variety of techniques now available has led to interesting new insights into the mechanisms of dormancy induction. Preliminary results obtained by the molecular biology approach have to be examined thoroughly and the role of the proposed "dormancy proteins" defined. The suggested correlation with the hormonal status has also to be confirmed. However, it must be noticed that both hormonal and molecular aspects were studied mainly in seeds whose dormancy is released by dry storage. The onset of a deep embryo dormancy requiring a long cold stratification to be released has now to be approached with the same tools.

REFERENCES

Ackerson, R.C. (1984). Abscisic acid and precocious germination in soybean. *Journal of Experimental Botany* 35: 414-421.

Amen, R.D. (1968). A model of seed dormancy. *Botanical Review* 34: 1-31.

Bartels, D., K. Engelhardt, R. Roncarati, K. Schneider, M. Rotter and F. Salamini. (1991). An ABA and GA modulated gene expressed in the barley embryo encodes an aldose reductase related protein. *The EMBO Journal* 10: 1037-1043.

Barthe, Ph., L.R. Hooge, S.R. Abrams, M. Th. Le Page-Degivry. (1993). Metabolism of (+)abscisic acid to dihydrophaseic acid-4'-β-D-glucopyranoside by sunflower embryos. *Phytochemistry* 34: 645-648.

Barthe, Ph., M. Th. Le Page-Degivry. (1993). Dual sources of abscisic acid in developing *Helianthus annuus embryo*. In: *Fourth International Workshop on Seeds: Basic and Applied Aspects of Seed Biology,* Vol. 2, eds. D. Côme and F. Corbineau, Angers 1992. Paris: ASFIS, pp. 671-675.

Basra, A.S., K.S. Gill, P.S. Bagga and H.S. Dhaliwal. (1993). Abscisic acid responsiveness and control of germinability in wheat genotypes. In *Pre-harvesting Sprouting in Cereals* 1992, eds. M.K. Walker-Simmons and J.L. Reid. St. Paul, Minnesota, USA: American Association of Cereal Chemists, pp. 83-90.

Bevington, J.M. and B.A. Hance. (1993). Protein synthesis in embryos of sugar maple (*Acer saccharum*) during stratification and emergence from dormancy. In: *Fourth International Workshop on Seeds: Basic and Applied Aspects of Seed Biology,* Vol. 2, eds. D. Côme and F. Corbineau, Angers 1992. Paris: ASFIS, pp. 659-664.

Bewley, J.D. and M. Black (1982). *Physiology of Biochemistry of Seeds,* Vol. 2. Berlin: Springer-Verlag.

Blanco, J., G. Garello, M. Th. Le Page-Degivry. (1994). Release of dormancy in sunflower embryos by dry storage: involvement of gibberellins and abscisic acid. *Seed Science Research,* 4: 57-62.

Black, M. (1983). Abscisic acid in seed germination. In: *Abscisic acid,* ed. F.T. Addicott. New York: Praeger Publishers. pp. 331-363.

Black, M. (1991). Involvement of ABA in the physiology of developing and mature seeds. In: *Abscisic Acid: Physiology and Biochemistry,* eds. W.J. Davies and H.G. Jones. Oxford: Bios Scientific Publishers, pp. 99-124.

Black, M., J. Butler and M. Hughes. (1987). Control and development of dormancy in cereals. In *Fourth International Symposium on Pre-harvest Sprouting in Cereals,* ed. D.J. Mares. Boulder: Westview Press, pp. 379-392.

Bonamy, P.A. and F.G. Dennis, Jr. (1977). Abscisic acid levels in seeds of peach. I. Changes during maturation and storage. *Journal of American Society of Horticultural Science* 102: 23-26.

Broquedis, M. (1983). Evolution de l'acide abscissique lié (abscissate de β-D-glucopyranose) et de l'acide abscissique libre au cours du développement de la baie de raisin. *Connaissance Vigne Vin* 17: 247-257.

Corbineau, F., Poljakoff-Mayber and Côme, D. (1991). Responsiveness to abscisic acid of embryos of dormant oat (*Avena sativa*) seeds. Involvement of ABA-inducible proteins. *Physiologia Plantarum* 83: 1-6.

De Vries, D.P. and L. Dubois. (1987). The effect of temperature on fruit set, seed set and seed germination in "SONIA" × "HADLEY" hybrid tea-rose crosses. *Euphytica* 36: 117-120.

Di Nola, L., C.F. Mischke and R.B. Taylorson. (1990). Changes in the composition and synthesis of proteins in cellular membranes of *Echinochloa crus-galli* (L.) Beauv. seeds during the transition from dormancy to germination. *Plant Physiology* 92: 427-433.

Dyer, W.E. (1993). Dormancy-associated embryonic mRNAs and proteins in imbibing *Avena fatua* caryopses. *Physiologia Plantarum* 88: 201-211.

Eeuwens, C.J. and W.W. Schwabe. (1975). Seed and pod wall development in *Pisum sativum* L. in relation to extracted and applied hormones. *Journal of Experimental Botany*, 26: 1-14.

Finkelstein, R.R., A.J. De Lisle, A.E. Simon and M.L. Crouch. (1987). Role of abscisic acid and restricted water uptake during embryogeny in *Brassica*. In: *Molecular Biology of Plant Growth Control*, New York: Alan R. Liss, pp. 73-84.

Gage, D.A., F. Fong and J.A.D. Zeevaart. (1989). Abscisic acid biosynthesis in isolated embryos of *Zea mays*. L. *Plant Physiology* 89: 1039-1041.

Galau, G.A., D.W. Hughes and L. Dure. (1986). Abscisic acid induction of cloned cotton late embryogenesis-abundant (Lea) RNAs. *Plant Molecular Biology* 7: 155-170.

Garcia-Maya, M., J.M. Chapman and M. Black. (1990) Regulation of α-amylase formation and gene expression in the developing wheat embryo. *Planta* 181: 296-303.

Giraudat, J., B.M. Hauge, C. Valon, J. Smalle, F. Parcy and H.M. Goodman. (1992). *Isolation of Arabidopsis ABI3* gene by positional cloning. *Plant Cell 4*: 1251-1261.

Goldmark, P.J., J. Curry, C.F. Morris and M.K. Walker-Simmons. (1992). Cloning and expression of an embryo-specific mRNA up-regulated in hydrated dormant seeds. *Plant Molecular Biology* 19: 433-441.

Groot, S.P.C. and C.M. Karssen. (1992). Dormancy and germination of abscisic acid-deficient tomato seeds. Studies with the *sitiens* mutant. *Plant Physiology* 99: 952-958.

Gudin, S., L. Arene, A. Chavagnat and C. Bulard. (1990). Influence of endocarp thickness on rose achene germination: genetic and environmental factors. *HortScience* 25: 786-788.

Hole, D.J., J.D. Smith and B.G. Cobb. (1989). Regulation of embryo dormancy by manipulation of abscisic acid in kernels and associated cob tissue of *Zea mays* L. cultured *in vitro*. *Plant Physiology* 91: 101-105.

Hong, B., R. Barg and T.H.D. Ho. (1992). Developmental and organ-specific expression of an ABA- and stress-induced protein in barley. *Plant Molecular Biology* 18: 663-674.

Jacobsen, S.E. and N.E. Olszewski. (1993). Mutations at the Spindly locus of *Arabidopsis* alter gibberellin signal transduction. *Plant Cell* 5: 887-896.

Jarvis, B.C., B. Frankland and J.H. Cherry. (1968). Increased nucleic-acid synthesis in relation to the breaking of dormancy of hazel seed by gibbrellic acid. *Planta* 83: 257-266.

Jarvis, B.C. and P.R.M. Shannon. (1981). Changes in poly (A) RNA metabolism in relation to storage and dormancy-breaking of hazel seed. *New Phytologist* 88: 31-40.

Juricic, S.S, S. Orlando and M. Th. Le Page-Degivry. (1995). Genetic and ontogenic changes in sensitivity to abscisic acid in *Brassica napus* seeds. *Plant Physiology and Biochemistry* 33: 593-598.

Karssen, C.M., D.L.C. Brinkhhorst-van Der Swan, A.E. Breekland and M. Koornneef. (1983). Induction of dormancy during seed development by endogenous abscisic acid: studies on abscisic acid deficient genotypes of *Arabidopsis thaliana* (L.) Heynh. *Planta* 157: 158-165.

Karssen, C.M., H.W.M. Hilhorst and M. Koornneef. (1990). The benefit of biosynthesis and response mutants to the study of the role of abscisic acid in plants. In *Plant Growth Substances 1988,* eds. R.P. Pharis and S.B. Rood. Berlin: Springer-Verlag, pp. 23-31.

Karssen, C.M. and E. Lacka. (1986). A revision of the hormone-balance theory of seed dormancy: Studies on gibberellin and/or abscisic acid deficient mutants of *Arabidopsis thaliana* L. In *Plant Growth Substances 1985* ed. M. Bopp. Heidelberg: Springer-Verlag, pp. 315-323.

Kawakami, N., C. Kawabata and K. Noda. (1992). Differential changes in levels of mRNAs during maturation of wheat seeds that are susceptible and resistant to preharvest-sprouting. *Plant & Cell Physiology* 33: 511-517.

Keith, K., M. Kraml, N.G. Dengler and P. McCourt. (1994). *fusca* 3, a heterochromic mutation affecting late embryo development in Arabidopsis. *Plant Cell* 6: 589-600.

King, R.W. (1982). Abscisic acid in seed development. In *The Physiology and Biochemistry of Seed Development, Dormancy and Germination,* ed. A.A. Khan. New York: Elsevier Biomedical Press, pp. 157-185.

Koornneef, M., G. Reuling, and C.M. Karssen. (1984). The isolation and characterization of abscisic acid-insensitive mutants of *Arabidopsis thaliana. Physiologia Plantarum* 61: 377-383.

Koornneef, M. and J.H. Van der Veen. (1980). Induction and analysis of gibberellin sensitive mutants in *Arabidopsis thaliana* (L.) Heynh. *Theoretical and Applied Genetics* 58: 257-263.

Le Page-Degivry, M. Th., Ph. Barthe and G. Garello. (1990). Involvement of endogenous abscisic acid in onset and release of *Helianthus annuus embryo dormancy. Plant Physiology* 92: 1164-1168.

Le Page-Degivry M. Th., Ph. Barthe, J. Blanco and G. Garello. (1993). Involvement of abscisic acid in hormonal regulation of sunflower embryo dormancy. In *Fourth International Workshop on Seeds: Basic and Applied Aspects of*

Seed Biology, Vol. 2, eds. D. Côme and F. Corbineau, Angers 1992. Paris: ASFIS, pp. 615-623.

Le Page-Degivry, M. Th. and C. Bulard. (1979). Acide abscissqieu lié et dormance embryonnaire chez *Pyrus malus. Physiologia Plantarum* 46: 115-120.

Le Page-Degivry, M. Th. and G. Garello. (1992). In situ abscisic acid synthesis. A requirement for induction of embryo dormancy in *Helianthus annuus. Plant Physiology* 98: 1386-1390.

Le Page-Degivry M. Th., Ph. Barthe, J. Blanco and G. Garello. (1996a). ABA involvement in the psychrolabile dormancy of *Fagus* embryo. In *5th International Workshop on Seeds,* ed. R. Ellis (in press).

Le Page-Degivry, M. Th., Ph. Barthe and G. Garello. (1996b). The physiological relevance of ABA conjugation at different stages of *Fagus* embryo culture. *Xth FESPP Congress,* Firenze, Abstract.

Le Page-Degivry M. Th., J. Blanco, Ph. Barthe and G. Garello. (1996c). Changes in hormone sensitivity in relation to onset and breaking of sunflower embryo dormancy. In *Plant Dormancy: Its Physiology, Biochemistry and Molecular Biology,* ed. G.A. Lang. Wallingford: CAB International (in press).

Leon-Kloosterziel, K.M., G.A. Van De Bunt, J.A.D. Zeevaart and M. Koornneef. (1996). *Arabidopsis* mutants with a reduced seed dormancy. *Plant Physiology* 110: 233-240

Li, B. and M.E. Foley. (1994). Differential polypeptide patterns in imbibed dormant and afterripened *Avena fatua* embryos. *Journal of Experimental Botany* 45: 275-279.

Li, B. and M.E. Foley. (1995). Cloning and characterization of differentially expressed genes in imbibed dormant and afterripened *Avena fatua* embryo. *Plant Molecular Biology* 29: 823-831.

Liu, R., O.A. Olsen, M. Kreis and N.G. Halford. (1992). Molecular cloning of a novel barley seed protein gene that is repressed by abscisic acid. *Plant Molecular Biology* 18: 1192-1195.

McCarty, D.R., T. Hattori, C.B. Carson, V. Vasil, M. Lazar and I.K. Vasil. (1991). The viviparous-1 developmental gene of maize encodes a novel transcriptional activator. *Cell 6:* 895-905.

Morris, C.F., R.J. Anderberg, P.J.Goldmark and M.K. Walker-Simmons. (1991). Molecular cloning and expression of abscisic acid-responsive genes in embryos of dormant wheat seeds. *Plant Physiology* 95: 814-821.

Morris, C.F., J.M. Moffat, R.G. Sears and G.M. Paulsen. (1989). Seed dormancy and responses of caryopses, embryos, and calli to abscisic acid in wheat. *Plant Physiology* 90: 643-647.

Parcy, F., C. Valon, M. Raynal, P. Gaubier-Comella, M. Delseny and J. Giraudat. (1994). Regulation of gene expression programs during *Arabidopsis* seed development: roles of the ABI 3 locus and of endogenous abscisic acid. *Plant Cell* 6: 1567-1582.

Phinney, B.O. (1984). Gibberellin A$_1$, dwarfism and the control of shoot elongation in higher plants. In *The Biosynthesis and Metabolism of Plant Hormones,* eds. A. Crozier and J.R. Hilmann. Cambridge: Cambridge University Press, pp.17-41.

Prévost, I. and M. Th. Le Page-Degivry. (1985). Inverse correlation between ABA content and germinabilty throughout the maturation and the *in vitro* culture of the embryo of *Phaseolus vulgaris*. *Journal of Experimental Botany* 36: 1457-1464.

Quebedeaux, B., P.B. Sweetser and J.C. Rowell. (1976). Abscisic acid levels in soybean reproductive structures during development. *Plant Physiology* 58: 363-366.

Ried, J.L. and M. Walker-Simmons. (1990). Synthesis of abscisic acid-responsive, heat-stable proteins in embryonic axes of dormant wheat grain. *Plant Physiology* 93: 662-667.

Ried, J.L. and M.K. Walker-Simmons. (1993) Group 3 late embryogenesis abundant proteins in desiccation-tolerant seedlings of wheat (*Triticum aestivum* L.). *Plant Physiology* 102: 125-131.

Rogers, J.C., D. Dena and G. Heck. (1985). Aleurain: a barley thiol protease closely related to mammalian cathepsin, H. *Proceeding National Academy of Sciences, U.S.A.* 82: 6512-6516.

Schöpfer, P. and C. Plachy. (1984). Control of seed germination by abscisic acid. II. Effect on embryo water uptake in *Brassica napus* L. *Plant Physiology* 76: 155-160.

Schuwrink, R.C., G. Van Duijn, and F. Heidekamp. (1993). Is abscisic acid involved in barley aleurone dormancy? In: *Pre-harvest Sprouting in Cereals, 1992*, eds. M.K. Walker-Simmons and J.L. Ried. St. Paul Minnesota, USA: American Association of Cereal Chemists, pp. 224-231.

Sembdner, G., R. Atzorn and G. Schneider. (1994). Plant hormone conjugation. *Plant Molecular Biology* 26: 1459-1481.

Skriver K. and J. Mundy. (1990). Gene expression in response to abscisic acid and osmotic stress. *Plant Cell* 2: 503-512.

Slater, R.J. and J.A. Bryant. (1987). RNA polymerase activity during breakage of seed dormancy by low temperature treatment of fruits of *Acer platanoides* (Norway Maple). *Journal of Experimental Botany* 38: 1026-1032.

Smith, J.D., F. Fong, C.W. Magill, B.G. Cobb and D.G. Bai. (1989). Hormones, genetic mutants and seed development. In: *Recent Advances in the Development and Germination of Seeds*, ed. R.B. Taylorson. New York: Plenum Press, pp. 57-69.

Strand, E. (1989). Studies on seed dormancy in small grain species. I. Barley. *Norwegian Journal of Agricultural Science* 3: 85-99.

Takahashi, N., B.O. Phinney and J. McMillan. (1991). Gibberellins. Heidelberg: Springer-Verlag.

Thévenot, C. and M. Durand. (1973) Mise en évidence d'une domrance chez les embryos de pommiers immatures. C.R. *Acad Agric Fr.* 59: 908-918.

Thomann, E.B., J. Sollnger, C. White and C.J. Rivin. (1992). Accumulation of group 3 late embryogenesis abundant proteins in *Zea mays* embryos. *Plant Physiology* 99: 607-614.

Tuan, D.Y.H. and J. Bonner. (1964). Dormancy associated with repression of genetic activity. *Plant Physiology* 39: 768-772.

Van Onckelen H., R. Caubergs, S. Horemans and J.A. De Greef. (1980) Metabolism of abscisic acid in developing seeds of *Phaseolus vulgaris* L. and its correlation to germination and α-amylase activity. *Journal of Experimental Botany* 31: 913-920.

Walbot, V., M. Clutter and I.M. Sussex. (1972). Reproductive development and embryogeny in *Phaseolus*. *Phytomorphology* 22: 59-68.

Walker-Simmons, M. (1987). ABA levels and sensitivity in developing wheat embryos of sprouting resistant and susceptible cultivars. *Plant Physiology* 84: 61-66.

Walker-Simmons, M. (1989). Dormancy in cereals–Levels of and response to abscisic acid. In *Plant Growth Substances 1988,* ed. R. Pharis. Heidelberg: Springer-Verlag, pp. 400-406.

Walker-Simmons, M., K.E. Crane and S. Yao. (1989). Synthesis of acid-soluble, ABA-inducible proteins in wheat embryos from dormant grain. In *Recent Advances in the Development and Germination of Seeds,* ed. R.B. Taylorson. New York: Plenum Press, pp. 47-55.

Walker-Simmons, M.K., R.J.Anderberg, P.A. Rose and S.R. Abrams. (1992). Optically pure abscisic acid analogs–tools for relating germination inhibition and gene expression in wheat embryos. *Plant Physiology* 9: 501-507.

Walton, D.C. (1980/1981). Does ABA play a role in seed germination? *Israel Journal of Botany* 29: 168-180.

Wang, M., S. Heimovaara-Dijkstra and B. Van Duijn. (1995). Modulation of germination of embryos isolated from dormant and non dormant barley grains by manipulation of endogenous abscisic acid. *Planta* 195: 586-592.

Wolswinkel, P. (1987). Assimilate transport in developing seeds of sunflower (*Helianthus annus*). *Journal of Plant Physiology* 127: 1-10.

SUBMITTED: 05/05/96
ACCEPTED: 11/04/96

Injuries to Reproductive Development Under Water Stress, and Their Consequences for Crop Productivity

Hargurdeep S. Saini
Sylvie Lalonde

SUMMARY. Reproductive development of plants, from meiosis to seed set, is highly vulnerable to water deficit. Two peaks of high sensitivity are encountered during this period. The first one occurs during meiosis in reproductive cells, and is common to all species studied. Water deficit at this stage causes pollen sterility, but usually affects female fertility only when the stress is severe. Pollen sterility does not result from a desiccation of the reproductive organs, but is an indirect consequence of water deficit in the vegetative parts, and may be mediated by a transportable sporocidal signal. The second peak of sensitivity occurs during flowering, and is conspicuous in

Hargurdeep S. Saini, Directeur général, Université de Montréal, Institut de recherche en biologie végétale, 4101 rue Sherbrooke est, Montreal, Qc., Canada H1X 2B2. Sylvie Lalonde, Université de Montréal, Institut de recherche en biologie végétale, 4101 rue Sherbrooke est, Montreal, Qc., Canada H1X 2B2.

Address correspondence to: Hargurdeep S. Saini, Université de Montréal, Institut de recherche en biologie végétale, 4101 rue Sherbrooke est, Montreal, Qc., Canada H1X 2B2 (E-mail: sainih@ere.umontreal.ca).

Major parts of the authors' research reported in this article were supported by grants to H.S.S. from the Natural Sciences and Engineering Research Council of Canada and The Rockefeller Foundation.

[Haworth co-indexing entry note]: "Injuries to Reproductive Development Under Water Stress, and Their Consequences for Crop Productivity." Saini, Hargurdeep S., and Sylvie Lalonde. Co-published simultaneously in *Journal of Crop Production* (The Food Products Press, an imprint of The Haworth Press, Inc.) Vol. 1, No. 1 (#1), 1998, pp. 223-248; and: *Crop Sciences: Recent Advances* (ed: Amarjit S. Basra) The Food Products Press, an imprint of The Haworth Press, Inc., 1998, pp. 223-248. Single or multiple copies of this article are available for a fee from The Haworth Document Delivery Service [1-800-342-9678, 9:00 a.m. - 5:00 p.m. (EST). E-mail address: getinfo@haworth.com].

rice, maize and some dicots. Depending on species, stress during this period can cause loss of pollen fertility, spikelet death or abortion of newly formed seed. These injuries, unlike those caused by the meiotic-stage stress, are associated with a decline in the water status of the reproductive structures. Changes in carbohydrate availability and metabolism appear to be involved in the effects of stress at both these stages. *[Article copies available for a fee from The Haworth Document Delivery Service: 1-800-342-9678. E-mail address: getinfo@haworth.com]*

KEYWORDS. Water stress, drought, sterility, anther, pollen, fertilization, grain set, grain abortion, carbohydrates, hormones

PREAMBLE

Living with some form of environmental stress is a daily fact of life for crop plants in most situations. Plants face a variety of abiotic and biotic stresses that can limit productivity and, in the extreme, even threaten the survival of a crop.

Water deficit probably limits global crop productivity more than any other stress (Fischer and Turner, 1978; Boyer, 1982). The nature and extent of damage, and the ability of a plant to recover from it, depend on the developmental stage at which a plant encounters water deficit. These considerations in turn influence the degree to which the economic yield of a crop is affected. For example, a transitory drought during the vegetative growth phase could affect the yield of a forage crop much more severely than that of a grain crop. In addition, certain developmental stages are more sensitive to water deficit than others. Among these is the reproductive development, which includes several water-stress-sensitive processes (Salter and Goode, 1967). Consequences of this high sensitivity are particularly important for crops in which products of sexual reproduction constitute economic yield, as in grain crops, which provide staple food for most of humanity. Heightened sensitivity during the reproductive phase assumes added importance when one considers that the risk of drought during this phase is relatively greater because transpiration at this late stage can be high and the soil moisture can be low, especially in the event of inadequate supplementary rains under rain-fed conditions.

Reproductive phase in the development of a plant starts with the transformation of a vegetative meristem into an inflorescence or flower primordium, and ends when the fruit or seed reaches maturity. This rather broad period covers several sub-stages, including floral initiation, differentiation of various parts of an inflorescence and/or flower, male and female meiosis, development of pollen and embryo sac, pollination, fertilization, and fruit and seed development. This review does not cover the entire spectrum of events within this phase. Instead, we focus on the period from just before the onset of meiosis in the microspore- and megaspore-mother-cells to the early post-fertilization events during the transition of embryo sac to seed. We arbitrarily refer to this period as "reproductive development." We have limited this discussion to the specific effects of stress on reproductive processes, without including general stress responses that affect reproduction, such as flower abscission. The literature covered in this review reflects the reality that a bulk of the work in this field has been done with cereals. Finally, owing to the greater accessibility of the male reproductive tissues, nearly all researchers time stress treatments according to male development. Hence, we use male development as the primary reference throughout this review, and refer to female development only where appropriate.

STAGES OF SENSITIVITY TO WATER STRESS

A comprehensive review of the early work by Salter and Goode (1967) shows that scientists have been aware since the turn of the century that adequate water availability during the reproductive phase is important for maintaining crop yield. The precise stress-sensitive stages of development within this broad phase vary among different species.

The stage of greatest sensitivity to drought in wheat, barley and oats is similar, if not identical. Numerous reports, particularly by Russian workers, have pointed out that grain yields of these crops was most sensitive to drought during and just before ear emergence (van der Paauw, 1949; Novikov, 1952; Skazkin, 1961 and the references cited therein; Aspinall, Nicholls, and May, 1964; Wells and Dubetz, 1966; Salter and Goode, 1967 and the references cited therein; Dubetz and Bole, 1973; Fischer, 1973). Udol'skaja (1936) was probably the first to link this sensitivity to approximately the period of reduction division in the sex organs of wheat. More recent and precise experiments have confirmed that the critical drought-sensitive stage is centered around the period from meiosis to tetrad break-up in anthers (Bingham, 1966; Saini and Aspinall, 1981; Dembinska, Lalonde and Saini, 1992). In the female tissue, this period corresponds to the meiosis in the megaspore mother cell and the subsequent degeneration of three redundant megaspores in the tetrad (Bennett et al., 1973). Although the work with barley and oats has been less detailed, the similarities between their responses and those of wheat also point to meiosis as the stage of maximal stress sensitivity (van der Paauw, 1949; Novikov, 1952; Skazkin and Zavadskaya, 1957; Zavadskaya and Skazkin, 1960; Salter and Goode, 1967).

Rice and maize grain yields are also highly vulnerable to water deficit during reproductive development (Robins and Domingo, 1953; Sato, 1954; Denmead and Shaw, 1960; Matsushima, 1962; Salter and Goode, 1967; O'Toole and Moya, 1981). Within this period, two distinct peaks of sensitivity are evident. Similar to wheat, barley and oats, the first peak occurs during meiosis and is quite conspicuous in rice (Sato, 1954; Namuco and O'Toole, 1986; Sheoran and Saini, 1996a). This peak has also been observed in maize, but it has not attracted as much attention as in other cereals (Kisselbach, 1950; Downey, 1969; Moss and Downey, 1971). In addition, both rice and maize display a second peak of extreme sensitivity to water deficit during anthesis and initial stages of grain development (Claassen and Shaw, 1970; Hsiao, 1982; O'Toole and Namuco, 1983; Schoper, Lambert, and Vasilas, 1986; Westgate and Boyer, 1986a; Ekanayake, De Datta, and Steponkus, 1989; Ekanayake, Steponkus, and De Datta, 1990). Some sensitivity at this latter stage, particularly when the stress is severe, has also been observed in wheat, barley and oats, but it is much less pronounced compared to rice or maize (Aspinall, Nicholls, and May, 1964; Wardlaw, 1971; Fischer, 1973; Sandhu and Horton, 1977; Brocklehurst, Moss, and Williams, 1978).

In other grasses such as rye, millet and sorghum, the broad period from flower development to just after fertilization is quite sensitive to drought, but the precise stages of sensitivity in reproductive organs are not clearly known (Salter and Goode, 1967; Lewis, Hiler, and Jordan, 1974; Mahalakshmi and Bidinger, 1985; Mahalakshmi, Bidinger, and Raju, 1987; Craufurd, Flower and Peacock, 1993). Certain dicot species are also sensitive to drought during anthesis (Turner, 1993; Westgate and Peterson, 1993 and the references cited therein).

In essence, two stages of development are critically sensitive to water stress: meiosis, which is common to all the cereals examined, and anthesis and grain initiation, which is conspicuous in rice, maize and dicots.

The nature of injury to the structure and function of reproductive organs, and the underlying mechanisms depend on the stage at which the plant experiences water stress. Accordingly, they can be divided into two broad categories:

Effects of Meiotic Stage Water Stress

The Nature of Injury

The most common effect of water stress at this stage is the induction of pollen sterility. This effect has been observed in wheat (Skazkin, 1961; Saini and Aspinall, 1981), barley (Skazkin and Zavadskaya, 1957;

Zavadskaya and Skazkin, 1960), oats (Novikov, 1952), rice (Sheoran and Saini, 1996a) and maize (Downey, 1969). In wheat, water stress results in small and shriveled anthers that do not dehisce (Saini and Aspinall, 1981). All pollen in such anthers, and a variable proportion of pollen in the apparently normal anthers are sterile. Water deficit causes similar abnormalities in rice anthers (Sheoran and Saini, 1996a). In addition, stress during meiosis shortens the panicle length in rice (Sheoran and Saini, 1996a), and results in the production of a large number of so called "blasted" spikelets, i.e., immature, small and discolored spikelets that desiccate rapidly after panicle emerges (Namuco and O'Toole, 1986). Stress of similar intensity does not cause the latter effect in wheat (Saini and Aspinall, 1981; Dorion, Lalonde, and Saini, 1996), but extreme drought can lead to spikelet death (Morgan, 1971). Slight shortening of spike length has also been reported in wheat (Bingham, 1966).

Sterile pollen grains of wheat and rice contain dilute cytoplasm and are devoid of starch, which is a conspicuous constituent of fertile pollen (Saini and Aspinall, 1981; Saini, Sedgley, and Aspinall, 1984; Dorion, Lalonde, and Saini, 1996; Sheoran and Saini, 1996a). In an Australian wheat cultivar 'Gabo', anther development following meiotic-stage stress proceeds normally until about the first pollen grain mitosis (PGM-1), when microspores lose contact with the tapetum, get dislodged from their normal peripheral location, and fail to develop further (Saini, Sedgley, and Aspinall, 1984). In about half of such anthers the filament degenerates at the same time. The disoriented pollen grains fail to accumulate starch, and have dilute cytoplasm. Their exine development is normal but intine is almost completely absent. A somewhat different pattern of development was recently observed in a Canadian wheat variety, where first symptoms of developmental disruption were observed at or soon after meiosis (Lalonde, Beebe, and Saini, 1997). These included degeneration of some meiocytes, loss of orientation of microspores and abnormal vacuolization of tapetal cells.

Only a few attempts have been made to find if water deficit also affects female fertility. Grain set upon reciprocal crosses between stressed and un-stressed plants showed that female fertility was not affected by a water stress treatment that caused complete male sterility in approximately 40% of the florets (Saini and Aspinall, 1981). This treatment lowered leaf water potential (Ψ_w) to approximately –2.3 MPa (control Ψ_w = –0.8 MPa), which borders on being a severe stress for wheat. A similar absence of effect on female fertility in wheat was also reported by Bingham (1966). Female fertility in oats was not affected even under severe drought (soil moisture = 13% of field capacity), unless the stress was prolonged (Skazkin and Lukomskaya, 1962). Under the latter conditions, the antipodal cells were deformed and gradually degenerated, and nucellar cells filled up the lumen of the embryo sac which degenerated to a withered strand lacking functional elements. Water stress during embryo-sac development in maize caused various abnormalities in embryo-sacs, including a complete suppression of development (Moss and Downey, 1971). Depending on the severity and length of stress, 15 to 43% of observed ovules were abnormal, compared to only 2.5% in well-watered plants. Grain set in these plants was severely reduced despite hand pollination with fertile pollen, indicating that the structurally abnormal embryo sacs were also sterile. Together, these observations, albeit limited in number, indicate that female fertility in cereals can be reduced by water stress, but is probably much less sensitive than male fertility. This generalization may extend to other stresses, such as heat or low temperature, which induce male sterility but have very little or no effect on female fertility (Hayase et al., 1969; Brooking, 1976; Satake and Yoshida, 1978; Saini and Aspinall, 1982a; Saini, Sedgley, and Aspinall, 1983). The greater stress-tolerance of female gametophyte is of potential adaptive significance. In the field, the risk of reduction in grain set due to pollen sterility could be offset by cross-pollination owing to an excess of pollen produced, and various degree to which male sterility is affected in different florets. However, the reduction in female fertility could not be overcome.

Tissue Water Status and Drought-Induced Sterility

In wheat plants subjected to water stress for 3 to 4 days between 15 and 5 days prior to ear emergence (period that includes meiosis), grain set was not affected until xylem Ψ_w fell to -1.2 MPa, below which it declined linearly with a decline in Ψ_w, and was virtually nil as xylem Ψ_w approached -2.4 MPa (Fischer, 1973). Decline in leaf relative water content and Ψ_w to 67% and -2.3 MPa compared to the control values of 93% and -0.8 MPa, respectively, reduced grain set to 36% from 55% in controls (Saini and Aspinall, 1981). A more rapid water stress of similar magnitude caused a slightly greater decline in grain set (Dorion, Lalonde, and Saini, 1996). In rice plants subjected to water stress for 18 to 24 hours during male meiosis, grain set declined linearly from 80% to 50% with a decline in leaf Ψ_w from -1.1 MPa (control) to -3.5 MPa (Namuco and O'Toole 1986). A much greater reduction in grain set was recently reported with less severe stress (leaf Ψ_w declined to -2.3 MPa) (Sheoran and Saini, 1996a). These differences are likely to be due to wide genotypic differences in the drought tolerance of rice to the water deficit during reproductive development (Garrity and O'Toole, 1994).

Interestingly, Ψ_w of the spikes, spikelets, anthers or other floral organs of water stressed wheat plants either does not change (Saini and Aspinall, 1981) or declines much less than that of the leaf (Morgan and King, 1984; Dorion, Lalonde, and Saini, 1996; Westgate, Passioura, and Munns, 1996). Relative water content of the spikelets also remains unchanged in water-stressed plants (Morgan, 1980). Similarly, Ψ_w of rice panicle changes little diurnally or in response to water deficit during meiosis, but changes markedly in concert with potential evapotranspiration after panicle emergence (Tsuda and Takami, 1993). This relative immunity of meiotic-stage inflorescence to water loss may be owed to its protected position; wheat and rice inflorescences at this stage are still enclosed within two or more leaf sheaths, and are approximately a week to 10 days from emergence (e.g., Saini and Aspinall, 1981; Sheoran and Saini, 1996a). Thus transpiration from the inflorescence parts would be minimal at this time. Moreover, a break in xylem between the floral stalk and the pericarp probably causes a hydraulic discontinuity at this point (Zee and O'Brien, 1970). A combined effect of these factors could cause a lag between the drop in Ψ_w of vegetative (e.g. leaf) and floral parts.

Even when the Ψ_w of spikelets, anthers or ovaries does decline in response to water stress, the decline is fully matched by a reduction in osmotic potential (Ψ_s), and hence, the spikelet turgor does not change despite a drop in leaf turgor to zero (Morgan, 1980; Morgan and King, 1984; Westgate, Passioura, and Munns, 1996). Thus, water-stress-induced reduction in grain set does not correlate with the water status of the reproductive structures. Further, the grain set declines only after the leaf turgor falls to zero (Morgan, 1980; Morgan and King, 1984).

Hormonal Regulation of Drought-Induced Sterility—
Role of a Transportable Sporocide

The above observations that the water status of reproductive organs of wheat changes little despite a substantial decline in leaf Ψ_w in response to meiotic-stage water stress, led to the hypothesis that male sterility in water stressed plants is not caused by desiccation of reproductive structures, but results from indirect consequences of a drop in Ψ_w or turgor elsewhere in the plant (Morgan, 1980; Saini and Aspinall, 1981). The fact that grain set declines only after leaf turgor falls to zero (Morgan, 1980), suggests the involvement of a turgor responsive phenomenon. Since a major consequence of a drop in cell turgor following water stress is the accumulation of the hormone abscisic acid (ABA) in various plant parts (Aspinall, 1980; Pierce and Raschke, 1980; Walton, 1980), much attention has been focused on the role of ABA in controlling fertility in water-stressed wheat.

Meiotic-stage spikelets and anthers of stressed wheat plants accumulate ABA despite no change in their turgor (Morgan, 1980; Saini and Aspinall, 1982b; Westgate, Passioura, and Munns, 1996), indicating that the hormone is transported from leaves or other vegetative tissues. Long distance transport of ABA to wheat spikes and in other species has been demonstrated (Goldbach and Goldbach, 1977; Wolf, Jescke, and Hartung, 1990). Application of exogenous ABA to spike, to leaf, or via a wick inserted below the spike, causes pollen sterility and loss of grain set in wheat (Morgan, 1980; Saini and Aspinall, 1982b; Zeng, King, and Morgan, 1985). ABA has a similar effect when applied to detached spikes cultured in a nutrient solution (Waters, Martin, and Lee, 1984). The concentration of ABA in the spikelets of plants that received exogenous ABA was slightly higher but within the same order of magnitude as that induced by a water stress treatment that produced a similar loss of grain set (Saini and Aspinall, 1982b). Recently, a tight correlation ($r > 0.99$) was shown between grain set and ABA levels in anthers, ovaries and glumes of water-stressed wheat plants (Westgate, Passioura, and Munns, 1996). The stage of sensitivity to applied ABA and water stress is remarkably similar; both have the maximal effect during meiosis (Morgan, 1980; Saini and Aspinall, 1981; Saini and Aspinall, 1982b; Morgan and King, 1984; Zeng, King, and Morgan, 1985). Like water stress, ABA application reduces grain set via the induction of male sterility without affecting female fertility (Saini and Aspinall, 1982b). Anthers and pollen grains affected by stress or ABA look morphologically similar at maturity (Morgan, 1980; Saini and Aspinall, 1981; Saini and Aspinall, 1982b). Differences in the seed set between two Australian cultivars of wheat under well-watered conditions were negatively correlated with the ABA content of the spike (Morgan and King, 1984). Similarly, the Cornerstone nuclear-male-sterile mutant of wheat had greater ABA levels in spike and leaves, and had a lower rate of ABA metabolism than the corresponding fertile plants (Zeng and King, 1986). Finally, distal florets within a spikelet set fewer grains and contained higher ABA levels, compared to the more fertile basal florets (Lee, Martin, and Bangerth, 1988). Taken together, these observations support the idea that ABA produced in water-deficient leaves or other vegetative parts is translocated to the reproductive organs, where it acts as a male sporocide. However, these observations are entirely correlative, and do not provide any direct evidence of a causal relationship between ABA and stress-induced male sterility. A major difficulty in addressing this issue has been the lack of a specific inhibitor of ABA biosynthesis or action. In an alternative approach, Dembinska et al. (1992) grew wheat plants with their root system split into two equal halves. When half the roots were subjected to water stress while the remainder kept wet, water uptake by the wet half maintained leaf Ψ_w at the normal level, whereas ABA produced in the dry roots was translocated to the spike. Thus, the spike ABA content increased to the same level as in plants with the entire root system stressed. However, the grain set declined only in the latter plants, indicating that increased spike ABA level could not be the sole regulator of fertility. This view is also supported by an earlier report that a majority of anthers that abort upon ABA treatment follow a sequence of developmental events dissimilar to that in any of the water stress-affected anthers (Saini, Sedgley, and Aspinall, 1984). It is, however, conceivable that ABA and some other consequence(s) of reduced Ψ_w act together to induce sterility. Indeed, in an analogous situation, current water status of the plant appears to influence stomatal response to ABA (Tardieu and Davies, 1992). It is also important to appreciate that the sporocidic effect of ABA is not unique, because a variety of synthetic and natural substances induce sterility in plants (McRae, 1985; Cross and Ladyman, 1991; Sawhney and Shukla, 1994). Moreover, stresses that do not affect tissue ABA levels can cause nearly complete male sterility and failure of grain set (e.g., Saini and Aspinall, 1982b). Hence, apparent similarities among the effects of applied chemicals or different stresses on pollen fertility may simply be a reflection of the limited range of vulnerable events during the reproductive development rather than a similarity in the underlying mechanisms.

The experiments with split-root system (Dembinska, Lalonde, and Saini, 1992) also indicate that, unlike some other responses such as stomatal closure (Davies and Zhang, 1991), male sterility is not caused by a signal from the drying roots but results from the effects of a decline in shoot Ψ_w. A similar conclusion was reached in a recent study in which pressurization of roots of water-stressed plants to maintain leaf Ψ_w at the level of well-watered plants improved grain set compared to un-pressurized stressed plants, indicating that shoot Ψ_w was more important determinant of grain set than the root Ψ_w (Westgate, Passioura, and Munns, 1996).

Could other plant hormones be involved in this response to drought?

Water stress has been claimed to promote ethylene production in plants (Yang and Hoffman, 1984), and treatments with ethylene or ethephon, an ethylene releasing compound, induce male sterility (Rowell and Miller, 1971; Fairey and Stoskopf, 1975; Verma and Kumar, 1978). Therefore, it has been speculated that ethylene may be involved in the male-sterilent effect of water stress (Morgan, 1980). However, this seems improbable because unphysiologically high ethylene concentrations are required to cause male sterility (Bennett and Hughes, 1972; Fairey and Stoskopf,

1975; Saini, 1982). Moreover, water stress does not enhance ethylene production by intact wheat and other plants (Morgan et al., 1990; Narayana, Lalonde, and Saini, 1991). Consistent with this, no difference in ethylene production was detected between fertile and sterile Cornerstone wheat described earlier (Zeng and King, 1986).

Involvement of a decrease in auxin content in abnormal embryo-sac development and production of male organs in female flowers of watered-stressed maize has also been hypothesized (Moss and Downey, 1971). The idea is supported by the observations that changes in auxin content are associated with the differentiation of female inflorescence (Sladky, 1969), auxin application promotes femaleness and induces male sterility in maize (Heslop-Harrison, 1961), and water stress can cause an increase in IAA-oxidase activity and a decrease in auxin content (Darbyshire, 1971; Aspinall, 1980). To our knowledge, this idea has not been pursued further; neither has the possibility that fertility could be influenced by a decrease in the level of cytokinins, possibly via an inhibition of their synthesis in roots and/or transport to shoot (Aspinall, 1980; Davies et al., 1986; Davies and Zhang, 1991). Application of cytokinins can decrease heat-induced kernel abortion in maize and improve flower production and fruit set in other plants (Carlson et al., 1987; Atkins and Pigeaire, 1993; Mosjidis et al., 1993; Cheikh and Jones, 1994). Consistent with the physiological effects of cytokinins, pressurizing the roots of water-stressed wheat plants prevented browning of apical spikelets and improved their fertility (Westgate, Passioura, and Munns, 1996). Thus, the possible involvement of cytokinins in the control of fertility in water-stressed plants merits further investigation through experimental manipulations and measurement of cytokinin content.

Cellular and Metabolic Bases for Water-Stress-Induced Male Sterility

Virtually no conclusive information is available on the cellular and metabolic events leading to male sterility under water stress conditions. Even the information on events accompanying this response is very limited.

Perhaps the first step in this direction was the observation by Skazkin and Zavadskaya (1957) of abnormalities in chromosomal pairing and separation during meiosis in pollen mother cells of water-stressed barley. An increase in univalents, lagging chromosomes, noncongression of bivalents in metaphase, and micronuclei formation, was also observed in water-stressed rice (Namuco and O'Toole, 1986). Under severe stress, the entire meiotic process is arrested in rice (Namuco and O'Toole, 1986). Aberrant chromosomal behavior could result in uneven distribution of chromosomes in daughter nuclei. This could, and certainly a complete arrest of meiotic division would, cause pollen sterility. No such abnormalities have been noticed in wheat (Saini, 1982), where meiotic division is nearly always completed (Saini, Sedgley, and Aspinall, 1984; Lalonde, Beebe, and Saini, 1997). The products of meiotic division, with some exceptions (Lalonde, Beebe, and Saini, 1997), continue to develop normally for several days before aborting (Saini, Sedgley, and Aspinall, 1984; Lalonde, Beebe, and Saini, 1997). This points to a more subtle lesion than abnormal or failed meiosis as the cause of male sterility.

Structural lesions in development can sometime give clues to the underlying metabolic causes of a response. Normal pollen grains accumulate starch during the later part of their development, which is subsequently utilized to support pollen germination and pollen tube growth (Miki-Hirosige and Nakamura, 1983; Pacini and Franchi, 1988; Clément et al., 1994). Cereal pollen grains rendered sterile by drought or other stresses lack starch (Ito, 1978; Saini and Aspinall, 1981; Saini and Aspinall, 1982a; Sheoran and Saini, 1996a). In water-stress-affected wheat anthers, meiosis proceeds normally but the subsequent pollen grain development is arrested, notably around the first pollen grain mitosis (Saini, Sedgley, and Aspinall, 1984). Concomitant with this, starch deposition is also inhibited (Saini, Sedgley, and Aspinall, 1984). This suggests that a disturbance in carbohydrate availability and/or metabolism may play a role in the failure of pollen development.

Grain number in detached wheat spikes cultured in nutrient solution is positively correlated with sucrose uptake (Waters, Martin, and Lee, 1984). Addition of ABA to the nutrient solution causes a decrease in grain set as well as sucrose uptake by the spikes, and increase in sucrose concentration in the medium reverses the decline in grain set (Waters, Martin, and Lee, 1984). In contrast, anthers of water-stressed wheat and rice plants accumulate sucrose and other sugars (Dorion, Lalonde, and Saini, 1996; Sheoran and Saini, 1996a), despite the expected inhibition of both the rate of photosynthesis and export of assimilate from the leaves (Boyer and McPherson, 1975; Hanson and Hitz, 1982). Thus, sugar starvation *per se* is not the likely cause of pollen abortion. Instead, certain enzymes of sugar metabolism and starch biosynthesis are selectively inhibited: In wheat, the activities of starch synthase and ADP-glucose phosphorylase, and the expression of ADP-glucose pyrophosphorylase gene are not appreciably affected by stress (Dorion, Lalonde, and Saini, 1996; Lalonde, 1996). In rice, activities of both these enzymes decline during meiotic-stage stress or soon thereafter (Sheoran and Saini, 1996a). The most dramatic effect of stress in both species is the decline in soluble acid invertase activity, which never recovers even after the stress is relieved (Dorion, Lalonde, and Saini, 1996; Sheoran and Saini, 1996a).

Invertase is the dominant enzyme of sucrose cleavage in the anthers of many species (Bryce and Nelsen, 1979; Nakamura, Sado, and Arai, 1980; Nakamura, Suzuki, and Suzuki, 1992; Dorion, Lalonde, and Saini, 1996; Sheoran and Saini, 1996a). Reduced invertase activity in anthers would curtail proper processing of incoming sucrose, and the resulting decline in carbon availability could jeopardize crucial metabolic and developmental processes. In wheat, where the developmental anatomy of stress-induced pollen abortion is known (Saini, Sedgley, and Aspinall, 1984; Lalonde, Beebe, and Saini, 1997), the decline in invertase activity precedes or coincides with the first anatomical signs of pollen abortion. This finding is significant because, in an area of research marked by a scarcity of clues about mechanisms, it identifies a metabolic lesion that occurs early enough to merit consideration among the potential causal events. At least, it defines a point in carbohydrate metabolism from which to work backward in search of a primary triggering event for water-stress-induced male sterility.

Deficiencies of nutrients, such as nitrogen, boron and copper, cause male sterility (Howlett, 1936; Skazkin and Rozkova, 1956; Skazkin and Zavadskaya, 1957; Graham, 1975; Sharma et al., 1987; Azouaou and Souvré, 1993). Sometimes applications of boron or nitrogen just before the critical developmental period can ameliorate the effects of water stress on fertility (Skazkin and Rozkova, 1956; Skazkin and Zavadskaya, 1957). The possibilities that water deficit could affect fertility via restricting uptake or delivery of certain nutrients to developing anthers, or that deficiency of certain nutrients could increase the sensitivity to stress, merit attention.

Effects of Flowering-Stage Water Stress

For the purpose of this review, we use the term 'flowering-stage' to include anthesis, pollination, fertilization and initial seed development from fertilized embryo-sac. As mentioned earlier, two crops, rice and maize, are particularly sensitive at this stage.

The Nature of Injury

Water stress during flowering in rice can reduce harvest index by as much as 60%, largely as a result of a reduction in grain set (O'Toole and Moya, 1981; Hsiao, 1982; Garrity and O'Toole, 1994). Grain set is also inhibited by desiccation caused by dry wind; humid wind does not have the same effect (Ebata and Ishikawa, 1989). Water stress during or just before flowering causes failure of panicle to fully exsert (emerge) from the flag leaf sheath, a delay in flowering, reduction in the percentage of spike-lets that open at anthesis, and severe desiccation and death of spikelets (O'Toole and Namuco, 1983; Ekanayake, De Datta, and Steponkus, 1989). Stress-affected lemma and palea turn white and anthers shrivel and dry (Ekanayake, De Datta, and Steponkus, 1989). Water stress also reduces the number of anthers that dehisce, and lowers the amount of pollen-shed and its *in vivo* germinability (Ekanayake, Steponkus, and De Datta, 1990). Presumably, these reproductive abnormalities result in a failure of fertil-ization. Grain abortion in some of the spikelets that are fertilized also accounts for a part of the reduction in grain number (O'Toole and Namuco, 1983). However, the relative importance of non-fertilization and seed abortion is not clear.

Grain number and, therefore, yield of maize are also severely reduced by water stress during flowering (Claassen and Shaw, 1970). This loss of grains can be attributed to one or more of the following causes: When stress occurs just prior to anthesis, silk growth relative to the tassel growth can be retarded by several days, causing asynchrony between pollen-shed-ding and silk emergence, and thus a failure of pollination (Kisselbach, 1950; Du Plessis, and Dijkhuis, 1967; Moss and Downey, 1971; Herrero and Johnson, 1981; Westgate and Boyer, 1985a). A failure of tassel emer-gence or anther exsertion can also have similar effects (Herrero and John-son, 1981). Water stress during anthesis does not affect pollen viability or its ability to effect fertilization (Herrero and Johnson, 1981; Hall et al., 1982; Schoper, Lambert, and Vasilas, 1986; Westgate and Boyer, 1986a). Water-stressed female plants do not set seed even when pollinated with normal pollen (Schoper, Lambert, and Vasilas, 1986; Westgate and Boyer, 1986a). This failure of seed set is not attributable to inhibition of pollen germination, pollen tube growth or fertilization, but results from the fail-ure of fertilized embryo-sac to develop beyond 2 or 3 days because of poor embryo and endosperm development and a lack of seed coat differenti-ation (Westgate and Boyer, 1986a).

Tissue Water Relations, Cell Growth, and the Effects of Flowering-Stage Water Stress on Fertility

Emerging and newly emerged rice panicles have very low diffusive resistance to water, and, unlike leaves, they have poor ability to prevent water loss even under severe desiccation (O'Toole, Hsiao, and Namuco, 1984; Ekanayake, De Datta, and Steponkus, 1993; Tsuda and Takami, 1993). Therefore, many of the deleterious effects of water stress during flowering can probably be attributed to the desiccation of the floral parts. Certain upland adapted cultivars of rice are better able to prevent water loss from panicle and suffer less sterility under stress (Ekanayake, De Datta, and Steponkus, 1993). However, reduction in transpiration by covering the panicles does not prevent sterility (Garrity, O'Toole, and Vidal, 1986), indicating that other adaptive features probably act in concert with reduced transpiration to limit sterility. Insufficient turgidity caused by excessive water loss could also be responsible for the inhibition of spikelet opening (Ekanayake, De Datta, and Steponkus, 1989), because this process is driven by the pressure generated by the turgidity of floral structures (Parmar, Siddiq, and Swaminathan, 1979).

One of the most conspicuous effects of water stress in rice is the failure of panicle to fully exsert from the flag leaf sheath (O'Toole and Namuco, 1983; Cruz and O'Toole, 1984). All spikelets that do not exsert are sterile, and approximately 25 to 30% of the water-stress-induced spikelet sterility is associated with poor panicle exsertion. The failure of seed set appears to be due to the inability of unexserted spikelets to complete anthesis and shed pollen, even when they are otherwise apparently healthy (O'Toole and Namuco, 1983). The reasons for the non-exsertion of panicle are not known, but may include inhibition of panicle elongation (Sheoran and Saini, 1996a). In this regard, a near total suppression of tubulin gene expression in panicles of plants subjected to brief meiotic-stage water deficit, and the inability of the expression to recover for several days after the plants are re-watered, is significant (Sheoran and Saini, 1996b). Rice panicle undergoes rapid elongation prior to panicle exsertion, and microtubules, and tubulin play a central role in cell division, cell elongation and cell wall deposition (Fosket and Morejohn, 1992).

Reproductive structures of maize are much more prone to water loss and a drop in Ψ_w at anthesis than at the later stages (Westgate and Thomson Grant, 1989). When soil water is depleted, Ψ_w of silks, leaf, root and stem declines (Westgate and Boyer, 1985b). In leaf, root and stem, Ψ_s declines enough in response to this drop in Ψ_w to maintain full turgor, but silks are unable to adequately lower their Ψ_s and maintain turgor (Westgate and Boyer, 1985b). Considering the importance of turgor for cell enlargement (Lockhart, 1965; Cosgrove, 1981; Cosgrove, 1993), this poor ability of silks to maintain turgor may be responsible for the delay in their growth and emergence under water stress (Moss and Downey, 1971; Herrero and Johnson, 1981). Pollen Ψ_w is quite low even in well-watered plants, and decline in water status of droughted plants does not affect pollen Ψ_w (Westgate and Boyer, 1986b). On the other hand, silk Ψ_w declines in concert with a decline in the Ψ_w of vegetative parts of the plant

(Westgate and Boyer, 1986b). However, this decline in silk Ψ_W does not prevent pollen germination because pollen Ψ_W is very low and always remains lower than that of silks, ensuring that pollen can draw water even from relatively dry silks (Westgate and Boyer, 1986a,b). This, combined with the maintenance of pollen viability even at very low Ψ_W (Barnabas and Rajki, 1981; Schoper, Lambert, and Vasilas, 1986; Westgate and Boyer, 1986a; Schoper et al., 1987), shows that pollen desiccation is not a factor in limiting seed set under water stress. As fertilization is apparently not inhibited under stress but further grain development is (Westgate and Boyer, 1986a), considerable effort has been devoted to identify the maternal factors that affect kernel abortion.

Carbohydrate Availability and Kernel Abortion Under Water Deficit

Since the level of leaf Ψ_W that causes kernel abortion in maize also completely inhibits photosynthesis and lowers carbohydrate reserves in stem (Westgate and Boyer, 1985a), it was hypothesized that the abortion may be attributable to limited carbohydrate availability during pollination (Westgate and Boyer, 1986a). When photosynthesis is inhibited by either water stress or varying light intensity, kernel number is reduced in proportion to the inhibition of photosynthesis (Schussler and Westgate, 1991a). Despite this inhibition, carbohydrates continue to accumulate in vegetative sinks, such as leaf and stalk, but their movement to reproductive sinks is severely reduced (Schussler and Westgate, 1991a,b). These data indicate that early kernel development is dependent on the supply of assimilate from concurrent photosynthesis, which cannot be substituted by remobilization of reserves stored in other tissues. Since sugar concentrations in the ovaries do not change, and sugar uptake by ovaries isolated from stressed plants is inhibited, it appears that the kernel set depends on the rate of movement of current assimilate to reproductive organs and not on the concentration of sugars *per se* (Schussler and Westgate, 1991b). In accord with this, experimental manipulations that either increase accumulation of carbohydrate reserves prior to anthesis or reduce sink size for a fixed availability of assimilates, do not diminish the extent of water-stress-induced kernel loss (Schussler and Westgate, 1994; Zinselmeier, Westgate, and Jones, 1995). Infusing liquid culture medium into the stem in a quantity sufficient to replace carbohydrates lost by the inhibition of photosynthesis during flowering-stage water deficit, partially or completely prevents kernel abortion (Boyle, Boyer, and Morgan, 1991). Since this treatment does not rehydrate the infused plants, and kernel abortion is not prevented by the infusion of water alone, the results further support the idea that kernel abortion is caused by limited assimilate supply.

Partitioning of carbon into reserves depends on assimilate supply and sink demand (Jenner, 1982). Because kernel abortion is not prevented by increase in assimilate supply through cultural or genetic manipulations (Schussler and Westgate 1991a,b; Schussler and Westgate, 1994; Zinselmeier, Westgate, and Jones, 1995), the ability of ovaries to utilize the available assimilate may also be impaired by water stress. This idea is also supported by reduced sucrose uptake by ovaries isolated from stressed plants, and lack of complete kernel set upon feeding culture medium to plants under water stress (Boyle, Boyer, and Morgan, 1991; Schussler and Westgate, 1991b). Direct evidence for this was recently furnished by Zinselmeier et al. (1995), who showed that acid invertase activity in the ovaries of water-stressed plants was strongly inhibited, parallel with a cessation of ovary growth, a massive accumulation of sucrose and a decrease in the level of reducing sugars. Maize ovaries induced to abort *in vitro* also have low acid invertase activity (Hanft and Jones, 1986). The decline in acid invertase activity in water-stress-affected kernels could be related to changes in sucrose delivery to the ovary during water deficit (Xu et al., 1996). Invertase is a well known marker of sink strength, and the maize mutant *miniature-1* lacking soluble and wall-bound acid invertase fails to produce normal kernels (Miller and Chourey, 1992). Thus, kernel abortion in water-stressed maize may be caused by the decline in acid invertase activity, and the resulting inability of the ovaries to maintain adequate sink strength. These metabolic events are remarkably similar to the inhibition of invertase activity and associated disruption of carbohydrate metabolism in water-stress-affected anthers of wheat and rice described earlier in this review. Whether this portends a similarity of mechanisms of responses to water stress of these rather different tissues, is an exciting possibility.

CONCLUDING REMARKS

While the high sensitivity of crop plants to water deficit during reproductive development, and its consequences for yield have been known since the turn of the century, serious enquiries into the physiological, biochemical and molecular causes of this response did not begin until the early 80s. The following generalizations can be extracted from the literature: Water stress during the meiotic-stage, which is sensitive in all the species examined, reduces grain set principally through an induction of pollen sterility. Flowering-stage stress has major effects on rice and maize;

it causes desiccation and functional abnormalities in floral organs, leading to a failure of pollination in rice, and kernel abortion within a few days after pollination in maize.

The idea that induction of sterility by meiotic-stage water stress involves an endogenous sporocide transported from vegetative to reproductive organs, has been explored in some detail. The focus has been almost exclusively on the role of ABA as such a sporocide. Despite considerable correlative evidence in favor of this idea, direct demonstration of a causal link between plant-produced ABA and sterility is still lacking. In fact, the most recent evidence contradicts this hypothesis. Further progress in this area is unlikely until specific inhibitors of ABA synthesis or action, or other means of manipulating ABA levels without causing widespread physiological dysfunctions are developed. In addition, more attention needs to be paid to the possibility that sporocidal 'signal' may also be a deficiency that originates outside the reproductive structures. Cytokinins and assimilate supply are potential candidates in this regard (Davies and Zhang, 1991; Xu et al., 1996). Poor ability of panicles to retain water appears to be a major cause of damage by flowering-stage water stress in rice. Hence, any attempts to understand the mechanisms underlying this response, and to improve plant tolerance at this stage must also consider the attributes that affect panicle transpiration and osmoregulation, as well as root characteristics, such as depth, that influence water delivery to the shoot (Garrity, O'Toole, and Vidal, 1986; Hanson et al., 1990). Impairment of assimilate supply and the ability of reproductive structures to use it, appear to be important factors, both in kernel abortion in maize and disruption of pollen development in wheat and rice (Zinselmeier et al., 1995; Dorion, Lalonde, and Saini, 1996; Sheoran and Saini, 1996a). The similarities between these two systems could imply similar regulatory mechanisms, and lessons from one could be useful in understanding the other. The next challenge is to determine if the observed changes are a part of the chain of events leading to developmental failure, or are simply manifestations of it.

Despite considerable effort over the last decade and a half, the mechanisms underlying the vast and important problem of reproductive failure during water stress remain unclear. However, as a result of the progress that has been made, we have moved beyond the search for the proverbial 'needle in the hay stack,' and certain well identified targets are now ready for critical scrutiny.

REFERENCES

Aspinall, D. (1980). Role of abscisic acid and other hormones in adaptation to water stress. In *Adaptation of Plants to Water and High Temperature Stress*, eds. N. C. Turner and P. J. Kramer. Brisbane: John Wiley & Sons, pp. 155-172.

Aspinall, D., P. B. Nicholls and L. H. May. (1964). Effect of soil moisture stress on the growth of barley. I. Vegetative development and grain yield, *Australian Journal of Agricultural Research* 15:729-745.

Atkins, C. A. and A. Pigeaire. (1993). Application of cytokinins to flowers to increase pod set in *Lupinus angustifolius* L. *Australian Journal of Agricultural Research* 44:1799-1819.

Azouaou, Z. and A. Souvré. (1993). Effects of copper deficiency on pollen fertility and nucleic acids in the durum wheat anther. *Sexual Plant Reproduction* 6:199-204.

Barnabas, B. and E. Rajki. (1981). Fertility of deep frozen maize (*Zea mays* L.) pollen. *Annals of Botany* 48:861-864.

Bennett, M. D. and W. G. Hughes. (1972). Additional mitosis in wheat pollen induced by ethrel. *Nature* 240:566-568.

Bennett, M. D., M. K. Rao, J. B. Smith and M. W. Bayliss. (1973). Cell development in the anther, the ovule and the young seed of *Triticum aestivum* L. var. Chinese Spring. *Philosophical Transactions of Royal Society of London* B266:39-81.

Bingham, J. (1966). Varietal response in wheat to water supply in the field, and male sterility caused by a period of drought in a glass house experiment. *Annals of Applied Biology* 57:365-377.

Boyer, J. S. (1982). Plant Productivity and Environment. *Science* 218:443-448.

Boyer, J. S. and H. G. McPherson. (1975). Physiology of water deficits in cereal crops. *Advances in Agronomy* 27:1-23.

Boyle, M. G., J. S. Boyer and P. W. Morgan. (1991). Stem infusion of liquid culture medium prevents reproductive failure of maize at low water potential. *Crop Science* 31:1246-1242.

Brocklehurst, P. A., J. P. Moss and W. Williams. (1978). Effect of irradiance and water supply on grain development in wheat. *Annals of Applied Biology* 90:265-276.

Brooking, I. R. (1976). Male sterility in *Sorghum bicolor* L. Moench induced by low night temperature. I. Timing of the stage of sensitivity. *Australian Journal of Plant Physiology* 3:589-596.

Bryce, W. H. and O. E. Nelsen. (1979). Starch-synthesizing enzymes in the endosperm and pollen of maize. *Plant Physiology* 63:312-317.

Carlson, D. R., D. J. Dyar, C. D. Cotterman and R. C. Durley. (1987). The physiological basis for cytokinin induced increases in pod set in IX93-100 soybeans. *Plant Physiology* 84:233-239.

Cheikh, N. and R. J. Jones. (1994). Disruption of maize kernel growth and development by heat stress. Role of cytokinin/abscisic acid balance. *Plant Physiology* 106:45-70.

Claassen, M. M. and R. H. Shaw. (1970). Water deficit effects on corn. II. Grain components. *Agronomy Journal* 62:652-655.

Clément, C., L. Chavant, M. Burrus and J. C. Audran. (1994). Anther starch variations in *Lilium* during pollen development. *Sexual Plant Reproduction* 7:347-356.

Cosgrove, D. J. (1981). Analysis of the dynamic and steady-state responses of growth rate and turgor pressure to changes in cell parameters. *Plant Physiology* 72:1439-1446.

Cosgrove, D. J. (1993). How do plant cell walls extend? *Plant Physiology* 102:1-6.

Craufurd, P. Q., D. J. Flower and J. M. Peacock. (1993). Effect of heat and drought stress on sorghum (*Sorghum bicolor*). I. Panicle development and leaf appearance. *Experimental Agriculture* 29:61-76.

Cross, J. W. and J. A. R. Ladyman. (1991). Chemical agents that inhibit pollen development–tools for research. *Sexual Plant Reproduction* 4:235-243.

Cruz, R. T. and J. C. O'Toole. (1984). Dryland rice response to an irrigation gradient at flowering stage. *Agronomy Journal* 76:178-183.

Darbyshire, B. (1971). The effect of water stress on activity of indoleacetic acid oxidase in pea plants. *Plant Physiology* 47:65-67.

Davies, W. J., A. R. de Costa, T. A. Lodge and J. Metcalfe. (1986). Plant growth substances and the regulation of growth under drought. *Australian Journal of Plant Physiology* 13:105-125.

Davies, W. J. and J. Zhang. (1991). Root signals and the regulation of growth and development of plants in drying soils. *Annual Review of Plant Physiology and Plant Molecular Biology* 42:55-76.

Dembinska, O., S. Lalonde and H. S. Saini. (1992). Evidence against the regulation of grain set by spikelet abscisic acid levels in water-stressed wheat. *Plant Physiology* 100:1599-1602.

Denmead, O. T. and R. H. Shaw. (1960). The effect of soil moisture stress at different stages of growth on the development and yield of corn. *Agronomy Journal* 52:272-274.

Dorion, S., S. Lalonde and H. S. Saini. (1996). Induction of male sterility in wheat by meiotic-stage water deficit is preceded by a decline in invertase activity and changes in carbohydrate metabolism in anthers. *Plant Physiology* 111: 137-145.

Downey, L. A. (1969). Crop Density and Water Use Studies with Irrigated Maize (*Zea mays* L.) on Sodic Clay Soils. M.Sc. Agr. Thesis, Sydney: Department of Agronomy, University of Sydney (cited by Moss & Downey, 1971).

Du Plessis, D. F. and F. J. Dijkhuis. (1967). The influence of the time lag between pollen shedding and silking on the yield of maize. *South African Journal of Agricultural Sciences* 10:667-674.

Dubetz, S. and J. B. Bole. (1973). Effect of moisture stress at early heading and of nitrogen fertilizer on three spring wheat cultivars. *Canadian Journal of Plant Science* 53:1-5.

Ebata, M. and M. Ishikawa. (1989). Effects of wind and rain on the fertilization, kernel development and kernel characters of rice plant. *Japanese Journal of Crop Science* 58:555-561.

Ekanayake, I. J., S. K. De Datta and P. L. Steponkus. (1989). Spikelet sterility and flowering response of rice to water stress at anthesis. *Annals of Botany* 63:257-264.

Ekanayake, I. J., S. K. De Datta and P. L. Steponkus. (1993). Effect of water deficit stress on diffusive resistance, transpiration, and spikelet desiccation of rice (*Oryza sativa* L.). *Annals of Botany* 72:73-80.

Ekanayake, I. J., P. L. Steponkus and S. K. De Datta. (1990). Sensitivity of pollination to water deficits at anthesis in upland rice. *Crop Science* 30:310-315.

Fairey, D. T. and N. C. Stoskopf. (1975). Effect of granular ethephon on male sterility in wheat. *Crop Science* 15:29-32.

Fischer, R. A. (1973). The effect of water stress at various stages of development on yield processes in wheat. In *Plant Response to Climatic Factors,* ed. R. O. Slatyer. Paris: UNESCO, pp. 233-241.

Fischer, R. A. and N. C. Turner. (1978). Plant Productivity in the arid and semiarid zone. *Annual Review of Plant Physiology* 29:277-317.

Fosket, D. E. and L. Morejohn. (1992). Structural and functional organization of tubulin. *Annual Review of Plant Physiology and Plant Molecular Biology* 43:201-240.

Garrity, D. P. and J. C. O'Toole. (1994). Screening rice for drought resistance at the reproductive phase. *Field Crops Research* 39:99-110.

Garrity, D. P., J. C. O'Toole and E. T. Vidal. (1986). Manipulating panicle transpiration resistance to increase rice spikelet fertility during flowering stage water-stress. *Crop Science* 26:789-795.

Goldbach, H. and E. Goldbach. (1977). Abscisic acid translocation and influence of water stress on grain abscisic acid content. *Journal of Experimental Botany* 28:1342-1350.

Graham, R. D. (1975). Male sterility in wheat plants deficient in copper. *Nature* 254:514-515.

Hall, A. J., F. Vilella, N. Trapani and C. Chimenti. (1982). The effects of water stress and genotype on the dynamics of pollen-shedding and silking in maize. *Field Crops Research* 5:349-363.

Hanft, J. M. and R. J. Jones. (1986). Kernel abortion in maize (*Zea mays* L.). I. Carbohydrate concentration patterns and acid invertase activity of maize kernels induced to abort *in vitro*. *Plant Physiology* 81:503-510.

Hanson, A. D. and W. D. Hitz. (1982). Metabolic responses of mesophytes to plant water deficits. 33:163-203.

Hanson, A. D., W. J. Peacock, L. T. Evans, C. J. Arntzen and G. S. Khush. (1990). Drought resistance in rice. *Nature* 345:26-27.

Hayase, H., T. Satake, I. Nashiyama and N. Ito. (1969). Male sterility caused by cooling treatment at the meiotic stage in rice plants. II. The most sensitive stage to cooling and fertilizing ability of pistils. *Proceedings of the Crop Science Society of Japan* 38:706-711.

Herrero, M. P. and R. R. Johnson. (1981). Drought stress and its effect on maize reproductive systems. *Crop Science* 21:105-110.

Heslop-Harrison, J. (1961). Experimental control of sexuality and inflorescence structure in *Zea mays* L. *Proceedings of the Linnean Society of London* 172:108-123.

Howlett, F. S. (1936). The effect of carbohydrate and nitrogen deficiency upon microsporogenesis and development of the male gametophyte in the tomato, *Lycopersicum esculentum* Mill. *Annals of Botany* 50:767-803.

Hsiao, T. C. (1982). The soil-plant-atmosphere continuum in relation to crop production. In *Drought Resistance in Crops with Emphasis on Rice.* Los Baños, Philippines: International Rice Research Institute, pp. 39-52.

Ito, N. (1978). Male sterility caused by cooling treatment at the young microspore stage in rice plants. XVI. Changes in carbohydrates, nitrogenous and phosphorous compounds in rice anthers after cooling treatment. *Japanese Journal of Crop Science* 17:318-323.

Jenner, C. F. (1982). Storage of starch. In *Plant Carbohydrates I. Intracellular Carbohydrates, Encyclopedia of Plant Physiology, New Series, 13A,* eds. F. A. Loewus and W. Tanner. Berlin: Springer-Verlag, pp. 707-747.

Kisselbach, T. A. (1950). Progressive development and seasonal variations of the corn crop. *Nebraska Agriculture Experimental Station Bulletin* 166:29.

Lalonde, S. (1996). Développement des anthères de blé *(Triticum aestivum L.)* assujetties à une contrainte hydrique avec une emphase sur les aspects biochimiques et moléculaires de la relation de l'ADP-glucose pyrophosphorylase et de l'accumulation d'amidon. Ph.D. Thesis, Montreal: Sciences biologiques, Université de Montréal.

Lalonde, S., D. Beebe and H. S. Saini. (1997). Early signs of disruption of wheat anther development, associated with the induction of male sterility by meiotic-stage water deficit. *Sexual Plant Reproduction* (in press).

Lee, B. T., P. Martin and F. Bangerth. (1988). Phytohormone levels in the florets of a single wheat spikelet during pre-anthesis development and relationships to grain set. *Journal of Experimental Botany* 39:927-933.

Lewis, R. B., E. A. Hiler and W. R. Jordan. (1974). Susceptibility of grain sorghum to water deficit at three growth stages. *Agronomy Journal* 66:589-591.

Lockhart, J. A. (1965). An analysis of irreversible cell extension. *Journal of Theoretical Biology* 8:264-275.

Mahalakshmi, V. and F. R. Bidinger. (1985). Flowering response of pearl-millet to water-stress during panicle development. *Annals of Applied Biology* 106:571-578.

Mahalakshmi, V., F. R. Bidinger and D. S. Raju. (1987). Effect of timing of water deficit on pearl millet (*Pennisetum americanum*). *Field Crops Research* 15:327-339.

Matsushima, S. (1962). Some experiments on soil-water-plant relationship in the cultivation of rice. *Proceedings of Crop Science Society of Japan* 31:115-121.

McRae, D. H. (1985). Advances in chemical hybridization. *Plant Breeding Reviews* 3:169-191.

Miki-Hirosige, H. and S. Nakamura. (1983). Growth and differentiation of amyloplasts during male gamete development in *Lilium longiflorum*. In *Pollen: Biology and Implications for Plant Breeding*, eds. D. L. Mulcahy and E. Ottaviano. New York: Elsevier Biomedical, pp. 141-147.

Miller, M. E. and P. S. Chourey. (1992). The maize invertase-deficient *miniature-1* seed mutation is associated with aberrant pedicel and endosperm development. *Plant Cell* 4:297-305.

Morgan, J. M. (1971). The death of spikelets in wheat due to water deficit. *Australian Journal of Experimental Agriculture and Animal Husbandry* 11:349-351.

Morgan, J. M. (1980). Possible role of abscisic acid in reducing seed set in water-stressed wheat plants. *Nature* 289:655-657.

Morgan, J. M. and R. W. King. (1984). Association between loss of leaf turgor, abscisic acid levels and seed set in two wheat cultivars. *Australian Journal of Plant Physiology* 11:143-150.

Morgan, P. W., C. J. He, J. A. De Greef and M. P. De Proft. (1990). Does water deficit stress promote ethylene synthesis by intact plants? *Plant Physiology* 94:1616-1624.

Mosjidis, C. O., C. M. Peterson, B. Truelove and R. R. Dute. (1993). Stimulation of pod and ovule growth of soybean, *Glycine max* (L.) Merr. by 6-benzylaminopurine. *Annals of Botany* 71:193-199.

Moss, G. I. and L. A. Downey. (1971). Influence of drought stress on female gametophyte development in corn (*Zea mays* L.) and subsequent grain yield. *Crop Science* 11:368-372.

Nakamura, N., M. Sado and Y. Arai. (1980). Sucrose metabolism during the growth of *Camellia japonica* pollen. *Phytochemistry* 19:205-209.

Nakamura, N., Y. Suzuki and H. Suzuki. (1992). Pyrophosphate-dependent phosphofructokinase from pollen: properties and possible role in sugar metabolism. *Physiologia Plantarum* 86:616-622.

Namuco, O. S. and J. C. O'Toole. (1986). Reproductive Stage Water-Stress and Sterility. 1. Effect of Stress During Meiosis. *Crop Science* 26:317-321.

Narayana, I., S. Lalonde and H. S. Saini. (1991). Water-stress-induced ethylene production in wheat: A fact or artifact? *Plant Physiology* 96:406-410.

Novikov, V. P. (1952). The effect of deficiency of water in the soil at different stages of development in oats. *Doklady Akademii Science SSSR* 82:641-643.

O'Toole, J. C., T. C. Hsiao and O. S. Namuco. (1984). Panicle water relations during water stress. *Plant Science Letters* 33:137-143.

O'Toole, J. C. and T. B. Moya. (1981). Water deficit and yield in upland rice. *Field Crops Research* 4:247-259.

O'Toole, J. C. and O. S. Namuco. (1983). Role of panicle exsertion in water-stress induced sterility. *Crop Science* 23:1093-1097.

Pacini, E. and G. G. Franchi. (1988). Amylogenesis and amylosis during pollen grain development. In *Sexual Reproduction in Higher Plants*, eds. M. Cresti, P. Gori and E. Pacini. Berlin: Springer-Verlag, pp. 181-186.

Parmar, K. S., E. A. Siddiq and M. S. Swaminathan. (1979). Variation in components of flowering behaviour of rice. *Indian Journal of Genetics and Plant Breeding* 39:542-550.

Pierce, M. and K. Raschke. (1980). Correlation between loss of turgor and accumulation of abscisic acid in detached leaves. *Planta* 148:174-182.

Robins, J. S. and C. E. Domingo. (1953). Some effects of severe moisture deficits at specific growth stages in corn. *Agronomy Journal* 45:618-621.

Rowell, P. L. and D. G. Miller. (1971). Induction of male sterility in wheat with 2-chloroethylphosphonic acid (ethrel). *Crop Science* 11:629-631.

Saini, H. S. (1982). Physiological Studies on Sterility Induced in Wheat by Heat and Water Deficit. PhD Thesis, Adelaide: Department of Plant Physiology, University of Adelaide.

Saini, H. S. and D. Aspinall. (1981). Effect of water deficit on sporogenesis in wheat (*Triticum aestivum* L.). *Annals of Botany* 48:623-633.

Saini, H. S. and D. Aspinall. (1982a). Abnormal sporogenesis in wheat (*Triticum aestivum* L.) induced by short periods of high temperature. *Annals of Botany* 49:835-846.

Saini, H. S. and D. Aspinall. (1982b). Sterility in wheat (*Triticum aestivum* L.) induced by water deficit or high temperature: Possible mediation by abscisic acid. *Australian Journal of Plant Physiology* 9:529-537.

Saini, H. S., M. Sedgley and D. Aspinall. (1983). Effect of heat stress during floral development on pollen tube growth and ovary anatomy in wheat (*Triticum aestivum* L.). *Australian Journal of Plant Physiology* 10:137-144.

Saini, H. S., M. Sedgley and D. Aspinall. (1984). Developmental anatomy in wheat of male sterility induced by heat stress, water deficit or abscisic acid. *Australian Journal of Plant Physiology* 11:243-253.

Salter, P. J. and J. E. Goode. (1967). *Crop Responses to Water at Different Stages of Growth. Research Review No. 2.* Farnham Royal, England: Commonwealth Agricultural Bureaux.

Sandhu, B. S. and M. L. Horton. (1977). Response of oats to water deficit. I. Physiological Characteristics. *Agronomy Journal* 69:357-360.

Satake, T. and S. Yoshida. (1978). High temperature induced sterility in indica rices at flowering. *Japanese Journal of Crop Science* 47:6-17.

Sato, K. (1954). Relationship between rice crop and water. *Japanese Journal of Breeding* 4:264-289.

Sawhney, V. K. and A. Shukla. (1994). Male sterility in flowering plants: Are plant growth substances involved? *American Journal of Botany* 81:1640-1647.

Schoper, J. B., R. J. Lambert, B. L. Vasilas, and M. E. Westgate. (1987). Plant factors controlling seed set in maize. The influence of silk, pollen, and ear-leaf water status and tessel heat treatment at pollination. *Plant Physiology* 83:121-125.

Schoper, J. B., R. J. Lambert, and B. L. Vasilas. (1986). Maize pollen viability and ear receptivity under water and high temperature stress. *Crop Science* 26:1029.

Schussler, J. R. and M. E. Westgate. (1991a). Maize kernel set at low water potential: I. Sensitivity to reduced assimilates during early kernel growth. *Crop Science* 31:1189-1195.

Schussler, J. R. and M. E. Westgate. (1991b). Maize kernel set at low water potential: II. Sensitivity to reduced assimilates at pollination. *Crop Science* 31:1196-1203.

Schussler, J. R. and M. E. Westgate. (1994). Increasing assimilate reserves does not prevent kernel abortion at low water potential in maize. *Crop Science* 34:1569-1576.

Sharma, P. N., C. Chatterjee, C. P. Sharma and S. C. Agarwala. (1987). Zinc deficiency and anther development in maize. *Plant and Cell Physiology* 28:11-18.

Sheoran, I. S. and H. S. Saini. (1996a). Drought-induced male sterility in rice: changes in carbohydrate levels and enzyme activities associated with the inhibition of starch accumulation in pollen. *Sexual Plant Reproduction* 9:161-169.

Sheoran, I. S. and H. S. Saini. (1996b). Molecular cloning and characterization of rice panicle tubulin, and its potential role in drought-induced reproductive failure. *14th International Congress of Sexual Plant Reproduction,* Lorne, Australia, p. 136.

Skazkin, F. D. (1961). The critical period in plants as regards insufficient water supply. *Timiryazevskie Chteniya Akademii Nauk. SSSR* 21:1-51.

Skazkin, F. D. and K. A. Lukomskaya. (1962). Effect of soil water deficiency on formation of female gametophyte. *Doklady Akademii Nauk. SSSR* 146: 1449-1451.

Skazkin, F. D. and V. G. Rozkova. (1956). The effect of boron on cereals growth with insufficient soil moisture at the critical period of development. *Doklady Akademii Nauk. SSSR* 108:962-964.

Skazkin, F. D. and I. G. Zavadskaya. (1957). On the influence of soil moisture deficiency and nitrogen nutrition on microsporogenesis in barley. *Doklady Akademii Nauk. SSSR* 117:240-242.

Sladky, Z. (1969). Role of growth regulators in differentiation process of maize (*Zea mays* L.) organs. *Biologia Plantarum* 11:208-215.

Tardieu, F. and W. J. Davies. (1992). Stomatal response to abscisic acid is a function of current water status. *Plant Physiology* 98:540-545.

Tsuda, M. and S. Takami. (1993). Changes of water potential in rice panicle under increasing drought stress at various stages. *Japanese Journal of Crop Science* 62:41-46.

Turner, L. B. (1993). The Effect of water-stress on floral characters, pollination and seed set in white clover (*Trifolium repens* L.). *Journal of Experimental Botany* 44:1155-1160.

Udol'skaja, N. L. (1936). Drought resistance of spring wheat varieties [in Russian]. *Omgiz, Omsk*: p. 124 (Cited by Salter and Goode, 1967).

van der Paauw, F. (1949). Water relations of oats with special attention to the influence of periods of drought. *Plant and Soil* 1:303-341.

Verma, M. M. and J. Kumar. (1978). Ethrel–a male gametocide that can replace the male sterility genes in barley. *Euphytica* 27:865-868.

Walton, D. C. (1980). Biochemistry and physiology of abscisic acid. *Annual Review of Plant Physiology* 31:453-489.

Wardlaw, I. F. (1971). Early stages of grain development in wheat: response to water stress in a single variety. *Australian Journal of Biological Sciences* 24:1047-55.

Waters, S. P., P. Martin, and B. T. Lee. (1984). Influence of sucrose and abscisic acid on the determination of grain number in wheat. *Journal of Experimental Botany* 35:829-840.

Wells, S. A. and S. Dubetz. (1966). Reaction of barley varieties to soil water stress. *Canadian Journal of Plant Science* 46:507-512.

Westgate, M. E. and J. S. Boyer. (1985a). Carbohydrate reserves and reproductive development at low leaf water potentials in maize. *Crop Science* 25:762-769.

Westgate, M. E. and J. S. Boyer. (1985b). Osmotic adjustment and the inhibition of leaf, root, stem and silk growth at low water potentials in maize. *Planta* 164:540-549.

Westgate, M. E. and J. S. Boyer. (1986a). Reproduction at low silk and pollen water potentials in maize. *Crop Science* 26:951-956.

Westgate, M. E. and J. S. Boyer. (1986b). Silk and pollen water potentials in maize. *Crop Science* 26:947-951.

Westgate, M. E., J. B. Passioura, and R. Munns. (1996). Water status and ABA content of floral organs in drought stressed wheat. *Australian Journal of Plant Physiology* (in press).

Westgate, M. E. and C. M. Peterson. (1993). Flower and pod development in water-deficient soybeans (*Glycine max* L. Merr.). *Journal of Experimental Botany* 44:109-117.

Westgate, M. E. and G. L. Thomson Grant. (1989). Water deficits and reproduction in maize. Responses of the reproductive tissue to water deficits at anthesis and mid-grain fill. *Plant Physiology* 91:862-867.

Wolf, O., W. D. Jescke, and W. Hartung. (1990). Long distance transport of abscisic acid in salt stressed *Lupinus albus* plants. *Journal of Experimental Botany* 41:593-560.

Xu, J., W. T. Avigne, D. R. McCarty, and K. E. Koch. (1996). A similar dichotomy of sugar modulation and developmental expression affects both paths of sucrose metabolism–evidence from maize invertase gene family. *Plant Cell* 8:1209-1220.

Yang, S. F. and N. E. Hoffman. (1984). Ethylene biosynthesis and its regulation in higher plants. *Annual Review of Plant Physiology* 35:155-89.

Zavadskaya, I. G. and F. D. Skazkin. (1960). On microsporogenesis in barley as affected by soil moisture deficiency and by application of nitrogen at various stages of development. *Doklady Akademii Nauk. SSSR* 131:692-694.

Zee, S. Y. and O'Brien. (1970). A special type of tracheary element associated with 'xylem discontinuity' in the floral axis of wheat. *Australian Journal of Biological Sciences* 23:783-791.

Zeng, Z. R. and R. W. King. (1986). Regulation of grain number in wheat–

changes in endogenous levels of abscisic acid. *Australian Journal of Plant Physiology* 13:347-352.

Zeng, Z. R., R. W. King, and J. M. Morgan. (1985). Regulation of grain number in wheat–genotype difference and responses to applied abscisic acid and to high-temperature. *Australian Journal of Plant Physiology* 12:609-619.

Zinselmeier, C., M. E. Westgate, and R. J. Jones. (1995). Kernel set at low water potential does not vary with source/sink ratio in maize. *Crop Science* 35:158-163.

Zinselmeier, C., M. E. Westgate, J. R. Schussler, and R. J. Jones. (1995). Low water potential disrupts carbohydrate metabolism in maize (*Zea mays* L.) ovaries. *Plant Physiology* 107:385-391.

SUBMITTED: 09/03/96
ACCEPTED: 12/22/96

Low Temperature Emergence in Crop Plants: Biochemical and Molecular Aspects of Germination and Early Seedling Growth

Cory L. Nykiforuk
Anne M. Johnson-Flanagan

SUMMARY. In Canada, seeding of agronomically important crops takes place during the early months of spring when temperatures are well below the optimum. Low temperature reduces the rate and success of germination. This in turn can result in slow asynchronous emergence from the soil and poor stand establishment. As emergence is a function of both germination and early seedling growth, the effect of low temperature on these developmental processes is of great interest. This review examines how low temperature affects germination and early seedling growth in relation to biochemical and molecular processes. As *Brassica napus* L. cv. Westar, canola, does not exhibit primary or secondary dormancy, it serves as an ideal species in which to study low temperature emergence. First, emergence is reviewed by independent sections covering germination and early seedling growth. Germination is dissected into the three phases of

---

Cory L. Nykiforuk, Department of Chemistry, University of Lethbridge, 4401 University Drive, Lethbridge, Alberta, Canada, T1K 3M4. Anne M. Johnson-Flanagan, Associate Professor, Department of Agriculture, Foods, and Nutritional Science, University of Alberta, Edmonton, Alberta, Canada, T6G 2R3.

Address correspondence to: Cory L. Nykiforuk, University of Lethbridge, 4401 University Drive, Lethbridge, Alberta, Canada, T1K 3M4 (E-mail: NYKICL @hg.uleth.ca).

[Haworth co-indexing entry note]: "Low Temperature Emergence in Crop Plants: Biochemical and Molecular Aspects of Germination and Early Seedling Growth." Nykiforuk, Cory L., and Anne M. Johnson-Flanagan. Co-published simultaneously in *Journal of Crop Production* (The Food Products Press, an imprint of The Haworth Press, Inc.) Vol. 1, No. 1 (#1), 1998, pp. 249-289; and: *Crop Sciences: Recent Advances* (ed: Amarjit S. Basra) The Food Products Press, an imprint of The Haworth Press, Inc., 1998, pp. 249-289. Single or multiple copies of this article are available for a fee from The Haworth Document Delivery Service [1-800-342-9678, 9:00 a.m. - 5:00 p.m. (EST). E-mail address: getinfo@haworth.com].

249

germination. Early seedling growth is broken into sections covering storage reserve mobilization, the expression of gene sets related to developmental stages, and the role of ABA. Next, comparisons between chilling injury in seeds and chilling injury in plants, oxidative stress, and seed priming are discussed. Finally, the current state of experimental evidence and results are used to resynthesize low temperature emergence in *Brassica napus* with final thoughts on future directions. *[Article copies available for a fee from The Haworth Document Delivery Service: 1-800-342-9678. E-mail address: getinfo@haworth. com]*

KEYWORDS. Low temperature emergence, germination, early seedling growth, chilling injury, *Brassica napus*

ABBREVIATIONS. ABA, abscisic acid; ICL, isocitrate lyase; MS, malate synthase; LEA, late embryogenesis abundant; ELISA, enzyme linked immunosorbent assay; SOD, superoxide dismutase; CAT, catalase; APX, ascorbate peroxidase; PEG, polyethylene glycol; PVA, polyvinyl alcohol

INTRODUCTION

In northern latitudes the length of the growing season, allowing for full sporophytic development of agronomically important crops, is reduced. Understanding the mechanisms involved in seedling emergence during exposure to low temperature is essential if we wish to improve low temperature emergence, thereby increasing the geographical distribution of crop plants and extending the growing season. Related reviews have covered chilling injury of plants (Lyons, 1973; Lyons, Raison, and Steponkus, 1979; Levitt, 1980; Graham and Patterson, 1982; Wang, 1982; Parkin et al., 1989) and chilling injury in seed (Simon, 1979; Wolk and Herner, 1982; Roberts, 1988; Bedi and Basra, 1993). However, there is no consensus as to how low temperature damages plant tissue and what inherent or phenotypic adaptations enable a plant to maintain growth at low nonfreezing temperatures. Low temperature emergence requires further attention because the emergence or vigor of the seedling can influence stand establishment (Stewart et al., 1990), growth, and the final yield (Pollock and Toole, 1966; Phillips and Youngman, 1971; Perry, 1972; Hobbs and Obendorf, 1972; Simon, 1979; Stewart et al., 1990).

The impact of low temperature on emergence is correlated with the plants' geographical distribution (Thompson, 1972; Lyons, 1973; Hallgren and Oquist, 1990). Generally, the seeds of the seeds of tropical and subtropical plants will germinate within the temperature range of 10-45°C (Monteith, 1981), and display metabolic dysfunctions in the temperature range of 10-12°C (Hallgren and Oquist, 1990). These, therefore, are chilling-sensitive. Temperate species can tolerate low, nonfreezing temperatures, germinating within the temperature range of 0-35°C, and are, therefore, considered to be chilling-resistant (Monteith, 1981; Ong and Monteith, 1985; Hallgren and Oquist, 1990). The lower threshold for germination is generally 2-3°C (Bedi and Basra, 1993). Susceptible plants include corn (*Zea mays*) (Cal and Obendorf, 1972; Cohn et al., 1979), cotton (*Gossypium herbaceum*) (Christiansen, 1969), lima bean (*Phaseolus lunatus*) (Pollock, 1969), field bean (*Phaseolus vulgaris*) (Pollock and Manalo, 1970), rice (*Oryza sativa*), cucumber (*Cucumis sativus*), tobacco (*Nicotiana tabacum*) (Markowski, 1989 a,b), and soybean (*Glycine max*) (Hobbs and Obendorf, 1972). Resistant crops include wheat (*Triticum aestivum*), barley (*Hordeum vulgare*), oats (*Avena sativum*) (Bodsworth and Bewley, 1981), mustard (*Sinapis alba*) (Simon et al., 1976), and canola (*Brassica napus* and *B. campestris*) (Acharya et al., 1983; Kondra et al., 1983; King et al., 1986; Barber et al., 1991; Wilson et al., 1992). A list of chilling-sensitive and -resistant species with their lower temperature limits for germination are given in Table 1. Chilling-sensitive and -resistant seed can be separated on the basis of the time required to attain 50% germination over a wide temperature range. Simon et al. (1976) demonstrated that the time required for 50% germination increases steadily as temperature is lowered in mustard (*Sinapis alba*), a resistant species, but discontinuities occurred at a specific temperature (around 12-13°C) in cucumber (*Cucumis sativus*) and mung bean (*Phaseolus aureus*), two sensitive species (Fig. 2 in Simon et al., 1976).

This review will focus on biochemical and molecular aspects during low temperature emergence in temperate species representing important agronomic crops, with the ultimate hope of giving the reader a broader appreciation for this area of study. First, an overview of emergence is required to place the processes involved in germination and early seedling growth in their proper context and concurrently explore the effects of low temperature on these developmental processes. This will be followed by sections interrelating current experimentation in the area of gene expression, the role of ABA, chilling injury, and oxidative stress with chilling injury in seed.

TABLE 1. Chilling sensitive and resistant species with their corresponding lower limit for germination

CHILLING SENSITIVE SPECIES

Species   Lower limit for germination Reference

Species	Lower limit for germination	Reference
Cucumis sativus	11-12°C	Simon et al., 1976
Cucumis melo	16-19°C	Mayer and Poljakoff-Mayber, 1989
Phaseolus aureus	11-12°C	Simon et al., 1976
Silene otites	10°C	Thompson, 1972
Ajuga reptans	16°C	Thompson, 1972
Lycopersicon esculentum	10°C	Liptay and Schöpfer, 1983; Scott and Jones, 1985; Dahal et al.,1990
Zea mays	8-10°C	Simon, 1979; Mayer and Poljakoff-Mayber, 1989
Oryza sativa	10-12°C	Simon, 1979; Mayer and Poljakoff-Mayber, 1989
Gossypium herbaceum	10-12°C	Simon, 1979; Roberts, 1988
Glycine max	10-12°C	Bramlage et al., 1979; Roberts, 1988
Nicotiana tabacum L.	14-16°C	Simon, 1979; Mayer and Poljakoff-Mayber, 1989; Mohapatra and Suggs, 1989
Solanum carolinense	20°C	Mayer and Poljakoff-Mayber, 1989
Lychnis flos-cuculi	10°C	Thompson, 1972

CHILLING RESISTANT SPECIES

Species   Lower limit for germination Reference

Species	Lower limit for germination	Reference
Sinapis alba	3-4°C	Simon et al., 1976
Brassica napus	2-3°C	Nykiforuk and Johnson-Flanagan, 1994
Brassica oleracea	2-3°C	Thompson, 1972
Gypsophila perfoliata	2°C	Thompson, 1972
Petrorhagia prolifera	3°C	Thompson, 1972
Agrostemma githago	3°C	Thompson, 1972
Lactuca sativa	2°C	Thompson, 1972
Deucus carota	2-3°C	Thompson, 1972; Hegarty, 1973
Allium ampeloprasum	2-3°C	Thompson, 1972

Hordeum vulgare	2-3°C	Simon, 1979
Hordeum sativum	3-5°C	Mayer and Poljakoff-Mayber, 1989
Secale cereale	3-5°C	Mayer and Poljakoff-Mayber, 1989
Triticum aestivum	2-5°C	Simon, 1979; Mayer and Poljakoff-Mayber, 1989; Addae and Pearson, 1992
Avena sativum	3-5°C	Simon, 1979; Mayer and Poljakoff-Mayber, 1989
Avena fatua	4°C	Sawhney, Quick, and Hsiao, 1985
Amarathus caudatus L. var. viridis	5°C	Gutterman, Corbineau, and Côme, 1992
Convolvulus arvensis	0.5-3°C	Mayer and Poljakoff-Mayber, 1989
Lepidium draba	0.5-3°C	Mayer and Poljakoff-Mayber, 1989
Helianthus annuus	1-5°C	Mwale et al., 1994

EMERGENCE

Emergence has been traditionally, but incorrectly, used to define germination (Perry, 1972). Emergence is comprised of two developmental processes; germination and early seedling growth (Nykiforuk and Johnson-Flanagan, 1994; Nykiforuk, 1996). Under proper conditions, successful germination establishes a vigorously growing seedling from a quiescent, mature, dry seed. Germination in the strictest sense occurs after hydration of the seed and ends with the developmental processes leading to the elongation of the embryonic axis, usually the radicle (Bewley and Black, 1994). Subsequent seedling growth relies upon the mobilization of the stored reserves to fuel cellular division followed by expansion basipetally in localized regions of the meristems (Finkelstein and Crouch, 1984). Successful emergence, resulting in the establishment of a young seedling, will only occur if both developmental processes occur in a timely manner. Although events associated with early seedling growth are independent from events resulting in germination, any impairment in germination will subsequently affect seedling growth.

Some seeds require specific environmental or chemical cues for germination (Mayer and Poljakoff-Mayber, 1989; Bewley and Black, 1994). Such seed are referred to as dormant. Dormancy can only be broken when the internal (the state of the seed itself) and external (environmental factors) controls are overcome (Ross, 1984; Tran and Cavanagh, 1984). When discussing the effects of low temperature on germination, it is important to exclude species that exhibit forms of dormancy. This is because germination in dormant seed will occur over a different temperature range than seed in which dormancy has been broken (Bewley and Black, 1994). In nondormant seed, species will germinate under the appropriate environmental conditions, and are referred to as quiescent. For example, in some annuals, such as *Brassica napus,* dormancy has been bred out to acquire a more uniform germination (Adler et al., 1993).

By focusing our attention on species that are nondormant we are able to exclude the complications arising from factors influencing dormancy. Canola (*B. napus* and *B. campestris*) are major agronomic crops that do not exhibit primary or secondary dormancy, and therefore serve as ideal species to study the effects of low temperature on germination and early seedling growth. A general overview of germination and early seedling growth using *B. napus* as a model species will aid the reader in understanding the complexities encountered and gain an appreciation for how little is really known about low temperature emergence.

Germination

Desiccation marks the end of embryo maturation and completes the embryogenesis program (Kermode, 1990). Germination is the developmental period that begins with water uptake and ends with radicle emergence through the seed coat (Bewley and Black, 1994). Therefore, germination is a physiological phenomenon preceding root growth (Côme and Tissaoui, 1972). The process of germination is subdivided into three phases (Tissaoui and Côme, 1975; Simon, 1984; Vertucci, 1989; Bewley and Black, 1994). During phase I, rapid hydration (imbibition) of the seed occurs and metabolism begins. During phase II, a lag follows while major metabolic events take place in preparation for radicle emergence. The final event, phase III, is characterized by radicle elongation, and marks the end of germination and the beginning of seedling growth. Often the early stages of germination are reversible to the end of phase II. Seed that are partially hydrated then lyophilized are still viable when hydrated again. This process can be repeated until the seed reaches phase III. At this stage the seed is committed to continuing the process of germination (Mayer and Marbach, 1981). Thereafter, drying down the seed will result in the death of the seed.

Factors influencing the success of germination include mucilage content (Eskin, 1992), testa thickness (Teran et al., 1994), genotype (Wilson et al., 1992), seedlot (Barber et al., 1991; Nykiforuk and Johnson-Flanagan, 1994), storage conditions (Frisbee et al., 1988; Vertucci, 1992), seed age (Roberts and Ellis, 1984), seed pre-treatments (Nelson et al., 1984; Basra et al., 1989; Coolbear and McGill, 1990; Todari and Negi, 1992), seed injury (Buntin et al., 1995), and conditions under which the seed matures (Gray and Thomas, 1982; Taylor et al., 1991). As these factors influence germination they should be considered when testing low temperature germination. The ability of a seed to germinate at low temperatures is dependent upon the actual temperature, the duration of the chilling exposure, and phase of germination when the chilling exposure occurs (Wolk and Herner, 1982; Bedi and Basra, 1993). Chilling injury in seed refers to injury incurred during germination (Harrington and Kihara, 1960), and is manifested as a loss of viability or reduced growth at low nonfreezing temperatures.

Generally, temperatures below the optimum result in progressively poorer germination (Liengsiri and Hellum, 1988; Leviatov et al., 1994; Nykiforuk and Johnson-Flanagan, 1994; O'Conner and Gusta, 1994). Low temperature slows the rate of imbibition (Vertucci, 1989), and may damage embryos, preventing germination (Pollock and Toole, 1966; Pollock, 1969; Bramlage et al., 1979; Woodstock, 1988).

In *B. napus,* low temperature germination studies have been limited to assessing germination rates under low temperature regimes (Acharya et al., 1983; Kondra et al., 1983; Livingstone and deJong, 1990; Barber et al., 1991; Mills, 1993). As with other species (Mohapatra and Suggs, 1989; Addae and Pearson, 1992), *Brassica* genotypes (cultivars) display differences in germination rates at low temperatures (Acharya et al., 1983; Wilson et al., 1992; Mills, 1993). Acharya et al. (1983) determined that 10°C was the best temperature to select for superior low temperature germination in the cultivar Westar; however, selection for rapid germination at this temperature produced inconsistent results. Similarly, King et al. (1986) found that such selections did not necessarily lead to improved performance. These results suggest complex control of germination and early seedling growth.

Our results on the study of low temperature emergence in *B. napus* indicate a temperature dependent response (Figure 1) (Nykiforuk, 1996). At 10°C, the overall success of germination is not affected, but the process of germination is delayed. At temperatures of 6°C and lower, both delays and reductions in germination were noted. The delay in germination at 10°C was determined to result from a thermal effect, whereas germination at 6°C reflected both thermal effects and developmental impairments.

Phase I of Germination

Oil seeds, such as *B. napus,* have a very low hydration level, generally less than 10%, and low metabolic activity (Kuras, 1984). In *B. napus,* the first phase of imbibition proceeds rapidly (Shaykewich and Williams,

1971; Shaykewich, 1973). The seed coat affords some protection during this time as it slows the rate of imbibition (Simon, 1984). As the uptake of water increases, the force of water from the external environment into the seed decreases logarithmically (Shaykewich and Williams, 1971; Shaykewich, 1973; Bewley and Black, 1994) until the uptake of water slows and enters a lag period. *B. napus* seed requires 60-75% relative moisture content between the temperatures of 2-25°C, in the presence of oxygen, to germinate (Canola Growers Manual, 1991).

Imbibition at low temperature results in leakage of cellular components (Pollock and Toole, 1966; Bramlage et al., 1979; Leopold and Musgrave, 1979; Leopold, 1980; Schmidt and Tracy, 1989), and the damage and leakiness displayed by susceptible species is believed to result from the physical stress caused during rapid water uptake (Spaeth, 1989). The uptake of water over the first 24 hours of imbibition in *B. napus* at 22, 10, and 6°C is shown in Figure 2. However, the slight lag does not directly translate into temporal delays observed in the onset of germination and developmental impairment in the later stages of germination (compare Figures 1 and 2). At both 10 and 6°C, phase I of seed germination is extended by approximately 6 hours, as opposed to a 2 day delay in the onset of germination at 10°C and a failure of the majority of the seed to complete germination at 6°C. Therefore, damage during imbibition is not believed to have played a major role in the delay or reduction observed in germination at low temperature in *B. napus*.

Damaged seed coats (Tully et al., 1981; Taylor and Dickson, 1987) or removal of the seed coat (Ashworth and Obendorf, 1980) can amplify or result in chilling injury in sensitive species during imbibition. However, in tomato (*Lycopersicon esculentum*), a chilling-sensitive species, cutting the seed coat increases germination percentage and rate at low temperature (Singer et al., 1989). Leviatov et al. (1994) determined that different seed components played a role in the ability of tomatoes to germinate at low temperatures between a sensitive and a resistant line. Removal of the testa reduced the time required for germination by 3 days, but did not affect the overall success of germination between the lines. In addition, removal of the seed coat and endosperm in front of the radicle tip eliminated the differences in germination rates (Leviatov et al., 1994). Further, endomannanase activity was higher in the cold tolerant line prior to germination at 12°C, especially in the micropylar endosperm cap (Leviatov et al., 1995). This may affect the initial radicle driving force during germination (Perry, 1972).

Respiration begins rapidly after imbibition commences and increases as the water potential rises in the seed (Simon, 1984). The initial rise is

FIGURE 1. Germination percentage rates at the optimal and suboptimal temperature conditions in *Brassica napus* cv. Westar over the course of imbibition (DAI). The temperatures tested were (♦) 22°C, (■) 10°C, (●) 6°C, and (▲) 2°C. Each value represents the mean of three independent replicates ± 1 SD. These values were obtained from Nykiforuk (1996).

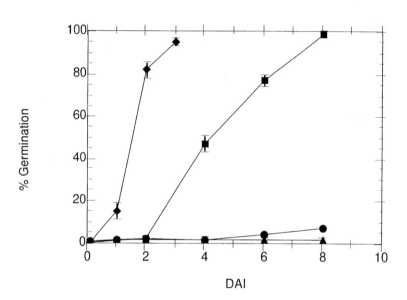

associated with hydration of the seed and activation of the mitochondrial enzymes (Mayer and Poljakoff-Mayber, 1989). In *B. napus,* respiration as measured by carbon dioxide evolution was lower at suboptimal temperatures in comparison to seeds imbibed at optimal temperatures (Nykiforuk, 1996).

Directly related to the increase in respiration, is the increase in energy charge as defined by Atkinson (1977). ATP production increases rapidly through both substrate level phosphorylation followed later by oxidative phosphorylation, providing energy to the metabolic machinery necessary to support germination (Mayer and Marbach, 1981; Simon, 1984). In a closely related crucifer, *Sinapis alba,* tritium from tritiated water was incorporated rapidly into oxo-acid intermediates of the Krebs cycle (Spedding and Wilson, 1968). In *B. napus,* starch grains appear within 3 hours of imbibition in the columella of the root cap (Kuras, 1986). This indicates a very early initial enzymatic transformation of storage lipids into carbo-

FIGURE 2. Changes in the fresh weight during the first 24 hours after imbibition (HAI) under different temperature conditions in *Brassica napus* cv. Westar. Otherwise as in Figure 1.

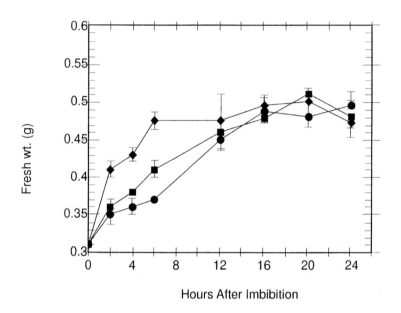

hydrates. Furthermore, starch is observed to accumulate in the embryonic cells just prior to germination (Kuras, 1986), providing the energy for intensified respiration during germination. These starch grains disappear soon after germination (Kuras, 1986). The effects of low temperature on these biochemical processes have not been assessed to date, but may contribute to the delays observed in the onset of germination.

Protein synthesis also commences at the start of imbibition (Bewley and Black, 1994). Shortly after the loss of residual mRNA during imbibition (Kermode, 1990), new messages are produced, some of which encode novel proteins involved in the mobilization of storage reserves after germination (Simon, 1984; Sanchez-Martinez et al., 1986; Lane, 1991). In *Brassica* seeds, RNA synthesis is detected in the outermost cells of the embryo two hours after imbibition, followed by synthesis in other cells (Payne et al., 1978).

Phase II of Germination

During the second phase of germination, respiration rates plateau (Simon, 1984; Bewley and Black, 1994), while RNA and protein synthesis proceed (Simon, 1984; Sanchez-Martinez et al., 1986; Kermode, 1990; Lane, 1991), and DNA synthesis (endoreduplication) lags behind, just prior to the start of cell division and after radicle elongation (Mayer and Marbach, 1981; Simon, 1984; Kuras, 1986). The speed of these events all play a role in controlling the length of the lag phase. These events may be delayed by factors such as inadequate product accumulation, lack of cell preparation (organellar synthesis and/or maturation, i.e., mitochondria and glyoxysomes), or lack of structural reorganization (cell wall loosening) all of which are necessary for radicle extension. In *B. napus,* the incorporation of methionine was reduced and delayed at suboptimal temperatures in comparison to seeds imbibed at optimal temperatures over the course of imbibition, but these differences were not noticeable when based upon equivalent degree days (Nykiforuk, 1996). This suggests that protein synthesis is one of the metabolic processes that is temperature dependent or constrained by thermal requirements.

In sensitive species, low temperature leads to an increase in the lag phase of water uptake (Simon et al., 1976; Vertucci and Leopold, 1983; Vertucci, 1989), reduced radicle growth (Christiansen, 1967; Cal and Obendorf, 1972; Cohn et al., 1979), and lower respiration rates (Simon et al., 1976; Leopold and Musgrave, 1979; Vertucci, 1989). The abnormalities in root growth occur because of the loss of apical control or abortion of the root tip (Bedi and Basra, 1993). Reductions in respiration rates have been correlated with abnormal mitochondrial development (Ilker et al., 1979; Chabot and Leopold, 1985; Hodson et al., 1987) at low temperature. During the switch from heterotrophic to autotrophic growth, chloroplast development is impaired in *Secale cereale* seedlings (Krol and Huner, 1984; Krol et al., 1987), a chilling-resistant species. Imbibition at 10 and 6°C in *B. napus* results in an increased lag during phase II of germination (shown in Figure 3; Nykiforuk, 1996), and is manifested as reduced radicle growth, reduced chlorophyll accumulation, and lower respiration rates in comparison to seeds imbibed at 22°C (Nykiforuk, 1996). Therefore, some of the developmental impairments displayed by susceptible species also occur in *B. napus* (a resistant species) during low temperature emergence. However, the threshold for these processes is lower in *B. napus,* occurring at temperatures below 10°C and ceasing around 6°C as compared to sensitive species which display impairments at around 10-12°C (Hallgren and Oquist, 1990). After germination, the baseline temperature for growth in *B. napus* seedlings is around 5°C (Morrison et

FIGURE 3. Changes in fresh weight during germination and early seedling growth at optimal and suboptimal temperatures in *Brassica napus* cv. Westar. Otherwise as in Figure 1.

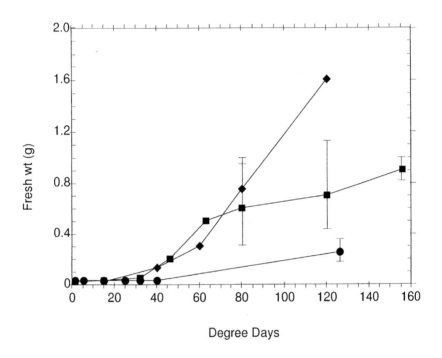

Degree Days

al., 1989). When the process of germination is followed on the basis of degree days, we observed slight increases in the lag phase and the onset of germination at 10°C and reductions and longer delays during phase II at 6°C (Nykiforuk, 1996). Further experimentation is required to study cellular processes during phase II of germination in order to quantify and observe what effects low temperature has within individual species before any general conclusions can be made. It is simply not correct to assume that temperature will affect all of these processes to the same degree.

The transition from phase II to phase III has been shown to be the most sensitive stage to temperature (Simon et al., 1976). The failure to germinate at low temperature is the result of an inability to complete phase II or an inability to enter phase III of germination (Simon et al., 1976). There are two hypotheses to explain this; (1) low temperature impairs proteins necessary for the process of germination to proceed and/or (2) the cellular membranes of the seed are disrupted and are unable to reconstitute properly at low temperature, resulting in the leakage of cellular components and restricting compartmentation of the organelles (Mayer and Marbach, 1981; Bewley and Black, 1994). These will be discussed herein later with respect to observations and speculations derived from chilling injury in plants.

Phase III of Germination

Only germinating seeds enter phase III (Bewley and Black, 1994), which is concurrent with radicle protrusion through the seed coat. Initially, the radicle elongates slowly by cell expansion, followed by rapid growth and cell division. Associated with cell division is a second increase in respiration (Simon, 1984). At lower temperatures (10°C and 6°C), an initial rise in carbon dioxide evolution, was followed by a lag, and a low rate of increase (Nykiforuk, 1996). Therefore, the length of the second phase was increased at low temperatures, as corroborated by reduced seedling growth, measured by fresh weight increases (Figure 3). Eventually, the radicle breaks through the testa to end germination (Kuras, 1986; Kuras, 1987; Bewley and Black, 1994). After germination, water uptake increases and the storage reserves are utilized to support seedling growth until photosynthesis and autotrophic growth can be maintained. The effect of low temperature on phase III can be best illustrated when the data are expressed on the basis of degree days (Table 2). At 22 and 10°C, similar rates in radicle growth were noted, while the rate at 6°C was lower. This indicates that phase III as well as phase II was affected at 6°C. The reduction in early seedling growth was also corroborated by differences in fresh weight increases (Figure 3; Nykiforuk, 1996). Thus, low temperature results in a temporal delay at 10°C, and a delay and reduction in germination at 6°C. This demonstrates that reductions in germination are not only because of thermal requirements, but biochemical and molecular processes may also be impaired at lower temperatures.

Early Seedling Growth

Following germination, seedling growth proceeds in either an epigeal or hypogeal fashion (Bewley and Black, 1994). Seedling growth in *B. napus* proceeds in an epigeal fashion, meaning that the cotyledons are raised by elongation of the hypocotyl. Thereafter, the cotyledons are converted to photosynthesizing leaf-like structures, and support the growth of the plumule (the growing meristem of the shoot) in the presence of light.

TABLE 2. Mean radicle lengths (mm) of seeds germinated at 22,10, and 6°C for days after imbibition (DAI). Mean radicle lengths were obtained by measuring radicle protrusion through the seed coat of germinated population. Degree days (DD) provides similar thermal timelines between each of the temperature conditions with respect to DAI. Each value represents the mean of at least three independent replicates ± 1 SD. Values in parentheses represent interpolated values calculated from the empirically measured values. Values in brackets represent DAI. This table was obtained from Nykiforuk (1996).

DD	22°C	[DAI]	10°C	[DAI]	6°C	[DAI]
0	0	[0]	0	[0]	0	[0]
4	(0.04)		(0.01)		0.02 ± 0.01	[1]
8	(0.1)		0.02 ± 0.01	[1]	0.02 ± 0.01	[2]
16	(0.18)		0.03 ± 0.01	[2]	0.04 ± 0.01	[4]
20	0.22 ± 0.1	[1]	(0.10)		(0.09)	
24	(0.60)		0.16 ± 0.1	[3]	0.14 ± 0.1	[6]
32	(1.35)		0.85 ± 0.3	[4]	0.49 ± 0.1	[8]
40	2.10 ± 0.6	[2]	(2.10)		0.89 ± 0.2	[10]
48	(3.31)		3.30 ± 0.5	[6]		
60	5.13 ± 1.0	[3]	(6.35)			
64	(5.75)		7.4 ± 1.6	[8]		
80	8.2 ± 0.3	[4]				

Further growth of the epicotyl results in the formation of true-leaf structures and establishes growth of the plant. The cotyledons at this time may remain photosynthetically active or senesce.

Storage Reserve Mobilization

Lipid Mobilization

The cells in mature *B. napus* seed are packed with oil bodies (Kuras, 1984). Germinating oil seeds rapidly mobilize storage lipids (Doman et al., 1982; Murphy et al., 1989a; Qouta et al., 1991) and convert them to sucrose, the primary nutrient during heterotrophic growth (Trelease,

1984). In *B. napus,* total lipids account for approximately 42% of the seed (Bell, 1993). Following germination the lipids steadily decrease between day 1 to day 10 in seedlings (Lin and Huang, 1983). Prior to this conversion, enzyme synthesis and organelle assembly is required to produce glyoxylate cycle enzymes, such as isocitrate lyase (ICL) and malate synthase (MS), and the glyoxysomes where these enzymes are localized (Trelease, 1984; Keller et al., 1991; Olsen and Harada, 1995). In addition, the lipid breakdown and increases in ICL activity are concurrent with decreases in storage protein (Doman et al., 1982; Qouta et al., 1991). As expected, increases in ICL activity and decreases in storage protein parallel the breakdown and mobilization of oil reserves at low temperature (Nykiforuk and Johnson-Flanagan, 1994; Nykiforuk, 1996). There has been little work on reserve mobilization with regard to low temperature emergence. Over the course of imbibition at low temperature, the mobilization of total lipids is reduced (Nykiforuk and Johnson-Flanagan, 1994). The rate of lipid breakdown is correlated to the rate of germination (Nykiforuk and Johnson-Flanagan, 1994).

ICL is normally considered to be a good indicator of the switch from germination to seedling growth (Harada et al., 1988), in addition to being indicative of lipid breakdown (Trelease, 1984). ICL (threo-D_s-isocitrate glyoxylate-lyase, E.C. 4.1.3.1) is a glyoxylate cycle enzyme responsible for converting isocitrate to glyoxylate and succinate (Vanni et al., 1990). ICL is suspected to be one of the regulatory enzymes in the mobilization of lipid reserves (Malhotra et al., 1984; Vanni et al., 1990) because the substrate isocitrate is partitioned between two competing cycles (Glyoxylate cycle and Krebs tricarboxylic acid cycle). In *B. napus,* ICL is active in the mature seed (Johnson-Flangan et al., 1992) and increases dramatically after germination during the first 3-4 days of seedling growth (Ettinger and Harada, 1990). Thereafter, ICL activity declines and is no longer detected in young leaves (Ettinger and Harada, 1990). Likewise, ICL transcripts are initially detected during late embryogeny 35-40 days post anthesis (DPA) and in the mature seed (Comai et al., 1989). Upon imbibition, the transcripts increase dramatically and remain high over the first six days of imbibition (Comai et al., 1989). Thereafter, the message declines and is no longer detected (Comai et al., 1989; Ettinger and Harada, 1990; Comai et al., 1992).

With respect to low temperature germination and early seedling growth, both lipid depletion and ICL activity at 10 and 6°C is delayed and reduced in *B. napus* (Nykiforuk and Johnson-Flanagan, 1994). Results from Western blots of ICL suggest this is the result of a temporal delay and lack of synthesis (Nykiforuk, 1996). Further insight into the effect of low temperature on the regulation of ICL was gained by comparing Western and Northern analyses (Nykiforuk, 1996). In comparison to the ratio of transcript to protein at 22°C, transcription rates were disproportionately high relative to the protein level at 10°C. This suggests that low temperature leads to dysfunctional post-transcriptional processes. At 6°C, both the transcript levels and accumulation of ICL were lower in comparison to seed at 22°C. These effects demonstrate that a lack of enzyme biosynthesis, rather than reduced activity results in poor lipid mobilization. Thus, ICL biosynthesis may contribute in part to lower seedling growth at 10°C, and very poor seedling growth at 6°C in *B. napus*.

Another possible cause of reduced lipid mobilization is decreased lipase activity. Lipase activity is concomitant with decreases in lipid reserves (Theimer and Rosnitschek, 1978; Lin and Huang, 1983; Hills and Murphy, 1988; Huang, 1992). Thus it is interesting to note that lipase is inhibited by ABA (Imeson et al., 1993), which has been observed to increase during low temperature imbibition (Nykiforuk, 1996). Another possibility is that metabolite transfer between organelles (glyoxysomes and mitochondria) and their transfer through the cytoplasm may be restricted or impaired at low temperature (Nykiforuk and Johnson-Flanagan, 1994).

Storage Protein Mobilization

According to Mikola (1983), the proteolysis of storage proteins occurs in three stages: (1) firstly, hydrolysis provides amino acids for the synthesis of hydrolytic enzymes required to break down the stored reserves (2) secondly, hydrolysis of the storage proteins provides amino acids to the growing seedling (3) and lastly, the mobilization system itself is broken down to provide the last ration of amino acids, just prior to the onset of autotrophic growth. Once the amino acids have been liberated from the storage proteins, they are translocated to areas of active growth (Bewley and Black, 1994). Translocation through the phloem is temperature dependent (Farrar, 1988), and differences in Arrhenius plots of translocation between chilling-sensitive and -resistant species show sharp declines at temperatures below 10 and 0°C, respectively (Giaquinta and Geiger, 1973).

In *B. napus*, the major storage proteins are cruciferin and napin (Lonnerdal and Janson, 1972; Finlayson, 1976; Crouch and Sussex, 1981), which are housed in protein bodies, and oleosin which is a component of the oil body (Huang, 1992). The breakdown of protein bodies occurs first in the radicle, then in the hypocotyl, and finally in the cotyledons (Kuras, 1987; Hoglund et al., 1992). One mechanism is believed to be involved in the mobilization of cruciferin and napin (Murphy et al., 1989b). Rapid mobilization of storage proteins occurs within the protein bodies two days after germination (Murphy et al., 1989b; Hoglund et al., 1992). By six days after germination, both cruciferin and napin are no longer detected (Murphy et al., 1989b). In *B. napus,* seeds imbibed and grown at low temperature exhibit delays in the mobilization of cruciferin (Nykiforuk, 1996). If this was strictly a thermal effect, developmentally similar seedlings should display comparable rates of protein mobilization, unless there is a loss of coordination of storage reserve breakdown. The results indicate that protein mobilization at $10°C$ is delayed because of thermal constraints, whereas at $6°C$ there is a loss of coordination (Nykiforuk, 1996).

Under optimum conditions, oleosins, which represent up to 10% of the total protein in the mature seed, rapidly disappear concomitant with lipid depletion (Huang, 1992). However, at low temperature the mobilization of oleosins does not correlate with decreases in lipid (Nykiforuk and Johnson-Flanagan, 1994) and is reduced and delayed (Nykiforuk, 1996). Again developmentally similar seedlings exhibited a loss of coordination between the breakdown of storage reserves and development at lower temperatures.

Developmental Switches: Gene Expression During Germination and Early Seedling Growth Under Low Temperature Conditions

During early sporophytic growth, a number of developmental stages have been delineated. In general, these are designated embryogenesis, germination and early seedling growth (post-germinative growth). Embryogenesis, the period encompassing the formation of the zygote until the development of the mature dry seed, has been described in detail by a number of researchers (Tykarska, 1976, 1979, 1980; Crouch and Sussex, 1981; Goldberg et al., 1989; Hughes and Galau, 1989, 1991; Fernandez et al., 1991; West and Harada, 1993; Jakobsen et al., 1994; Goldberg et al., 1994; Parcy et al., 1994). After a nondormant seed enters quiescence, there is a developmental switch in the programs from embryogenesis to germination (Harada et al., 1988; Kermode, 1990) initiated by imbibition. Germination follows (Kuras, 1984, 1986, 1987; Harada et al., 1988) and there is another developmental switch to processes associated with early seedling growth that lead to autotrophic growth and development (Harada et al., 1988; Dietrich et al., 1989).

The switches in development are associated with induction and suppression of gene sets (Goldberg et al., 1989; Hughes and Galau, 1989; Fernandez et al., 1991; Galau and Hughes, 1991; Jakobsen et al., 1994). The switch from embryogenesis to the onset of germination is associated with the loss of the LEA (late embryogenesis abundant) messages. LEA type messages are highly conserved across species including *Gossypium herbaceum* (Galau et al., 1986; Baker et al., 1988), *Daucus carota* (Choi et al., 1987), *Hordeum vulgare* (Hong et al., 1988), and *Oryza sativa* (Mundy and Chua, 1988). These messages have been temporally characterized in *B. napus* (Harada et al., 1989; Raynal et al., 1989; Jakobsen et al., 1994) and are present during late embryogenesis and degraded rapidly upon imbibition within 24 hours. Two other messages, COT 44 and IL1, which encode a putative peptidase and ICL, respectively, are present in low amounts during late embryogeny and accumulate to high levels during post-germinative growth in the seedlings (Harada et al., 1988; Comai, 1989; Dietrich et al., 1989). COT44 is present in low amounts during late embryogeny, increases during the first 24 hours after imbibition, and accumulates to high levels during post-germinative growth in seedlings (Harada et al., 1988; Dietrich et al., 1989). In a similar manner, ICL message is present during late embryogeny, and increases dramatically to high levels over the first 6 days of imbibition (Comai et al., 1989). Therefore, these messages signal the onset of germination and post-germinative growth.

In *B. napus,* the expression of these messages during low temperature germination and early seedling growth reflected the delays observed in the onset of germination, and the delays and reductions in seedling growth (Figure 4). Two LEA messages, LEA 76 and B86 transiently increase then decrease when seeds were imbibed at low temperature. Further, LEA 76 was still present in seedlings at 1 mm radicle length at 10 and 6°C (Figure 5). The eventual decrease in LEA type messages was correlated with the onset of germination at the suboptimal temperatures (Nykiforuk and Johnson-Flangan, 1994; Nykiforuk, 1996). The presence of LEA transcripts in seedlings may reflect an inability of the seed to switch completely from embryogenesis to germination program, while the transient increase in LEA messages may be associated with the temporal delay in development. The changes in LEA type messages were also correlated to the endogenous levels of ABA during low temperature germination and early seedling growth (Nykiforuk, 1996). This suggests that the delays observed in the developmental switch from embryogenesis to germination and subsequent post-germinative growth may involve ABA. With regard to COT44 and IL1, the accumulation of these messages is also delayed (Figure 4). Furthermore, during seedling growth the message levels were similar at 10°C and 22°C, but lower in seedlings grown at 6°C (Figure 5). Thus, quantitative differences correlate with reduced seedling growth at 6°C (Nykiforuk and Johnson-Flanagan, 1994).

FIGURE 4. Northern blot analysis of the developmental molecular markers LEA 76, B86, IL1, and COT 44 over the course of germination and early seedling growth at 22, 10, and 6°C. Each lane represents the total extracted RNA (20 µg) from one hundred seed collected on each DAI in each temperature condition from *Brassica napus* cv. Westar seed. Sd denotes total RNA extracted from mature seed. RNA extraction, blotting, and hybridization conditions were the same for all probes as outlined in Nykiforuk (1996).

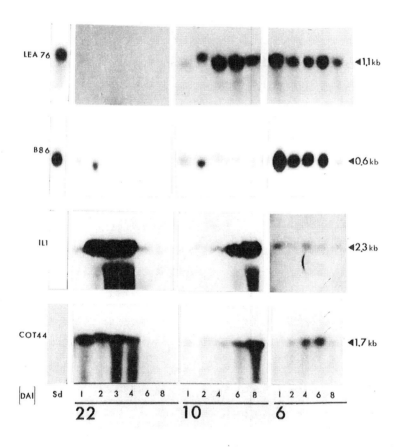

FIGURE 5. Northern blot analysis of the developmental molecular markers LEA 76, B86, IL1, and COT 44 during specific stages after germination as measured by radicle length (mm) at 22, 10, and 6°C. Each lane represents the total extracted RNA (20 µg) from 20-30 seedlings collected from each temperature condition from *Brassica napus* cv. Westar seed. Sd denotes total RNA extracted from mature seed; le denotes total RNA extracted from leaf tissue. RNA extraction, blotting, and hybridization conditions were the same for all probes as outlined in Nykiforuk (1996).

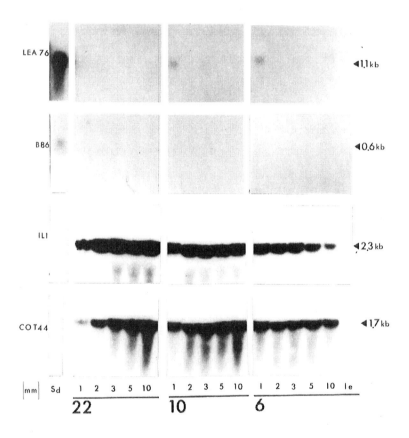

Abscisic Acid

ABA has been implicated as a mediator of the developmental switches during embryogenesis (Kermode, 1990). ABA levels increase during the cell expansion phase of embryogenesis and decrease thereafter to low levels in the mature *B. napus* seed (Finkelstein et al., 1985; Kermode, 1990; Singh and Sawhney, 1992). Not surprisingly, ABA has been shown to prevent precocious germination in immature *Brassica* embryos (Finkelstein et al., 1985). It also inhibits and reduces germination (Sharma et al., 1992). Exogenous ABA concentrations as little as 10 µM (Schöpfer and Plachy, 1984) will inhibit germination of *B. napus* seed, and concentrations ten fold higher will maintain seed in a quiescent state for long periods of time (Schöpfer and Plachy, 1985). It has been hypothesized that the accumulation of endogenous ABA leads to the inhibition of some developmental processes and the induction of others (Xin and Li, 1992; Prasad et al., 1994), thereby determining the rate at which germination and/or seedling growth proceeds. Thus it is interesting to note that chilling temperatures result in increased endogenous ABA levels in sensitive species including *Cucumis sativus* (Capell and Dorffling, 1989), *Lycopersicon esculentum* (Daie and Campbell, 1981), *Zea mays* (Capell and Dorffling, 1993), and *Phaseolus vulgaris* (Eamus and Wilson, 1983; Eze et al., 1983).

In our studies, a transient increase in endogenous ABA occurs over the course of imbibition at low temperature (Nykiforuk, 1996). Unfortunately, the experiment employed could not distinguish whether the secondary peak in ABA arose from a re-induction of ABA synthesis or from conjugated storage forms of ABA present in the dry seed, as the ELISA procedure employed will only detect free cis, trans (+) ABA (Martens et al., 1983). However, this experiment determined that a threshold in ABA may exist in the range of 65 to 81 pmol/g fresh wt, which in turn inhibits germination or inhibits processes allowing germination to proceed (Nykiforuk, 1996). Radicle growth was minimal concurrent with the secondary increase in endogenous ABA at low temperature (Nykiforuk, 1996). However, once endogenous ABA levels were below the putative threshold value radicle growth proceeded rapidly at 10°C and slowly at 6°C.

To further examine the role of ABA in germination, we imbibed seed in the presence of ABA or fluridone. If ABA was inhibitory for germination at low temperature, then fluridone (an inhibitor of ABA synthesis) (Gamble and Mullet, 1986), would be expected to overcome the inhibition. Exogenous ABA reduced the rate and success of germination, and at lower temperatures these effects were amplified (Nykiforuk, 1996). The results clearly showed that fluridone increased the rate and overall percent germination at 6°C (Nykiforuk, 1996).

CHILLING INJURY

As can be seen from the above, very little is known about the biochemical and molecular consequences of germination at low temperature. Therefore, our greatest source of information is derived from the literature available on chilling injury. From these, speculations can be made on the biochemical and molecular effects of chilling injury in seed. This, however, is a leap of faith and therefore should be used only to develop hypotheses.

Membrane structure has been shown to be sensitive to low temperature (Gordon-Kamm and Steponkus, 1984; Steponkus, 1984). One of the first hypotheses put forth by Lyons (1973) involves changes in bulk membrane lipids but this is no longer accepted. This theory has now been refined to acknowledge that membrane lipids can exist in heterogeneous states (Benga and Holmes, 1984; Storch and Kleinfeld, 1985). During exposure to low temperature, the domains of lipids undergo continuous transitions at a critical temperature from liquid crystal to gel phase in localized regions. Upon rewarming this would result in irreversible damage from structural perturbations, leading to a loss of semipermeability and membrane protein aggregation (Quinn, 1985). In addition, chilling in the light results in the formation of activated oxygen (Wise and Naylor, 1987), which in turn results in enhanced leakage of the cytoplasmic solutes (Kendall and McKersie, 1989). Indicators of oxidative stress and lipid degradation were observed in cucumber seedlings within 1-2 days of chilling (Hariyadi and Parkin, 1993). Irreversible increases in electrolyte leakage in cucumber seedlings subjected to chilling temperatures have also been observed and suggests that increases in passive permeability of seedling plasma membranes is likely a secondary response to chilling (Hariyadi and Parkin, 1993).

The failure to germinate upon exposure to low nonfreezing temperatures in the range of chilling injury may also result because of extraordinarily high activation energies of proteins essential to the germination process or inactivation of "housekeeping" proteins (Christophersen, 1973; Simon et al., 1976). Soluble enzymes are inactivated by a low temperature-induced weakening of the hydrophobic bonds, leading to the unfolding and/or dissociation of large enzymes (Levitt, 1980). This theory has been further expanded by Graham and Patterson (1982) to involve changes in enzyme levels, alterations in kinetic properties, and/or cold liability of these proteins. Low temperature may inactivate enzymatic pathways by disrupting protein-protein and/or protein-lipid interactions (Parkin et al., 1989). For example, plasma membrane ATPase activity is influenced by low temperature influencing protein-lipid interactions and the protein itself (Wright et al., 1982).

Soluble enzymes that are affected by temperature include phenylalanine ammonia lyase (PAL) synthesis (Tan, 1980). PEP carboxylase is cold sensitive in maize (Uedan and Sugiyama, 1976) and exhibits a discontinuity in activity at around 14°C (Grierson et al., 1982). PEP carboxylase exhibits an increase in the K_m at temperatures below 10°C in sensitive plants, but remains constant in resistant plants (Graham et al., 1979). Pyruvate kinase (Sugiyama et al., 1979) and phosphofructokinase (Dixon et al., 1981), both dissociate from tetramers to dimers at low temperatures and become inactive. Many other enzymes are chilling labile or freezing sensitive (Guy, 1990). The impaired function of some of these enzymes at low temperature could account for the reductions in respiration and the energy required to sustain growth during heterotrophic growth and subsequent autotrophic growth during low temperature germination and early seedling growth.

OXIDATIVE STRESS

Activated oxygen species including superoxide radicals, hydrogen peroxide and hydroxyl radicals (Saruyama and Tanida, 1995) are produced via electron transport systems in mitochondria and chloroplasts (Bowler et al., 1992) as a normal consequence of metabolism. These in turn can damage cellular components including the denaturing of proteins and causing peroxidation of membrane lipids (Michalski and Kaniuga, 1981; Elstner, 1982). The products of lipid peroxidation in turn can produce volatile compounds (hydrocarbons, aldehydes, and alcohols) that may further damage the cell (Benson, 1990). In order to reduce these harmful products, antioxidant enzymes (e.g., superoxide dismutase and catalase), lipid soluble membrane-associated antioxidants (e.g., α-tocopherol and β-carotene), and water soluble reductants (e.g. glutathione and ascorbate) are produced, which quench or convert the radicals to less harmful byproducts (Bowler et al., 1992; Walker and McKersie, 1993 and references therein). Superoxide dismutase (SOD; EC 1.15.1.1) reacts with superoxide radicals and converts them to oxygen and hydrogen peroxide (Bowler et al., 1992). Superoxide radicals can also interact with hydrogen peroxide to form highly reactive hydroxyl radicals, and therefore the dismutation of the superoxide radicals is an important defense mechanism in the cell (Prasad et al., 1994). The hydrogen peroxide in turn is converted to water and oxygen by the enzyme catalase (CAT; EC 1.11.1.6) and/or ascorbate peroxidase (APX; EC 1.11.1.11) (Bowler et al., 1992).

Activated oxygen is believed to be responsible for cold injury in plants (Wise and Naylor, 1987; Kendall and McKersie, 1989; Hodgson and Raison, 1991; Okuda et al., 1991; Hariyadi and Parkin, 1993; Walker and McKersie, 1993; Prasad et al., 1994). Therefore, the stability and/or synthesis of these enzymes may be linked to resistance or sensitivity to low temperature.

The extent of oxidative stress encountered in seed during germination and early seedling growth at low temperature has not been quantified to date. However, there is some evidence to suggest that oxidative stress is a problem. Increases in hydrogen peroxide have been reported in rice (Saruyama and Tanida, 1995) and wheat seedlings (Okuda et al., 1991) during chilling treatments. Lower hypocotyl length, protein content and higher peroxidase activity were observed in root seedlings from a cold-sensitive *B. napus* comparison to a cold-tolerant *B. campestris* line (Nam and Park, 1992). Wilson and McDonald (1986) report that free radical damage encountered during seed storage may be enhanced upon germination, and therefore low temperature may actually exacerbate this effect.

Investigation into the possibility that during low temperature germination there is an increase in activated oxygen leading to peroxidation of membrane lipids and/or denaturation of proteins deserves merit. Perhaps resistant species exhibit differences in SOD, CAT or APX activities in comparison to sensitive species. Catalase has been found to be cold labile in rice, pea, cucumber, maize, rye and wheat (Taylor et al., 1974; Omran, 1980; Feirabend et al., 1992; Saruyama and Tanida, 1995). In the case of rice at the germination stage, SOD increased, CAT was found to be more stable in a tolerant cultivar in comparison to a sensitive cultivar, and APX was believed to compensate for the losses of activity in CAT, remaining the same or higher (Saruyama and Tanida, 1995). Perhaps the overexpression of CAT targeted to the glyoxysome (Olsen et al., 1993), or cytosolic SOD, and/or mitochondrial SOD targeted to the mitochondria (Bowler et al., 1992) could alleviate oxidative stress encountered during low temperature germination. Sen Gupta et al. (1993) demonstrated that photosynthetic rates were 20% higher in transgenic tobacco overexpressing Cu/Zn-SOD (Bowler et al., 1992) in comparison to nontransformed tobacco plants when subjected to chilling temperatures.

SEED PRIMING

One popular method used to overcome low temperature effects on germination is priming seeds with PEG, PVA, and/or water. Seed priming has been shown to result in more uniform and rapid low temperature germination (Rao and Dao, 1987; Zheng et al., 1994) in a wide variety of plants including wheat, oats (Akalehiywot and Bewley, 1977), soybean (Knypl and Khan, 1981), parsley (Akers et al., 1987), maize (Basra et al., 1988), carrot, celery and onion (Szafirowska et al., 1981; Brocklehurst and Dearman, 1983a, 1983b) and canola (Rao and Dao, 1987; Zheng et al., 1994). Hydropriming of *B. oleracea* seed greatly improved the rate of germination at 10°C (Fujikura et al., 1993). In these studies, the hydration level of the seed was raised under controlled conditions, followed by dehydrating the seed to original moisture levels (Heydecker and Gibbons, 1978; Heydecker and Coolbear, 1977; Brocklehurst and Dearman, 1984; Fujikura et al., 1993). The priming treatment maintains the seed in phase II of germination.

Yan et al. (1989) showed that seed priming in *Glycine max* seed protected oxidative phosphorylation from low temperature imbibitional stress. Yang et al. (1992) found priming in the presence of PVA and PEG at low temperature assisted in the duplication and differentiation of mitochondria. Therefore, mitochondrial development during low temperature imbibition plays an important role in seed vigor and resistance to chilling injury. These effects may be mediated by decreased water uptake rates as PVA pretreatment reduces the water absorption rate of soybean seeds (Guo et al., 1989).

From these and other studies, primed seed had enhanced metabolic activity (respiration rates) (Yan, 1987), enzymes involved in the mobilization of seed reserves are activated (Khan et al., 1978), there is improved synthesis of RNA and protein (Khan et al., 1978; Knypl et al., 1980) and membrane integrity is restored, repaired, or synthesized (Burgess and Powell, 1984; Basra et al., 1988). The priming of seed initiates processes leading to germination by maintaining the seed in phase II. Subsequent hydration then enables these processes to carry on from where they left off.

Priming in *B. napus* is believed to improve low temperature germination by initiating metabolic events and/or the leaching of germination inhibitors (Zheng et al., 1994). The presence of a water soluble germination inhibitor in *B. napus* modulated by low temperature and light has been postulated (Bazanska and Lewak, 1986). The identity of this compound is unknown, however, possible candidates include ABA (Schöpfer and Plachy, 1984, 1985; Sharma et al., 1992) and/or phenolics (Nandakumar and Rangaswamy, 1985).

LOW TEMPERATURE EMERGENCE IN Brassica napus

From our studies, we have determined that reduced seedling emergence at low temperature in *B. napus* is a function of both germination and early seedling growth (Nykiforuk, 1996). When studying the effects of low temperature on emergence, it is essential to use degree days as a basis to compare biochemical processes in order to know if these processes are delayed or reduced because of thermal requirements or if the delay is also because of developmental impairments. From our studies, some processes appear to be affected by temperature because of thermal requirements. For instance at 10°C, germination rates slow without reducing the germination success. This appears to be strictly a thermal effect. On the other hand, it is apparent that the switch from embryogenesis to germination and post-germinative growth is delayed because of developmental impairments. This was determined by examination of molecular events in developmentally similar seedlings. There is also a large reduction in seedling growth. This appears to be a combination of both thermal requirements and developmental impairments resulting in the observed delays and reductions in specific processes. The reduced seedling growth is associated with lower rates of CO_2 evolution, lower rates of chlorophyll accumulation, lower rates of ICL synthesis, lower lipid mobilization, and developmental delays in oleosin depletion. These effects do not appear to be caused by decreased storage protein degradation. In addition, low rates of fresh weight increase may indicate a developmental effect. Seed at 6°C is again delayed because of thermal requirements, but developmental impairments have a more significant effect resulting in poor germination and seedling growth. A proportion of the seed is unable to make the transition from phase II to phase III of germination. Further, seedling growth was also affected by a loss of developmental coordination, i.e., a loss of coordination between the breakdown of storage reserves (cruciferin; and lipids via ICL synthesis and oleosin degradation) and their subsequent utilization. We speculated that the lack of coordination between the mobilization of storage proteins and the *de novo* synthesis of proteins prevents some seed from germinating and proceeding with growth at 6°C.

FUTURE DIRECTIONS

The purpose of this review was to gain insight into some of the underlying biochemical and molecular events occurring during low temperature germination and early seedling growth. Further research into the endogenous levels of hormones, their functions, and interactions during germination and early seedling growth need to be explored. Then alterations of endogenous levels via the application of hormone synthesis inhibitors or mutant seeds over- or under-expressing vital hormones could be pursued.

It is unclear whether ABA should be increased or decreased. On one hand ABA inhibits germination (Schöpfer and Plachy, 1984, 1985; Nykiforuk, 1996), while on the other, treatment of plants with exogenous ABA can reduce chilling injury in plants (Rikin and Richmond, 1976; Markhart III, 1986; Pardossi et al., 1992) and increase acclimation against chilling in maize seedlings (Anderson et al., 1994). Wilen et al. (1994) using ABA analogs increased freezing tolerance, fresh and dry weights in *B. napus* and *B. campestris* in plants grown at 10°C in comparison to acetone-treated or untreated seedlings. Based on these and other studies, Capell and Dörffling (1993) suggested that the ability to increase endogenous ABA levels as a protectant is a prerequisite against chilling injury in seedlings and plants. The mechanism of ABA protection is not understood, but ABA has been observed to protect mitochondria from oxidative damage in maize seedlings subjected to chilling temperatures by inducing antioxidant enzymes (Prasad et al., 1994). Janowiak and Dörffling (1995) observed higher and faster accumulation of ABA in more resistant chilled maize seedlings. Therefore, the influence of ABA on germination may differ from the protective role it may play during seedling growth.

It is obvious that a better understanding of the complex interactions and coordination of the developmental processes during germination and early seedling growth is needed. This in turn may enable researchers to discover the rate limiting steps in rapid and synchronous low temperature emergence. Unfortunately, any means attempted to improve emergence at low temperature may be met with resistance. Nature has been working on this problem, and evolution has selected for the best possible adaptations to the environment. Indeed, for any given instant the rate of germination will be governed by the slowest biological process involved, and therefore the entire temperature range over which germination will occur in any given species is not simply characterized by temperature coefficients because germination is a complex process and lower temperature will affect each constituent step individually (Mayer and Poljakoff-Mayber, 1989). Therefore, any observations in a low temperature germination study merely reflect the effect of temperature on the overall resultant effect of that process (Hegarty, 1973). Each process and developmental stage needs to be studied independently keeping the overall scheme of emergence in perspective to dissect which processes may be altered to alleviate the effects of low temperature and enhance germination and early seedling growth.

REFERENCES

Acharya, S.N., J. Dueck and R.K. Downey. (1983). Selection and heritability studies on canola/rapeseed for low temperature germination. *Canadian Journal of Plant Science* 63: 377-384.

Addae, P.C. and C.J. Pearson. (1992). Thermal requirements for germination and seedling growth of wheat. *Australian Journal of Agricultural Research* 43: 585-594.

Adler, L.S., K. Wikler, F.S. Wyndham, C.R. Linder and J. Schmitt. (1993). Potential for persistence of genes escaped from canola: germination cues in crop, wild, and crop-wild hybrid *Brassica rapa*. *Functional Ecology* 7: 736-745.

Akalehiywot, T. and J.D. Bewley. (1977). Promotion and synchronization of cereal grain germination by osmotic pre-treatment with polyethylene glycol. *Journal of Agricultural Science* 89: 503-506.

Akers, S.W., G.A. Berkowitz and J. Rabin. (1987). Germination of parsley seeds primed in aerated solutions of polyethylene glycol. *HortScience* 22: 250-252.

Anderson, M.D., T.K. Prasad, B.A. Martin and C.R. Steward. (1994). Differential gene expression in chilling-acclimated maize seedlings and evidence for the involvement of abscisic acid in chilling tolerance. *Plant Physiology* 105: 331-339.

Ashworth, E.N. and R.L. Obendorf. (1980). Imbibitional chilling injury in soybean axes: relationships to stelar lesions and seasonal environments. *Agronomy Journal* 72: 923-928.

Atkinson, D.E. (1977). *Cellular Energy Metabolism and Its Regulation*. New York, NY: Academic Press.

Baker, J.C., C. Steele, and L.S. Dure III. (1988). Sequence and characterization of 6 Lea proteins and their genes from cotton. *Plant Molecular Biology* 11: 277-291.

Barber, S.J., G. Rakow and R.K. Downey. (1991). Laboratory and growth room seed vigor testing of certified canola seed. *GCIRC Rapeseed Congress*, C-09, pp. 727-733.

Basra, A.S., S. Bedi and C.P. Malik. (1988). Accelerated germination of maize seeds under chilling stress by osmotic priming and associated changes in the embryo phopholipids. *Annals of Botany* 61: 653-659.

Basra, A.S., R. Dhillon and C.P. Malik. (1989). Influence of seed pre-treatments with plant growth regulators on metabolic alterations of germinating maize embryos under stressing temperature regimes. *Annals of Botany* 64: 37-41.

Bazanska, J. and S. Lewak. (1986). Light inhibits germination of rape seeds at unfavourable temperatures. *Acta Physiologiae Plantarum* 8: 145-149.

Bedi, S. and A.S. Basra. (1993). Chilling injury in germinating seeds: basic mechanisms and agricultural implications. *Seed Science Research* 3: 219-229.

Bell, J.M. (1993). Composition of canola meal. In *Canola Meal Feed Industry Guide*. Winnipeg, Canada: Canola Council of Canada.

Benga, G. and R.P. Holmes. (1984). Interactions between components in biological membranes and their implications for membrane function. *Progress in Biophysics and Molecular Biology* 43: 195-257.

Benson, E.E. (1990). Free radical damage in stored plant germplasm. *International Board for Plant Genetic Resources, Rome*, pp. 37-66.

Bewley, J.D. and M. Black, eds. (1994). *Seeds: Physiology of Development and Germination*. 2nd edition. New York, NY: Plenum Press.

Bodsworth, S. and J.D. Bewley. (1981). Osmotic priming of crop species with polyethylene glycol as a means of enhancing early and synchronous germination at cool temperatures. *Canadian Journal of Botany* 59:672-676.

Bowler, C., M. Van Montagu and D. Inze. (1992). Superoxide dismutase and stress tolerance. *Annual Review of Plant Physiology and Plant Molecular Biology* 43: 83-116.

Bramlage, W.J., A.C. Leopold and J.E. Specht. (1979). Imbibitional chilling sensitivity among soybean cultivars. *Crop Science* 19: 811-814.

Brocklehurst, P.A. and J. Dearman. (1983a). Interactions between seed priming treatments and nine seed lots of carrot, celery and onion. I. Laboratory germination. *Annals of Applied Biology* 102: 577-584.

Brocklehurst, P.A. and J. Dearman. (1983b). Interactions between seed priming treatments and nine seed lots of carrot, celery and onion. II. Seedling emergence and plant growth. *Annals of Applied Biology* 102: 585-593.

Brocklehurst, P.A. and J. Dearman. (1984). A comparison of different chemicals for osmotic pre-treatment of vegetable seeds. *Annals of Applied Biology* 105: 391-398.

Buntin, G.D., J.P. McCafferey, P.L. Raymer and J. Romero. (1995). Quality and germination of rapeseed and canola seed damaged by adult cabbage seedpod beevil, *Ceutorhynchus assimilis* (Paykull) (*Coleoptera: Curculionidae*). *Canadian Journal of Plant Science* 75: 539-541.

Burgess, R.W. and A.K. Powell. (1984). Evidence for repair processes in invigoration of seeds by hydration. *Annals of Botany* 55: 753-757.

Cal, J.P. and R.L. Obendorf. (1972). Imbibitional chilling injury in *Zea Mays* L. altered by initial kernel moisture and maternal parent. *Crop Science* 12: 369-373.

Canola Growers Manual. (1991). *Canola Council of Canada.* Winnipeg, Canada: Canola Council of Canada.

Capell, B. and K. Dörffling. (1989). Low temperature-induced changes of abscisic acid contents in barley and cucumber leaves in relation to their water status. *Journal of Plant Physiology* 135: 571-575.

Capell, B. and K. Dörffling. (1993). Genotype-specific differences in chilling tolerance of maize in relation to chilling-induced changes in water status and abscisic acid accumulation. *Physiologia Plantarum* 88: 638-647.

Chabot, J.P. and A.C. Leopold. (1985). Ultrastructural aspects of chilling injury in soybean seed radicle. *American Journal of Botany* 72: 1120-1126.

Choi, J.H., L. Liu, C. Borkird and Z.R. Sung. (1987). Cloning of genes developmentally regulated during plant embryogenesis *Proceedings of the National Academy of Science USA* 84: 1906-1910.

Christiansen, M.N. (1967). Period of sensitivity to chilling in germinating cotton. *Plant Physiology* 42: 520-522.

Christiansen, M.N. (1969). Seed moisture content and chilling injury to imbibing cotton seed. *Crop Science* 9: 672-673.

Christophersen, J. (1973). Basic aspects of temperature action on microorganisms. In *Temperature and Life,* eds. H. Precht, J. Christophersen, H. Hensel, and W. Larcher. Berlin: Springer-Verlag, p 3.

Cohn, M.A., R.L. Obendorf and G.T. Rytko. (1979). Relationship of stelar lesions to radicle growth in corn seedlings. *Agronomy Journal* 71: 954-958.

Comai, L., R.A. Dietrich, D.J. Maslyar, C.S. Baden and J.J. Harada. (1989). Coordinate expression of transcriptionally regulated isocitrate lyase and malate synthase genes in *Brassica napus* L. *The Plant Cell* 1: 293-300.

Comai, L., K.L. Matsudaira, R.C. Heupel, R.A. Dietrich and J.J. Harada. (1992). Expression of a *Brassica napus* malate synthase gene in transgenic tomato plants during the transition from late embryogeny to germination. *Plant Physiology* 98: 53-61.

Côme, D. and T. Tissaoui. (1972). Interrelated effects of imbibition, temperature and oxygen on seed germination. In *Seed Ecology,* ed. W. Heydecker. London: Butterworths, pp. 157-167.

Coolbear, P. and C.R. McGill. (1990). Effects of a low-temperature pre-sowing treatment on the germination of tomato seed under temperature and osmotic stress. *Scientia Horticulturae* 44: 43-54.

Crouch M.L. and I.M. Sussex. (1981). Development and storage-protein synthesis in *Brassica napus* L. embryos *in vivo* and *in vitro*. *Planta* 153: 64-74.

Dahal, P., K.J. Bradford and R.A. Jones. (1990). Effects of priming and endosperm integrity on seed germination rates of tomato genotypes. I. Germination at suboptimal temperatures. *Journal of Experimental Botany* 41: 1431-1439.

Daie, J. and W.F. Campbell. (1981). Response of tomato plants to stressful temperatures. *Plant Physiology* 67: 26-29.

Dietrich, R.A., D.J. Maslyar, R.C. Heupel and J.J. Harada. (1989). Spatial patterns of gene expression in *Brassica napus* seedlings: identification of a cortex-specific gene and localization of mRNAs encoding isocitrate lyase and a polypeptide homologous to proteinases. *The Plant Cell* 1: 73-80.

Dixon, W.L., F. Franks and T. Aprees. (1981). Cold lability of phosphofructokinase from potato tubers. *Phytochemistry* 20: 969-972.

Doman, D.C., J.C. Walker, R.N. Trelease and B.D. Moore. (1982). Metabolism of carbohydrate and lipid reserves in germinated cotton seeds. *Planta* 155: 502-510.

Eamus, D. and J.M. Wilson. (1983). ABA levels and effects in chilled and hardened *Phaseolus vulgaris*. *Journal of Experimental Botany* 34: 1000-1006.

Elstner, E.F. (1982). Oxygen activation and oxygen toxicity. *Annual Review of Plant Physiology* 33:73-96.

Eskin, N.A.M. (1992). Effect of variety and geographical location of the incidence of mucilage in canola seeds. *Canadian Journal of Plant Science* 72: 1223-1225.

Ettinger, W.F. and J.J. Harada. (1990). Translational or post-translational processes affect differentially the accumulation of isocitrate lyase and malate synthase proteins and enzyme activities in embryos and seedlings of *Brassica napus*. *Archives of Biochemistry and Biophysics* 281: 139-143.

Eze, J.M.O., E.B. Dumbroff and J.E. Thompson. (1983). Effects of temperature and moisture stress on the accumulation of abscisic acid in bean. *Physiologia Plantarum* 58: 179-183.

Farrar, J.F. (1988). Temperature and the partioning and translocation of carbon. In

Plants and Temperature, eds. S.P. Long and F.I. Woodward. Cambridge, England: The Company of Biologists Limited. Symp. Soc. Exp. Biol. Vol. 42: 203-236.

Feirabend, J., C. Schaan and B. Hertwig. (1992). Photoinactivation of catalase occurs under both high- and low-temperature stress conditions and accompanies photoinhibition of photosystem II. *Plant Physiology* 100:1554-1561.

Fernandez, D.E., F.R. Turner and M.L. Crouch. (1991). *In situ* localization of storage protein mRNAs in developing meristems of *Brassica napus* embryos. *Development* 111: 299-313.

Finkelstein, R.R. and M.L. Crouch. (1984). Precociously germinating rapeseed embryos retain characteristics of embryogeny. *Planta* 162: 125-131.

Finkelstein, R.R., K.M. Tenbarge, J.E. Shumway and M.L. Crouch. (1985). Role of ABA in maturation of rapeseed embryos. *Plant Physiology* 78: 630-636.

Finlayson, A.J. (1976). The seed protein contents of some *Cruciferae*. In *The Biology and Chemistry of the Cruciferae,* eds. J.G. Vaughan, J.G. McLeod, B.M. Jones. New York, NY: Academic Press, pp. 279-306.

Frisbee, C.C, C.W. Smith, L.E. Wiesner and R.H. Lockerman. (1988). Short term storage effects on dormancy and germination of chickpea (*Cicer arietinum*). *Journal of Seed Technology* 12: 16-23.

Fujikura, Y., H.L. Kraak, A.S. Basra and C.M. Karssen. (1993). Hydropriming, a simple and inexpensive priming method. *Seed Science and Technology* 21: 639-642.

Galau, G.A., D.W. Hughes, and L. III Dure. (1986). Abscisic acid induction of cloned cotton late embryogenesis-abundant (*Lea*) mRNAs. *Plant Molecular Biology* 7: 155-170.

Galau, G.A., K.S. Jakobsen and D.W. Hughes. (1991). The controls of late dicot embryogenesis and early germination. *Physiologia Plantarum* 81: 280-288.

Gamble, P.E. and J.E. Mullet. (1986). Inhibition of carotenoid accumulation and abscisic acid biosynthesis in fluridone-treated dark-grown barley *European Journal of Biochemistry* 160: 117-121.

Giaquinta, R.T. and D.R. Geiger. (1973). Mechanism of inhibition of translocation by localized chilling. *Plant Physiology* 51: 372-377.

Goldberg, R.B., S.J. Barker and L. Perez-Grau. (1989). Regulation of gene expression during plant embryogenesis. *Cell* 56: 149-160.

Goldberg, R.B., G. de Paiva and R. Yadegari. (1994). Plant embryogenesis: body plan elaboration and preparation for life after the seed. *Science* 266: 605-614.

Gordon-Kamm, W.J. and P.L. Steponkus. (1984). Lamellar-to-hexagonal$_{II}$ phase transitions in the plasma membrane of isolated protoplasts after freeze-induced dehydration. *Proceedings of the National Academy of Sciences USA.* 81: 6373-6377.

Graham, D., D.G. Hockley and B.D. Patterson. (1979). Temperature effects on phophoenol pyruvate carboxylase from chilling-sensitive and chilling-resistant plants. In *Low Temperature Stress in Crop Plants. The Role of the Membrane,* eds. J.M. Lyons, D. Graham, and J.K. Raison. New York, NY: Academic Press, pp. 453-461.

Graham, D. and B.D. Patterson. (1982). Responses of plants to low, nonfreezing temperatures: proteins, metabolism, and acclimation. *Annual Review of Plant Physiology* 33: 347-372.

Gray, D. and T.H. Thomas. (1982). In *The Physiology and Biochemistry of Seed Development, Dormancy and Germination,* ed. A.A. Khan. New York, NY: Elsevier Biomedical Press.

Grierson, W., J. Soule and K. Kawada. (1982). Physiological aspects of physiological stress. *Hort Review* 4: 247-271.

Guo, J.Q., W.T. Chen and Y.Z. Yang. (1989). Effects of polyvinyl alcohol pretreatment on the increase of soybean seed vigor and its protection against chilling injury. *Acta Phytophysiologica Sinica* 15: 251-256.

Gutterman, Y., F. Corbineau and D. Côme. (1992). Interrelated effects of temperature, light, and oxygen on *Amaranthus caudatus* L. seed germination. *Weed Research* 32: 111-117.

Guy, C.L. (1990). Cold acclimation and freezing stress tolerance: role of protein metabolism. *Annual Review of Plant Physiology and Plant Molecular Biology* 41: 187-223.

Hallgren, J-E. and G. Oquist. (1990). Adaptations to low temperatures. In *Stress Responses in Plants: Adaptation and Acclimation Mechanisms.* Wiley-Liss Inc, pp. 265-293.

Harada, J.J., C.S. Baden and L. Comai. (1988). Spatially regulated genes expressed during seed germination and postgerminative development are activated during embryogeny. *Molecular and General Genetics* 212: 466-473.

Harada, J.J., A.J. DeLisle, C.S. Baden and M.L. Crouch. (1989). Unusual sequence of an abscisic acid-inducible mRNA which accumulates late in *Brassica napus* seed development. *Plant Molecular Biology* 12: 395-401.

Hariyadi, P. and K.L. Parkin. (1993). Chilling-induced oxidative stress in cucumber (*Cucumis sativus* L. Cv. Calypso) seedlings. *Journal of Plant Physiology* 141: 733-738.

Harrington, J.F. and G.M. Kihara. (1960). Chilling injury of germinating muskmelon and pepper seeds. *Proceedings of the American Society of Horticultural Science* 75: 485-489.

Hegarty, T.W. (1973). Temperature coefficient (Q10), seed germination and other biological processes. *Nature* 243: 305-306.

Heydecker, W. and P. Coolbear. (1977). Seed treatment for improved performance-survey and attempted prognosis. *Seed Science and Technology* 5: 353-425.

Heydecker, W. and M. Gibbins. (1978). The priming of seeds. *Acta Horticulturae* 83: 213-223.

Hills, M.J. and D.J. Murphy. (1988). Characterization of lipases from the lipid bodies and microsomal membranes of erucic acid-free oilseed-rape (*Brassica napus*) cotyledons. *Biochemical Journal* 249:687-693.

Hobbs, P.R. and R.L. Obendorf. (1972). Interaction of initial seed moisture and imbibitional temperature on germination and productivity of soybean. *Crop Science* 12: 664-667.

Hodgson, R.A.J. and J.K. Raison. (1991). Superoxide production by thylakoids during chilling and its implications in the susceptibility of plants to chilling-induced photoinhibition. *Planta* 183: 222-228.

Hodson, M.J., L. Nola, Di and A.M. Mayer. (1987). The effect of changing temperatures during imbibition on ultrastructure in germinating pea embryonic radicles. *Journal of Experimental Botany* 38: 525-534.

Hoglund, A-S., J. Rodin, E. Larsson and L. Rask. (1992). Distribution of napin and cruciferin in developing rape seed embryos. *Plant Physiology* 98: 509-515.

Hong, B., S.J. Uknes and T-H.D. Ho. (1988). Cloning and characterization of a cDNA encoding a mRNA rapidly induced by ABA in barley aleurone layers. *Plant Molecular Biology* 11: 495-506

Huang, A.H.C. (1992). Oil bodies and oleosins in seeds. *Annual Review of Plant Physiology and Plant Molecular Biology* 43: 177-200.

Hughes, D.W. and G.A. Galau. (1989). Temporally modular gene expression during cotyledon development. *Genes Development* 3: 358-369.

Hughes, D.W. and G.G. Galau. (1991). Developmental and environmental induction of Lea and LeaA mRNAs and the postabscission program during embryo culture. *The Plant Cell* 3: 605-618.

Ilker, R., R.W. Briedenbach and J.M. Lyons. (1979). Sequence of ultrastructural changes in tomato cotyledons during chilling. In *Low Temperature Stress in Crop Plants. The Role of the Membrane,* eds. J.M. Lyons, D. Graham, and J.K. Raison. New York, NY: Academic Press, pp. 97-113.

Imeson, H.C., S.B. Rood, J. Weselake, K.P. Zanewich, B.A. Bullock, K.E. Stobbs and M.G. Kocsis. (1993). Hormonal control of lipase activity in oilseed rape germinants. *Physiologia Plantarum* 89: 476-482.

Jakobsen, K.S., D.W. Hughes and G.A. Galau. (1994). Simultaneous induction of postabscission and germination mRNAs in cultured dicotyledonous embryos. *Planta* 192: 384-394.

Janowiak, F. and K. Dörffling. (1995). Chilling-induced changes in the contents of 1-aminocyclopropane-1-carboxylic acid (ACC) and its N-malonyl conjugate (MACC) in seedlings of two maize inbreds differing in chilling tolerance. *Journal of Plant Physiology* 147: 257-262.

Johnson-Flanagan, A.M., Z. Huiwen, X-M. Geng, D.C.W. Brown, C.L. Nykiforuk and J. Singh. (1992). Frost, abscisic acid and desiccation hasten embryo development in *Brassica napus. Plant Physiology* 99: 700-706.

Keller, G-A., S. Krisans, S.J. Gould, J.M. Sommer, C.C. Wang, W. Schliebs, W. Kunau, S. Brody and S. Subrmani. (1991). Evolutionary conservation of a microbody targeting signal that targets proteins to peroxisomes, glyoxysomes and glycosomes. *The Journal of Cell Biology* 114: 893-904.

Kendall, E.J. and B.D. McKersie. (1989). Free radical and freezing injury to cell membranes of winter wheat. *Physiologia Plantarum* 76: 86-94.

Kermode, A.R. (1990). Regulatory mechanisms involved in the transition from seed development to germination. *Critical Reviews in Plant Sciences* 9(2): 155-195.

Khan, A.A., K.L Tao, J.S. Knypl, B. Borkowska and L.E. Powell. (1978). Osmotic conditioning of seeds: physiological and biochemical changes *Acta Horticulturae* 83: 267-278.

King, R.J., Z.P. Kondra and M.R. Thiagarajah. (1986). "Selection for fast germination in rapeseed (*Brassica napus* L. and *B. campestris* L.)," *Euphytica* 35: 835-842.

Kondra, Z.P., D.C. Campbell and J.R. King. (1983). Temperature effects on germination of rapeseed (*Brassica napus* L. and *B. campestris* L.). *Canadian Journal of Plant Science*. 63: 1063-1065.

Knypl, J.S. and A.A. Khan. (1981). Osmoconditioning of soybean seeds to improve performance at suboptimal temperatures. *Agronomy Journal* 73: 112-116.

Knypl, J.S., K.M. Janas and A. Radziwonowska-Jozwiak. (1980). Is enhanced vigour in soybean (*Glycine max*) dependent on activation of protein turnover during controlled hydration of seeds? *Physiologie Vegetale* 18: 157-161.

Krol, M. and N.P.A. Huner. (1984). Pigment accumulation during growth and development at low temperature. *Advances in Photosynthesis Research* IV(4): 463-466.

Krol, M., N.P.A. Huner and A. McIntosh. (1987). Choloplast biogenesis at cold-hardening temperatures. Development of photosystem I and photosystem II activities in relation to pigment accumulation. *Photosynthesis Research* 14: 97-112.

Kuras, M. (1984). Activation of rape (*Brassica napus* L.) embryo during seed germination. III. Ultrastructure of dry embryo axis. *Acta Societatis Botanicorum Poloniae* 53: 171-186.

Kuras, M. (1986). Activation of rape (*Brassica napus* L.) embryo during seed germination. IV. Germinating embryo. The first zones of mitoses, starch and DNA synthesis and their expansion pattern. *Acta Societatis Botanicorum Poloniae* 55: 539-563.

Kuras, M. (1987). Activation of rape (*Brassica napus* L.) embryo during seed germination. V. The first zones of ultrastructural changes and their expansion. *Acta Societatis Botanicorum Poloniae* 56: 77-91.

Lane, B. (1991). Cellular desiccation and hydration: developmentally regulated proteins, and the maturation and germination of seed embryos. *Biogenesis* 5: 2893-2901.

Leopold, A.C. (1980). Temperature effects on soybean imbibition and leakage. *Plant Physiology* 65: 1096-1098.

Leopold, A.C. and M.E. Musgrave. (1979). Respiratory changes with chilling injury of soybeans. *Plant Physiology* 64: 702-705.

Leviatov, S., O. Shoseyov and S. Wolf. (1994). Roles of different seed components in controlling tomato seed germination at low temperature. *Scientia Horticulturae* 56: 197-206.

Leviatov, S., O. Shoseyov and S. Wolf. (1995). Involvement of endomannase in the control of tomato seed germination under low temperature conditions. *Annals of Botany* 76: 1-6.

Levitt, J., ed. (1980). *Responses of Plants to Environmental Stresses. Chilling, Freezing and High Temperature Stresses.* Vol. 1. New York, NY: Academic Press.

Liengsiri, C. and A.K. Hellum. (1988). Effects of temperature on seed germination in *Pterocarpus macrocarpus*. *Journal of Seed Technology* 12: 66-75.

Lin, Y-H. and A.H.C. Huang. (1983). Lipase in lipid bodies of cotyledons of rape and mustard seedlings. *Archives of Biochemistry and Biophysics* 225: 360-369.

Liptay, A., and P. Schöpfer. (1983). Effect of water stress, seed coat restraint, and abscisic acid upon different germination capabilities of two tomato lines at low temperature. *Plant Physiology* 73: 935-938.

Livingstone, N.J. and E. deJong. (1990). Matric and osmotic potential effects on seedling emergence at different temperatures. *Agronomy Journal* 82: 995-998.

Lonnerdal, B. and J.C. Janson. (1972). Studies on *Brassica* seed proteins. 1. The low molecular weight proteins in rapeseed. Isolation and characterization. *Biochemical and Biophysical Acta* 278: 175-183.

Lyons, J.M. (1973). Chilling injury in plants. *Annual Review of Plant Physiology* 24: 445-466.

Lyons, J.M., J.K. Raison and P.L. Steponkus. (1979). The plant membrane in response to low temperature: an overview. In *Low Temperature Stress in Crop Plants. The Role of the Membrane*, eds. J.M. Lyons, D. Graham, and J.K. Raison. New York, NY: Academic Press, pp. 1-24.

Malhotra, O.P, U.N. Dwivedi and P.K. Srivastava. (1984). Steady state kinetics and negative cooperativity in the action of isocitrate lyase. *Indian Journal of Biochemistry and Biophysics* 21: 99-105.

Markhart III, A.H. (1986). Chilling injury: a review of possible causes. *HortScience* 21: 1329-1333.

Markowski, A. (1989a). Sensitivity of different species of field crops to chilling temperature. 1. Interaction of initial seed moisture and imbibition temperature. *Acta Physiologiae Plantarum* 10: 265-274.

Markowski, A. (1989b). Sensitivity of different species of field crops to chilling temperature. 2. Germination, growth and injuries of seedlings *Acta Physiologiae Plantarum* 10: 275-280.

Martens, R., B. Deus-Neumann and E.W. Weiler. (1983). Monoclonal antibodies and the detection and quantification of the endogenous plant growth regulator abscisic acid. *Federation of European Biochemical Society Letters* 160: 269-272.

Mayer, A.M. and I. Marbach. (1981). In *Progress in Phytochemistry* Volume 7, eds. L. Reinhold, J.B. Harborne, and T. Swain. New York, NY: Pergamon Press, pp. 95-136.

Mayer, A.M. and A. Poljakoff-Mayber, eds. (1989). *The Germination of Seeds.* 4th Edition, New York, NY: Pergamon Press.

Michalski, W.P. and Z. Kaniuga. (1981). Photosynthetic apparatus of chilling-sensitive plants. X. Relationship between superoxide dismutase acitvity and photoperoxidation of chloroplast lipids. *Biochemical and Biophysical Acta* 637: 159-167.

Mikola, J. (1983). In *Seed Proteins,* eds. J. Paussant, J. Mosse and S.G. Vaughan. London: Academic Press.

Mills, P.F. (1993). The effects of low temperatures on the germination and emergence of canola. *Project: 83-0036, Alberta Agriculture Farming for the Future*. Beaverlodge, Alberta: Alberta Agriculture.

Mohapatra, S.C. and C.W. Suggs. (1989). Cultivar differences in tobacco seed response to germination temperature. *Seed Science and Technology* 17: 639-647.

Monteith, J.L. (1981). Climatic variation and the growth of crops. *Quarterly Journal of the Royal Meteorological Society* 107: 749-774.

Morrison, M.J., P.B.E. McVetty and C.F. Shankewich. (1989). The determination and verification of a baseline temperature for the growth of westar summer rape. *Canadian Journal of Plant Science* 69: 455-464.

Mundy, J. and N.H. Chua. (1988). Abscisic acid and water stress induce the expression of a novel rice gene. *European Molecular Biology Organization Journal* 7: 2279-2286.

Murphy, D.J., I. Cummins and A.S. Kang. (1989a). Immunological investigation of lipases in germinating oilseed rape, *Brassica napus. Journal of the Science of Food and Agriculture* 47: 21-31.

Murphy, D.J., I. Cummins and A.J. Ryan. (1989b). Immunocytochemical and biochemical study of the biosynthesis and mobilisation of the major seed storage proteins of *Brassica napus. Plant Physiology and Biochemistry* 27: 647-657.

Mwale, S.S., S.N. Azam-Ali, J.A. Clark, R.G. Bradley and M.R. Chatha. (1994). Effect of temperature on the germination of sunflower (*Helianthus annuus* L.). *Seed Science and Technology* 22: 565-571.

Nam, M.H. and W.C. Park. (1992). Biochemical analysis of screening cold tolerance in *Brassica* seedling. *Research Reports of the Rural Development Administration* (Suweon) 34: 15-20.

Nandakumar, L. and N.S. Rangaswamy. (1985). Effect of some flavonoids and phenolic acids on seed germination and rooting. *Journal of Experimental Botany* 36: 1313-1319.

Nelson, J.M., A. Jenkins and G.C. Sharples. (1984). Soaking and other seed pretreatment effects on germination and emergence of sugarbeets at high temperature. *Journal of Seed Technology* 9: 79-86.

Nykiforuk, C.L. (1996). The enhancement of low temperature germination and early seedling growth in *Brassica napus* cv. Westar. Ph.D. Thesis, University of Alberta, Edmonton, Canada.

Nykiforuk, C.L. and A.M. Johnson-Flanagan. (1994). Germination and early seedling development under low temperature in canola. *Crop Science* 34: 1047-1054.

O'Conner, B.J. and L.V. Gusta. (1994). Effect of low temperature and seeding depth on the germination and emergence of seven flax (*Linum usitatissimum* L.) cultivars. *Canadian Journal of Plant Science* 74: 247-253.

Okuda, T., Y. Matsuda, A. Yamanaka and S. Sagisaka. (1991). Abrupt increase in the level of hydrogen peroxide in leaves of winter wheat is caused by cold treatment. *Plant Physiology* 97: 1265-1267.

Olsen, L.J., W.F. Ettinger, B. Damsz, K. Matsudaira, M.A. Webb and J.J. Harada.

(1993). Targeting of glyoxysomal proteins to peroxisomes in leaves and roots of a higher plant. *The Plant Cell* 5: 941-952.

Olsen, L.J. and J.J. Harada. (1995). Peroxisomes and their assembly in higher plants. *Annual Review of Plant Physiology and Plant Molecular Biology* 46: 123-146.

Omran, R.G. (1980). Peroxide levels and the activity of catalase, peroxidase, and idoleacetic acid oxidase during and after chilling cucumber seedlings. *Plant Physiology* 65: 407-408.

Ong, C.K. and J.L. Monteith. (1985). Response of pearl millet to light and temperature. *Field Crops Research* 11: 141-160.

Parcy, F., C. Valon, M. Raynal, P. Gaubier-Comella, M. Delseny and J. Giraudat. (1994). Regulation of gene expression programs during *Arabidopsis* seed development: Roles of the ABI3 locus and of endogenous abscisic acid. *The Plant Cell* 6: 1567-1582.

Pardossi, A., P. Vernieri and F. Tognoni. (1992). Involvement of abscisic acid in regulating water status in *Phaseolus vulgaris* L. during chilling. *Plant Physiology* 100: 1243-1250.

Parkin, K.L., A. Marangoni, R.L. Jackman, R.Y. Yada and D.W. Stanley. (1989). Chilling injury. A review of possible mechanisms. *Journal of Food Biochemistry* 13: 127-153.

Payne, P.I., M. Dobrzanska, P.W. Barlow and M.E. Gordon. (1978). The synthesis of RNA in imbibing seed of rape (*Brassica napus*) prior to the onset of germination: A biochemical and cytological study. *Journal of Experimental Botany* 29: 77-88.

Perry, D.A. (1972). Seed vigour and field establishment. *Horticultural Abstracts* 42: 334-342.

Phillips, J.C. and V.E. Youngman. (1971). Effect of initial seed moisture content on emergence and yield of grain sorghum. *Crop Science* 11: 354-357.

Pollock, B.M. (1969). Imbibition temperature sensitivity of lima bean seeds controlled by initial seed moisture. *Plant Physiology* 44: 907-911.

Pollock, B.M. and J.R. Manalo. (1970). Stimulated mechanical damage to garden beans during germination. *Journal of the American Society for Horticultural Science* 95: 415-417.

Pollock, B.M. and V.K. Toole. (1966). Imbibition period as the critical temperature-sensitive stage in germination of lima bean seeds. *Plant Physiology* 41: 221-229.

Prasad, T.K., M.D. Anderson, B.A. Martin and C.R. Stewart. (1994). Evidence for chilling-induced oxidative stress in maize seedlings and a regulatory role for hydrogen peroxide. *The Plant Cell* 6: 65-74.

Qouta, L.A., K.W. Waldron, E.A.H. Baydoun and C.T. Brett. (1991). Changes in seed reserves and cell wall composition of component organs during germination of cabbage (*Brassica oleracea*) seeds. *Journal of Plant Physiology* 138: 700-707.

Quinn, P.J. (1985). A lipid-phase separation model of low-temperature damage to biological membranes. *Cryobiology* 22: 128-146.

Rao, S.C. and T.H. Dao. (1987). Soil water effects on low-temperature seedling emergence of five *Brassica* cultivars. *Agronomy Journal* 79: 517-519.

Raynal, M., D. Depigny, R. Cooke, and M. Delseny. (1989). Characterization of a radish nuclear gene expressed during late seed maturation. *Plant Physiology* 91: 829-836.

Rikin, A. and A.E. Richmond. (1976). Amelioration of chilling injuries in cucumber seedlings by abscisic acid. *Physiologia Plantarum* 38: 95-97.

Roberts, E.H. (1988). Temperature and seed germination. In *Plants and Temperature*, eds. S.P. Long and F.I. Woodward. Cambridge: The Company of Biologists Limited., pp.109-132.

Roberts, E.H. and R.H. Ellis. (1984). The implications of the deterioration of orthodox seeds during storage for genetic resources conservation. In *Crop Genetic Resources Conservation and Evaluation*, eds. J.H.W. Holden and J.T. Williams. London: George, Allen and Unwin, pp. 18-36.

Ross, J.D. (1984). Metabolite aspects of dormancy. In *Seed Physiology. Volume 2. Germination and Reserve Mobilization*, ed. D.R. Murray. New York: Academic Press, pp. 45-76.

Sanchez-Martinez, D., P. Puigdomenech and M. Pages. (1986). Regulation of gene expression in developing *Zea mays* embryos. Protein synthesis during embryogenesis and early germination of maize. *Plant Physiology* 82: 543-549.

Saruyama, H. and M. Tanida. (1995). Effect of chilling on activated oxygen-scavenging enzymes in low temperature-sensitive and -tolerant cultivars of rice (*Oryza sativa* L.). *Plant Science* 109: 105-113.

Sawhney, R., W.A. Quick and A.I. Hsiao. (1985). The effect of temperature during parental vegetative growth on seed germination of wild oats (*Avena fatua* L.). *Annals of Botany* 55: 25-28.

Schmidt, D.H. and W.F. Tracy. (1989). Duration of imbibition affects seed leachate conductivity in sweet corn. *HortScience* 24: 346-347.

Schöpfer, P. and C. Plachy. (1984). Contol of seed germination by abscisic acid. II. Effect on embryo water uptake in *Brassica napus* L. *Plant Physiology* 76: 155-160.

Schöpfer, P. and C. Plachy. (1985). Contol of seed germination by abscisic acid. III. Effect on embryo growth potential (minimum turgor pressure) and growth coefficient (cell wall extensibility) in *Brassica napus* L. *Plant Physiology* 77: 676-686.

Scott, S.J. and R.A. Jones. (1985). Quantifying seed germination response to low temperature variation among *Lycopersicon* spp. *Experimental and Environmental Botany* 25: 129-137.

Sen Gupta, A., J.L. Heinen, A.S. Holoday, J.J. Burke and R.D. Allen. (1993). Increased resistance to oxidative stress in transgenic plants that overexpress chloroplastic Cu/Zn superoxide dismutase. *Proceedings of the National Academy of Sciences USA* 90: 1629-1633.

Sharma, S.S., S. Sharma and V.K. Rai. (1992). The effect of EGTA, calcium channel blockers (lanthanum chloride and nifedipine) and their interaction with abscisic acid on seed germination of *Brassica juncea* cv. RLM-198. *Annals of Botany* 70: 295-299.

Shaykewich, C.F. (1973). Proposed method for measuring swelling pressure of seeds prior to germination. *Journal of Experimental Botany* 24: 1056-1061.

Shaykewich, C.F. and J. Williams. (1971). Resistance to water absorption in germinating rapeseed (*Brassica napus* L.). *Journal of Experimental Botany* 22: 19-24.

Simon, E.W., A. Minchin, M.M. McMenamin and J.M. Smith. (1976). The low temperature limit for seed germination. *New Phytologist* 77: 301-311.

Simon, E.W. (1979). Seed germination at low temperatures. In *Low Temperature Stress in Crop Plants. The Role of the Membrane*, eds. J.M. Lyons, D. Graham, and J.K. Raison. New York: Academic Press, pp. 37-46.

Simon, E.W. (1984). Early events in germination. In *Seed Physiology. Volume 2. Germination and Reserve Mobilization,* ed. D.R. Murray, New York: Academic Press, pp. 77-115.

Singer, S.M., S.O. El-Abd, H.M.E. Saied and A.F. Abou-Hadid. (1989). Influence of tomato seed coat restraints on germination performance under suboptimal temperatures. *Egyptian Journal of Horticulture* 16(2): 127-132.

Singh, S. and V.K. Sawhney. (1992). Endogenous hormones in seeds, germination behaviour and early seedling characteristics in a normal and *ogura* cytoplasmic male sterile line of rapeseed (*Brassica napus* L.). *Journal of Experimental Botany* 43: 1497-1505.

Spaeth, S.C. (1989). Extrusion of protoplasm and protein bodies through pores in cell walls of pea, bean and faba bean cotyledons during imbibition. *Crop Science* 29: 452-459.

Spedding, D.J. and A.T. Wilson. (1968). Studies of the early reactions in the germination of *Sinapis alba* seeds. *Phytochemistry* 7: 897-901.

Steponkus, P.L. (1984). Role of the plasma membrane in freezing injury and cold acclimation. *Annual Review of Plant Physiology* 35: 543-584.

Stewart, C.R., B.A. Martin, L. Redding and S. Cerwick. (1990). Seedling growth, mitochondrial characteristics and alternative respiratory capacity of corn genotypes differing in cold tolerance. *Plant Physiology* 92: 761-766.

Storch, J. and A.M. Kleinfeld. (1985). The lipid structure of biological membranes. *Trends in Biochemical Sciences* 10: 418-421.

Sugiyama, T., M.R. Schmitt, S.B. Ku and G.E. Edwards. (1979). Differences in cold lability of pyruvate Pi dikinase among C-4 species. *Plant and Cell Physiology* 20: 965-971.

Szafirowska, A., A.A. Khan and N.M. Peck. (1981). Osmo-conditioning of carrot seeds to improve seedling establishment and yield in cotton soil. *Agronomy Journal* 73: 549-556.

Tan, S.C. (1980). Phenylalanine ammonia lyase and the phenylalanine ammonia lyase inactivating system-effects of light, temperature and mineral deficiencies. *Australian Journal of Plant Physiology* 7: 159-168.

Taylor, A.G. and M.H. Dickson. (1987). Seed coat permeability in semi-hard snap bean seeds: its influence on imbibitional chilling injury. *Journal of Horticultural Science* 62: 183-189.

Taylor, A.J., C.J. Smith and I.B. Wilson. (1991). Effect of irrigation and nitrogen fertilizer on yield, oil content, nitrogen accumulation and water use of canola (*Brassica napus* L.). *Fertilizer Research* 29: 249-260.

Taylor, A.O., C.R. Slack and H.G. McPherson. (1974). Plants under climatic stress. VI. Chilling and light effects on photosynthetic enzymes of sorghum and maize. *Plant Physiology* 54: 696-701..

Theimer, R.R. and I. Rosnitschek. (1978). Development and intracellular localization of lipase activity in rapeseed (*Brassica napus* L.) cotyledons. *Planta* 139: 249-256.

Teran, S.G., R.K. Maiti and J.L. Hernandez-Pinero. (1994). Seed ultrastructure of six horticultural species in relation to their germination capacity. *Phyton* 55: 123-128.

Thompson, P.A. (1972). Geographical adaptation of seeds. In *Seed Ecology*, ed. W. Heydecker. London: Butterworths, pp. 31-58.

Tissaoui, T. and D. Côme. (1975). Mise en evidence de trois phases physiologiaues defferentes au cours de la germination de l'embryon de pommier nondormant, grace a la mesure de l'activite respiratoire. *Physiologie Végétal* 13: 95-102.

Todari, N.P. and A.K. Negi. (1992). Pretreatment of some Indian *Cassia* seeds to improve their germination. *Seed Science and Technology* 20: 582-588.

Tran, V.W. and A.K. Cavanagh. (1984). Structural aspects of dormancy. In *Seed Physiology, Volume 2, Germination and Reserve Mobilization*, ed. D.R. Murray. New York: Academic Press, pp. 1-44.

Trelease, R.N. (1984). Biogenesis of glyoxysomes. *Annual Review of Plant Physiology* 35: 321-347.

Tully, R.E., M.E. Musgrave and A.C. Leopold. (1981). The seed coat as a control of imbibitional chilling injury. *Crop Science* 21: 312-315.

Tykarska, T. (1976). Rape embryogenesis. I. The proembryo development. *Acta Societatis Botanicorum Poloniae* 45: 3-16.

Tykarska, T. (1979). Rape embryogenesis. II. Development of embryo proper. *Acta Societatis Botanicorum Poloniae* 48: 391-421.

Tykarska, T. (1980). Rape embryogenesis. III. Embryo development in time. *Acta Societatis Botanicorum Poloniae* 49: 369-385.

Uedan, K. And T. Sugiyama. (1976). Purification and characterization of phosphoenolpyruvate carboxylase from maize leaves. *Plant Physiology* 57: 906-910.

Vanni, P., E. Giachetti, G. Pinzauti and B.A. Mcfadden. (1990). Comparative structure, function and regulation of isocitrate lyase an important assimilatory enzyme. *Comparative Biochemistry and Physiology* 3: 431-458.

Vertucci, C.W. (1989). In *Seed Moisture*. C.S.S.A. Special Publication Number 14, pp. 93-115.

Vertucci, C.W. (1992). A calorimetric study of the changes in lipids during seed storage under dry conditions. *Plant Physiology* 99: 310-316.

Vertucci, C.W. and A.C. Leopold. (1983). Dynamics of imbibition of soybean embryos. *Plant Physiology* 72: 190-193.

Walker, M.A. and B.D. McKersie. (1993). Role of the ascorbate-glutathione antioxidant system in chilling resistance of tomato. *Journal of Plant Physiology* 141: 234-239.

Wang, C.Y. (1982). Physiological and biochemical responses of plants to chilling stress. *HortScience* 17: 173-186.

West, M. and J.J. Harada (1993). Embryogenesis. *The Plant Cell* 5: 1361-1369.

Wilen, R.W., L.V. Gusta, B. Lei, S.R. Abrams and B.E. Ewan. (1994). Effects of abscisic acid (ABA) and ABA analogs on freezing tolerance, low temperature growth, and flowering in rapeseed. *Journal of Plant Growth Regulation* 13: 235-241.

Wilson, D.O. Jr. and M.B. McDonald. (1986). The lipid peroxidation model of seed ageing. *Seed Science and Technology* 14: 269-300.

Wilson, R.E, E.H. Jensen and G.C.J. Fernandez. (1992). Seed germination response to eleven forage cultivars of *Brassica* to temperature. *Agronomy Journal* 84: 200-202.

Wise, R.R. and A.W. Naylor. (1987). Chilling-enhanced photooxidation-evidence for the role of singlet oxygen and superoxide in the breakdown of pigments and endogenous antioxidants. *Plant Physiology* 83: 278-282.

Wolk, W.D. and R.C. Herner. (1982). Chilling injury of germinating seeds and seedlings. *HortScience* 17: 169-173.

Woodstock, L.W. (1988). Seed imbibition: a critical period for successful germination. *Journal of Seed Technology* 12: 1-15.

Wright, L.C., E.J. McMurchie, M.K. Pomeroy and J.K. Raison. (1982). Thermal behavior and lipid composition of cauliflower plasma membranes in relation to ATPase activity and chilling sensitivity. *Plant Physiology* 69: 1356-1360.

Xin, A. And P.H. Li. (1992). Abscisic acid-induced chilling tolerance of maize suspension-cultured cells. *Plant Physiology* 99: 707-711.

Yan, Y.T. (1987). Effect of PEG priming in preventing imbibitional chilling injury in soybean seeds. *Plant Physiology Communications* 4: 24-27.

Yan, Y.T., Z. Liang, G.H. Zheng and P.S. Tang. (1989). Effect of low temperature imbibition on mitochondrion respiration and phosphorylation of PEG primed soybean seed. *Acta Botanica Sinica* 31: 441-448.

Yang, Y.Z., W.T. Chen and J.Q. Guo. (1992). Effects of PVA and PEG pretreatment on development and ultrastructure of plumular root mitochondria in soybean seed during low temperature imbibition process. *Acta Botanica Sinica* 34: 432-436.

Zheng, G.H., R.W. Wilen, A.E. Slinkard and L.V. Gusta. (1994). Enhancement of canola seed germination and seedling emergence at low temperature by priming. *Crop Science* 34: 1589-1593.

SUBMITTED: 09/10/96
ACCEPTED: 11/13/96

The Changing Global Environment and World Crop Production

Sylvan H. Wittwer

SUMMARY. During a period of a presumed world food crisis, the importance of climate and weather, and the rising level of atmospheric carbon dioxide are highlighted as important changes in the global environment. There is a dual and simultaneous effect of the rising level of atmospheric carbon dioxide on first, global warming and second, on the enhancement of crop productivity as reflected by an increased photosynthetic capacity, greater water use efficiency and alleviation of other crop stresses. Climate variability has a greater impact on agricultural productivity than does climate change. The rising level of atmospheric carbon dioxide is a universally free subsidy, gaining in magnitude with time, on which all can reckon when it comes to crop productivity. *[Article copies available for a fee from The Haworth Document Delivery Service: 1-800-342-9678. E-mail address: getinfo@haworth.com]*

KEYWORDS. Climate, carbon dioxide, food, crop production, global warming, climate change

---

Sylvan H. Wittwer, Director Emeritus of the Michigan State University Agricultural Experiment Station, Professor Emeritus of Horticulture, Chairman of the Board of Agriculture (1973-1977) and member of the Climate Research Board (1977-1980) of the National Research Council, National Academy of Sciences, USA.

Address correspondence to: Sylvan H. Wittwer, 1590 Wittwer Avenue, P.O. Box 1169, Logandale, NV 89021, USA.

[Haworth co-indexing entry note]: "The Changing Global Environment and World Crop Production." Wittwer, Sylvan H. Co-published simultaneously in *Journal of Crop Production* (The Food Products Press, an imprint of The Haworth Press, Inc.) Vol. 1, No. 1 (#1), 1998, pp. 291-299; and: *Crop Sciences: Recent Advances* (ed: Amarjit S. Basra) The Food Products Press, an imprint of The Haworth Press, Inc., 1998, pp. 291-299. Single or multiple copies of this article are available for a fee from The Haworth Document Delivery Service [1-800-342-9678, 9:00 a.m. - 5:00 p.m. (EST). E-mail address: getinfo@haworth.com].

INTRODUCTION

Food production is a global concern and changes in the global environment affect food production. Every few years the public is bombarded with dire stories about an approaching world food crises. So it is today. Beginning with this decade, the popular media is again leading the world to believe that global environmental changes of climate, (the so-called "global warming") and increases in atmospheric carbon dioxide, are leading us to disaster. Beneath the rhetoric, however, is evidence that, first, current increases in the rising levels of atmospheric carbon dioxide are very favorable for the most essential of human activities, namely, the production of food. Secondly, a century's worth of weather records provide no evidence of catastrophic increases in global temperatures. In fact, the increase of 0.4°C occurred during the past century before human activity generated the rising level of atmospheric carbon dioxide and other greenhouse gases. Whether there is any global warming that can be attributed to human activity, is still highly debatable (Weiss, 1996).

Food, climate, and the rising levels of atmospheric carbon dioxide are uniquely interrelated (Wittwer, 1995). *Food production* is a critical resource, perhaps the world's most important renewable resource. The production of this renewable resource, upon which all life depends, is possible only through photosynthesis, the world's most important biochemical process. An essential raw material, almost always in short supply, is the atmospheric carbon dioxide. For example, an acre corn crop must process over 40,000 tons of air to produce the record yield of 130 bushels per acre recorded in the U.S. for 1994.

Globally, some 25 crops stand between people and starvation. The largest single food group is the cereal grains, of which rice, wheat and corn are leading members. Others include oats, barley and rye. They provide approximately 60% of the calories and 50% of the protein consumed by the human race. The legumes (soybeans, field beans, cowpeas, chickpeas, pigeonpeas) provide about 20% of the world's protein. The balance of calories, protein and essential vitamins and minerals is obtained from tuber (potato) and root crops (sweet potato, cassava, yams, cocoyams) and various fruits, nuts and vegetables. Food animals, deriving their food either directly or indirectly from plants, provide 20% of the protein with 5% coming from fish.

CLIMATE AND WEATHER

The most determinant factor in agricultural (food) production is weather or climate. For agriculture, climate must be managed both, as a resource to be used wisely on the one hand, or a hazard to be dealt with on the other. Food production is very much a function of climate, which in itself is unpredictable. In fact, the principal characteristic of climate is variability. From the perspective of food security, the stability of agricultural production is as important, if not more so, than the magnitude of output. Climate variability has a greater input on agricultural productivity—both its magnitude and stability—than does climate change. Extremes in weather, rather than averages, affect agriculture. Both crops and livestock are sensitive to weather over relatively short periods of time. Annual averages of temperature and rainfall do not convey short term deficiencies. Such deficiencies impact both the volume and stability of food output. History reveals that for food production, warming is better than cooling.

Of all the natural climatic hazards, drought is that which farmers fear most. The lack of water is a serious impediment to plant growth and global food production. This is illustrated by the fact that today, the 17% of the world's cropland that is irrigated produces one-third of the agricultural output. For the United States, the 12% of the cultivated farm land that is irrigated accounts for 37% of crop production. Agriculture in the United States consumes, mostly through irrigation, 80 to 85% of the nation's fresh water resources. For the world it is over 65%.

The most readily identifiable potential climatic impact of significant magnitude on future living standards of the human race is in agriculture (food) production. With few exceptions, crops are continuously exposed to the weather. The availability of the water resources and the efficiency of its use, will be a major key to future food security.

Carbon Dioxide

We now introduce the impacts of the rising level of atmospheric carbon dioxide. First, we have the effect on climate change, and second on food production. The climate change impact is characterized by the widely publicized global warming or the so-called "greenhouse effect." Presumably, this also is causing an increased frequency of extreme or hazardous events. Conversely, elevated levels of atmospheric carbon dioxide have a decidedly beneficial effect on crop production through an enhancement of photosynthetic capacity and an increase in water use efficiency. Arrhenius (1896), the author of the concept of "global warming" first suggested these possibilities over 100 years ago. He favored the view that a slow warming with a doubling of the earth's atmosphere of CO_2 would result in better living conditions and higher crop yields. Additionally, hundreds of experiments with crop plants now show partial alleviation of the harmful effects of both marginally low and high temperatures, air pollutants, and a lessening of the environmental stresses imposed by drought, alkalinity, and mineral stresses–both excesses and deficiencies–, low light intensities and UV-B radiation.

Concerning changes in levels of atmospheric carbon dioxide, there are some well-known facts. *First,* there is a documented increase. The isolated test site at Mauna Loa in Hawaii shows more than a 12% increase in the mean annual concentration, from 316 ppm by volume of dry air in 1959 to the 1996 level of 360 ppm. The current annual rate of increase is about 0.5% or 1.6 ppm. Carbon dioxide source sink models predict that the current level of atmospheric CO_2 will be doubled by the latter part of the 21st century.

Second, the increase is truly global. The earth's atmosphere is very effective in dispensing emissions from whatever the source, be it natural or man-made.

Third, with the average level of CO_2 rising, there is an annual oscillation of the earth's atmospheric CO_2. The earth's atmospheric CO_2 level begins to fall in the spring and continues through the summer months as it appears to be sequestered by the vegetation of the northern hemisphere. In late autumn, there is a resurgence of CO_2 into the atmosphere. This results in new heights by mid-winter. With the amplitude increasing by about 0.5% each year, it appears the concentration amount of the earth's biomass is either increasing or is steady. It is not decreasing.

Global Warming

Thus, there are two ongoing global experiments inadvertently being conducted by the world's people. The outcomes of either we do not know. *First* is the so-called global warming resulting from increasing amounts of atmospheric CO_2 and other radioactively active trace gases. *Second* is the magnitude of the stimulative effect of the earth's atmospheric CO_2 enrichment on improved photosynthetic capacity, and its effects on plant growth and development. This in turn increases food production, forestry output, and global biological productivity with an improvement in water use efficiency.

Meanwhile, these two experiments will likely continue well into the 21st century with the final results not fully realized. The topics of food security, the magnitude of climate change (global warming), and the beneficial biological effects of the rising levels of atmospheric carbon dioxide are rent with both political and scientific controversy. There are those that advocate immediate action with accompanying costs of billions of dollars for reducing the world's output of CO_2. Global initiatives concerning such were promoted at the Rio Earth Summit in 1992, and again in the recent Berlin Assembly in March 1995.

To date, our knowledge of the climate effects of the rising CO_2 content and other greenhouse gases in the atmosphere is inadequate for initiating any global attempt to change the climate (Mitchell et al., 1995). A wise alternative, however, will be to sustain an international research effort in global climate modeling. If the climate does change, some warming could be tolerated, and may even be beneficial with no reductions in food production. A warming trend would increase the lengths of the growing seasons, encourage farmer adaptations, and favor the introduction of new technologies and cultural practices. The result would be crops and food animals more resistant to environmental stresses. The prospects of climate change from increasing levels of atmospheric carbon dioxide do not frighten many agriculturists, farmers or foresters. Early model projections, leading to a warming of the planet ranging from 1.9 to 5.2°C above pre-industrial levels when there is a doubling of the present CO_2 level, have been progressively lowered by the United Nations Inter-governmental Panel on Climate Change (Mitchell et al., 1995). There is no evidence that climate variability or hazardous events (floods, hurricanes, tornadoes, heat waves, frost) would be more frequent as atmospheric carbon dioxide increases. Marked inter-annual variations have always been with us. The most recent for the grain belt of the United States was the hot-dry summer of 1989. This resulted in the first scare tactics of a global warming initiated by scientists. The cold-wet summer of 1992 followed. Again, 1995 and perhaps 1996 may be partial analogs of that experienced in the summer of 1989.

Food, Climate, and Carbon Dioxide

Food security, climate change and variability, and the rising level of carbon dioxide are all resources vital to the people of the earth. Of these resources, the rising levels of atmospheric carbon dioxide must not be viewed as USA Vice President Albert Gore has declared, and many others concur, in his best selling book *Earth in the Balance,* 1992:

> the process of filling the atmosphere with CO_2 and other pollutants–is a wilful expansion of our dysfunctional civilization into vulnerable parts of the world.

Such pronouncements are too often accompanied by projections of melting icecaps, coastal flooding, mega hurricanes, drought, disease, and famine.

Unlike other national resources (land, water, energy) essential for food production, which are costly and progressively in shorter supply, the rising level of atmospheric carbon dioxide is a universally free premium, gaining in magnitude with time, on which all can reckon for the future.

For the present, the direct effects of an increasing atmospheric CO_2 on food production and the outputs of range-lands and forests are much more important than any effects thus far manifest for climate (Ciais et al., 1995; Wittwer, 1995). A recent review of over 1000 individual experiments with 475 plant crop varieties, published in 342 peer-reviewed scientific formal articles, and authored by 454 scientists in 29 countries has shown an average growth enhancement of 52%, with a doubling of the current level of atmospheric carbon dioxide. Globally, it is estimated that overall crop productivity has been increased by 10% and may account for much of which has been attributed to the green revolution. Meanwhile, changes in climate in specific fields where crops actually grow and are cultivated remain defiantly uncertain. Conversely, the effects of an enriched CO_2 atmosphere on crop productivity, in large measure, are positive, and leave little doubt as to the benefits for global food security.

Projected global shortages in food supplies and concerns related to up-coming shortages of land, water, energy and mineral resources related to crop production must be cushioned by the realism that the resource base changes with time and technology. The efficiency of land use for crop production can be greatly enhanced by no-till practices, and the overall global resource of arable land that could be cultivated, if needed, exceeds one billion hectares. The efficiency of water use for crop production could be increased from the current global level, 35-40% to near the 85% now being achieved in Israel. New and more efficient sources of energy will be developed, and additional geological resources will be discovered with still greater efficiency in energy use. Innovative uses for mineral resources will be forthcoming and substitutes, such as fiber optics to replace copper. The genetic resource base for all food crops will continue to expand.

Climate and weather can be proclaimed as the most determinate factors for plant growth, crop productivity and food production. The climate resource base will also change with time and technology. Climate can be both a resource to be used wisely or a hazard to be dealt with. Crops are being developed to tolerate and partially overcome the adverse effects of drought, heat, cold and flooding.

Protected Cultivation

Protected cultivation, including windbreaks, crop irrigation, plastic soil mulches and plant covers along with both plastic and glass greenhouses show continual expansion world-wide now extending from temperate cropping zones to the tropics (Table 1). Often within these innovative crop

TABLE 1. Global Distribution of Protected Cultivation
(Modified after Wittwer and Castilla, 1995)

Structures	Geographical Areas (Hectares)				
	Orient	Mediterranean	N.&S. America	North Europe	Total
Plastic soil mulches	5,090,000[x]	191,000	85,000[z]	15,000	5,381,000
Direct cover (floating types)	5,500	10,300	1,500[z]	27,000	44,300
Low tunnels (row covers)	143,400	90,500	20,000z	3,300	257,200
High tunnels		27,600[y]			27,600
Plastic greenhouses	138,200	67,700	15,600	16,700	238,200
Glass Greenhouses	3,000	7,900	4,000	25,800	40,700

[x]Predominantly cotton, corn, rice, peanuts and rice seed beds in China and elsewhere in the Orient.

[y]High tunnels are often computed together with plastic greenhouses in countries other than Mediterranean. France is included in the Mediterranean group.

[z]Crude estimates only.

protection structures, atmospheric levels of carbon dioxide are artificially increased to enhance crop production. For example, today in Holland notably in "The Westlands" all crops, vegetable and ornamental, grown in all the glass greenhouses, covering over 10,000 hectares and with an annual return of over $3 billion are enriched with atmospheric CO_2 at a level of 1,000 ppm during daylight hours, and during the entire year when crops are produced. Increases related to enriched atmospheric levels of CO_2 for marketable yields of tomato, cucumbers, sweet pepper, eggplant and ornamental, range from 20 to 40% (Nederhoff, 1994). This could be a harbinger for the future.

By far, the most extensive means of protected cultivation is through irrigation. Deserts have been reclaimed and arid lands made productive. Many of the hazards of drought have been overcome. Rates of increase for irrigated land have been precipitous. From 1950 to the early 1980s, the crop area irrigated in the world increased from 94 to 250 million hectares. This increased agricultural output by 40 to 50%. Although the expansion of irrigation in the late 1980s and early 1990s has slowed dramatically in the United States and elsewhere, some nations, such as China, anticipate expansions of great magnitude with the harnessing of the Yangzi river. Over two-thirds of all the food in China is produced on cropland that is irrigated, and more than half of the food produced in India, Indonesia and Japan is grown in irrigated fields. Without irrigation, there would essentially be no food produced in Egypt, the Sudan, Israel, Jordan and Saudi Arabia. For the world's five most populated countries, China, India, Russia, the United States and Indonesia, water supplies for agriculture are increasingly becoming more critical. This is true for most other countries of the world, except those in northern Europe where drought seldom occurs and irrigation is infrequent. Further crop production is now limited by short water supplies in much of the United States, Russia, China, India, Pakistan, Spain, Italy, Turkey, Mexico, Chile, Australia, and most all Mediterranean, Arabian Gulf and African countries. Making crop irrigation more efficient must become a top future priority. This will come with time and technology. One option is the increased use of micro-irrigation which gives up to 95% efficiency (Smajstria, 1993).

REFERENCES

Arrhenius, S. (1896). On the influence of carbonic acid in the air upon the temperature of the ground. *Philosophical Magazine* 41: 237.

Ciais, P., M. Tans, M. Trolier, J.W.C. White and R.C. Francey. (1995). A large northern hemisphere terrestrial CO_2 sink indicated by the $^{13}C/^{12}C$ ratio of atmospheric CO_2. *Science* 269: 1099-1102.

Gore, A. (1992). *Earth in the Balance.* New York: Houghton-Mifflin.

Mitchell, J.F.B., T.C. Jokus, J.M. Gregory and S.F.B. Tett. (1995). Climate response to increasing levels of greenhouse gases and sulphate aerosols. *Nature* 376: 501-550.

Nederhoff, E.M. (1994). *Effects of CO_2 Concentration on Photosynthesis, Transpiration and Production of Greenhouse Fruit Vegetable Crops.* Naaldwijk: The Netherlands Glasshouse Crop Research Station.

Smajstria, A.G. (1993). Microirrigation for Citrus in Florida. *HortScience* 28: 295-298.

Weiss, P. (1996). Industry group assails climate chapter. *Science* 272: 1724.

Wittwer, S.H. (1995). *Food, Climate and Carbon Dioxide.* Boca Raton, Florida: CRC Press.

Wittwer, S.H. and N. Castilla. (1995). Protected cultivation of horticultural crop–wordwide. *HortTechnology.* 4.5: 6-24.

SUBMITTED: 06/26/96
ACCEPTED: 10/22/96

Crop Productivity
and Sustainable Development in China

Lianzheng Wang

SUMMARY. The current situation of crop production in China is reviewed, along with strategies to meet future food needs of its population. *[Article copies available for a fee from The Haworth Document Delivery Service: 1-800-342-9678. E-mail address: getinfo@haworth.com]*

KEYWORDS. China, crop productivity, grain production, sustainable development, agricultural resources

THE CURRENT SITUATION OF CROP PRODUCTION IN CHINA

Total Grain Production

In recent years, domestic and foreign experts, scientists, economists and officials have expressed grave concern about grain production in China. Questions like "Who will feed China?" are being asked (Brown, 1994). Therefore, I would like to take stock of the current situation of grain production in China and suggest ways to solve the grain problem (see also Wang, 1989, 1995, 1996).

Lianzheng Wang, Professor, Chinese Academy of Agricultural Sciences (CASS), and President of China Crop Science Society (CCSS), 30 Baishiqiao Road, Haidian District, Beijing 100081, China.

[Haworth co-indexing entry note]: "Crop Productivity and Sustainable Development in China." Wang, Lianzheng. Co-published simultaneously in *Journal of Crop Production* (The Food Products Press, an imprint of The Haworth Press, Inc.) Vol. 1, No. 1 (#1), 1998, pp. 301-308; and: *Crop Sciences: Recent Advances* (ed: Amarjit S. Basra) The Food Products Press, an imprint of The Haworth Press, Inc., 1998, pp. 301-308. Single or multiple copies of this article are available for a fee from The Haworth Document Delivery Service [1-800-342-9678, 9:00 a.m. - 5:00 p.m. (EST). E-mail address: getinfo@haworth.com].

301

From 1949 to 1995, the total grain production in China increased from 113.18 million tonnes to 466.62 million tonnes, and averaging 448.73 million tonnes from 1990 to 1995 (Table 1). In 1996, the total grain production rose to 480 million tonnes on a cultivated land area of about 112 million hectares which increased over two million hectares in comparison with 1995. The average total grain production during 1994-1996 was 463.9 million tonnes. It means, during the past 18 years (1978-1996), total grain production in China has increased by 973.5 million tonnes per year, recording an annual increase of 3.1%.

Certainly, China is showing the potential to feed itself. In the past ten years (1986-1995), an average annual import of 4.24 million tonnes is not over 5% of whole grain production in China (Table 2).

Since 1978, China has carried out reforms and an open door policy. In countryside, it has carried out a production responsibility system. As a result, the total grain production witnessed a rapid increase of 50 million tonnes from 1978 to 1982 (Table 3). A further increase of 50 million tonnes followed in just two years (1983-1984). The rate of increase slowed down after 1985, but total grain production has continued to increase in recent years (Table 3).

The reasons of rapid development of grain production in China are attributable to: (1) reform and good policies (including production responsibility system, increasing grain prices several times, (2) enlarging

TABLE 1. Grain production from 1949 to 1996 in China (10,000 tonnes)

Year	Total production	Year	Total production	Year	Total production
1949	11318	1965	19453	1981	32502
1950	13213	1966	21400	1982	35450
1951	14369	1967	21782	1983	38728
1952	16392	1968	20906	1984	40731
1953	16638	1969	21097	1985	37911
1954	16952	1970	23996	1986	39151
1955	18394	1971	25014	1987	40298
1956	19275	1972	24048	1988	39408
1957	19505	1973	26494	1989	40755
1958	20000	1974	27527	1990	44624
1959	17000	1975	28452	1991	43529
1960	14350	1976	28631	1992	44266
1961	14750	1977	28273	1993	45649
1962	16000	1978	30477	1994	44510
1963	17000	1979	33212	1995	46662
1964	18750	1980	32056	1996	48000

TABLE 2. Import and export of grain during 1986-1995 in China (10,000 tonnes)

Year	Import	Export
1986	773	942
1987	1628	737
1988	1533	717
1989	1658	656
1990	1372	583
1991	1345	1086
1992	1175	1364
1993	752	1535
1994	915	1263
1995	2040	64

TABLE 3. The number of years that the national total grain production was increased by 50 million tonnes each time

Year	Planting area (10,000 mu)	Total production (10,000 tonnes)	Total production (1 billion Jin) (1Jin = 0.5kg)	The number of years that the total production was increased by 50 million tonnes
1949	164.94	11318	226.36	
1952	185.97	16392	327.84	3
1958	191.42	20000	400.00	6
1971	181.27	25014	500.28	13
1978	180.60	30477	609.54	7
1982	170.19	35450	709.00	4
1984	169.33	40731	814.62	2
1993	165.76	45648	912.96	9

application of science and technology in agriculture and (3) increasing use of agriculture inputs.

Yield per Hectare

From 1949 to 1978, the yield per hectare was increased by 49.95 kg annually, while from 1978 to 1984, it increased by 180 kg annually, recording the fastest increase (Table 4). From 1984 to 1993, the yield per mu increased by 56.55 kg annually.

TABLE 4. Yield per hectare in different years (1 hectare = 15 mu)

Year	1949	1952	1958	1971	1978	1982	1984	1993
Yield/ha (kg)	1035	1320	1375	2070	2535	3135	3615	4125

Average per Capita Possession (kg/person)

The average per capita possession of grain which was only 209 kg in 1949 rose to 306 kg in 1957, 363 kg in 1982, 397 kg in 1984, 368 kg in 1994 and 402 kg in 1995 (Table 5). Recent estimates reveal that per capita meat consumption is 41 kg, that of aquatic products 21 kg, eggs 14 kg, fruits 35 kg and vegetables 198 kg.

Main Crops

In China, the number one crop is rice providing 44.89% of total grain production in 1978 and 38.9% in 1993. Second important crop is wheat accounting for 17.7% of total grain production in 1978 and about 23.3% in 1993. Corn occupied third place with figures of 18.4% in 1978 and about 22.5% in 1993. For 1993, the soybean contribution was 3.4% and potato and sweet potato 7.0%. Other crops including millet, sorghum and others were 6.1% in 1978 and 11.2% in 1993.

Rice got highest production in 1992 reaching 186.22 million tonnes. Corn production in 1995 was 111.99 million tonnes, surpassing wheat production for the first time. Soybean production in 1994 touched 15.6 million tonnes. The acreage and sown areas under main crops in China are given in Tables 6 and 7.

Yield Realization of Different Crops

In China, the yield of rice per hectare is the highest than other crops. In 1995, it was 6025 kg for rice, 4917 kg for corn and 3542 kg for wheat (Table 8).

WAYS TO INCREASE GRAIN PRODUCTION IN CHINA

China has excellent potential to increase grain production by adopting the following strategies.

TABLE 5. Average per capita possession of grain (kg/person)

Year	1949	1952	1957	1978	1982	1984
kg/per capita	209	288	306	324	363	397
Year	1990	1991	1992	1993	1994	1995
kg/per capita	397	397	378	379	368	402

TABLE 6. The acreage of main crops in China (10,000 hectares)

Year	Rice	Wheat	Corn	Soybean	Potato	Others
1993	3035.5	3023.4	2069.4	945.4	922.0	1055.1

TABLE 7. Sown areas of main crops in China (1,000 hectares)

	Total sown area	Rice	Wheat	Corn	Soybean
1986	110933	32266	29616	19124	8295
1987	111268	32193	28798	20212	8445
1988	110123	31987	28785	19692	8120
1989	112205	32700	29841	20353	8057
1990	113466	33064	30753	21401	7560
1991	112314	32590	30948	21574	7041
1992	110560	32090	30496	21044	7221
1993	110509	30355	30235	20694	9454
1994	109544	30171	28981	21152	9222
1995	110060	30744	28860	22776	8127

Increasing the Yield per Unit Area

1. Release of crop hybrids and good cultivars

From 1991-1995, 301 new hybrids and cultivars of crops in China were released. Rice cultivar Xieyou-415, hybrid between Japonica and Indica, was released with a yield of 8250-9750 kg per hectare which is higher than the yield of standard cultivar Shanyou 63 by about 11.9%. Five rice hybrids with two lines were released:70-you No.9, 70 you-04, Lianyou-

TABLE 8. The yield per hectare of different crops (kg)

Year	Rice	Wheat	Corn	Soybean
1952	2415	735	1350	825
1978	3975	1845	2805	1065
1982	4886	2449	3268	1073
1984	5373	2969	3960	1331
1993	5854	3519	4963	1619
1994	5831	3426	4693	1735
1995	6025	3542	4917	1661

Peite; Er-zha No.1 and Huazha No.1. The planting areas of rice hybrids reached 15.6 million hectares in 1994. Hybrid rice and cultivars are a technological breakthrough in increasing crop productivity.

New wheat cultivar Yangmai 158 was released. With an increase of about 14.27% over the standard cultivar, it occupied about one million hectares. Yumai 18 occupied the biggest planting area–1.4 million hectares. Lumai 14 and 15 occupied third and fourth place.

Corn hybrid Danyu 13 occupied the biggest planting area (1.9 million hectares) in 1994, followed by Yedan No.13 (1.6 million hectares), Yedan No.2 (1.56 million hectares) and Zhongdan No.2 (1.43 million hectares). In recent years, some QPM hybrids were released, for example, Zhongdan 3850 and Luyou 13, with lysine contents of about 0.4-0.44.

New soybean cultivars Hefeng 25, Heinong 35 and Heinong 34 and also high yielding and high protein content cultivars Ludou No.4, Jidou 7, Youdou 8, Suxie 1, Zhongdou 19.24, Zhonghuang 4,6 have been released.

2. *Increasing double planting index from 151% to 154%:* To increase 1% index is to increase 1 million hectares.
3. *Controlling insects, disease and weeds by chemical, biocontrol and IPM:* Biocontrol is desirable for reducing pollution and improving environmental sustainability.
4. *Improving field management and cultural practices.*
5. *Rational tillage and rotation practices.*

Integrated Nutrient Management

In China, the current dosage of chemical fertilizers per hectare is about 300 kg. At the same time, there is increasing use of organic manure including green manure, algae (Table 9).

TABLE 9. Application of chemical fertilizers

Year	Dosage of chemical fertilizers per hectare(kg)	Total dosage of chemical fertilizers in China (10,000 tonnes)
1952	0.75	7.8
1978	89.25	884
1982	153.75	1513
1984	177.75	1740
1993	285.00	3152
1994	300.00	3318
1995		3594

Increasing Irrigation Area

In 1979, the irrigated area was 45 million hectares which rose to 49.3 million hectares in 1995, registering an increase of 4.3 million hectares. In the next five years, we will set up 300 counties as a model saving irrigation water. It will save 6 billion M^3 water and can increase 8 billion kg grain output.

Enhancing Agricultural Mechanization and Power

There were 7848 (10,000 k.w.) small tractors in use in 1995 as compared with 1615 in 1980. There were 879.5 (10,000 k.w.) small trucks in 1980 and 6060.9 (1,000 k.w.) in 1995. The utilization of electricity in countryside was 28.3 billion (KWH) in 1979 and 165.5 billion (KWH) in 1995.

Enlarging the Land Use

China has a lot of wasteland–about 14.7 million hectares, which can be used for planting crops. In the coming decades, reclamation of new lands is planned.

Strengthening Agricultural Research and Education

The key problem in China is to increase crop productivity using advanced technology. China has already released rice, corn and rapeseed hybrids and we are actively working on wheat heterosis.

Rational Utilization of Natural Resources

China has grasslands of 390 million hectares. We have a big potential for development of animal husbandry. At the same time, we can use mountain areas for wooden foods, walnut, Chinese chestnut, oil tea, fruits production, etc.

Grain Saving and Value-Adding

We must develop wine production and decrease grain consumption on spirits.

Thus, we believe that China can solve its grain problem and feed itself!

REFERENCES

Brown, L.R. (1994). Who will feed China? *World Watch.* September/October, pp. 10-19.

Wang, L. (1989). Characteristics of agricultural production and the measures to achieve high and stable yields. *Scientia Agriculture Sinica* 22: 1-5

Wang, L. (1995) Grain production in China and its stable increase. *China Bulletin of Agronomy* 11: 1-4.

Wang, L. (1996). Current situation of grain in China and its ways of solution. *Theory Consultation,* No. 2: 22-26.

SUBMITTED: 12/08/96
ACCEPTED: 01/04/97

Index

Abscisic acid, 203,207-16,229-31,
    239,250-51,264,266,269,
    274-75
ADP-glucose pyrophosphorylase,
    233
Aegilops, 1-18
Agricultural mechanization, 307
Allelochemicals, 171,185-90
Allelopathy
    defined, 169-70
    agronomic crops, 173,179-80
    fruit crops, 178,183-84
    trees, 173,181-82
    weeds, 172,174-78,184
Allopolyploid, 2,5
Amphiploid, 8,11-15
AMP-PCR, 124-26
Amylase, 6,213
Assimilate partitioning, 36-38
ATP, 257
ATPase, 270
Autotoxicity, 170
Auxin, 232
Awns, 38

Backcrossing, 9,15-18
Bioeconomic model, 159-61

Canopy nitrogen and yield potential,
    47-70
Carbon assimilation and partitioning,
    36-38
Chemical fertilizers, 306-07
Chickpea, 119-20,123,125,128,
    131-32
Chilling

injury, 250-51,255-56,270,275
labile enzymes, 271
resistant species, 251-53,259,264
sensitive species, 251-53,259,
    264,269
Chromosome
    aberrations, 232
    doubling, 3
Climate change and crop production,
    291-98
Climate variability, 291-93,296
Cluster analysis, 87
Coefficient of nitrogen allocation,
    62-63
Cotton taxonomy, 80-81
Cotyledons, 261-62
Crop biomass production, 48-51,
    59,69
Crop hybrids, 305-07
Crop production in China, 301-08
Crop rotation, 153-54,185-86
Cytokinins, 232,239

Decision aids, 159-61
Degree days, 260,273
Dendrogram, 87
DNA
    methylation 105-07
    synthesis, 259
Dormancy, 249,253
Dormancy proteins, 203,213-14,216
Dwarfing genes, 27-41

Endomannanase, 256
Environmental stress and crops,
    224-25
Epigeal, 261

   309

Haworth
DOCUMENT DELIVERY
SERVICE

This valuable service provides a single-article order form for any article from a Haworth journal.

- *Time Saving:* No running around from library to library to find a specific article.
- *Cost Effective:* All costs are kept down to a minimum.
- *Fast Delivery:* Choose from several options, including same-day FAX.
- *No Copyright Hassles:* You will be supplied by the original publisher.
- *Easy Payment:* Choose from several easy payment methods.

Open Accounts Welcome for . . .
- Library Interlibrary Loan Departments
- Library Network/Consortia Wishing to Provide Single-Article Services
- Indexing/Abstracting Services with Single Article Provision Services
- Document Provision Brokers and Freelance Information Service Providers

---

MAIL or *FAX* THIS ENTIRE ORDER FORM TO:

Haworth Document Delivery Service
The Haworth Press, Inc.
10 Alice Street
Binghamton, NY 13904-1580

or FAX: 1-800-895-0582
or CALL: 1-800-342-9678
9am-5pm EST

PLEASE SEND ME PHOTOCOPIES OF THE FOLLOWING SINGLE ARTICLES:

1) Journal Title: _____
   Vol/Issue/Year:_____Starting & Ending Pages:_____
   Article Title:_____
   _____

2) Journal Title: _____
   Vol/Issue/Year:_____Starting & Ending Pages:_____
   Article Title:_____
   _____

3) Journal Title: _____
   Vol/Issue/Year:_____Starting & Ending Pages:_____
   Article Title:_____
   _____

4) Journal Title: _____
   Vol/Issue/Year:_____Starting & Ending Pages:_____
   Article Title:_____
   _____

(See other side for Costs and Payment Information)

COSTS: Please figure your cost to order quality copies of an article.

1. Set-up charge per article: $8.00
   ($8.00 × number of separate articles) _____

2. Photocopying charge for each article:

   1-10 pages: $1.00 _____

   11-19 pages: $3.00 _____

   20-29 pages: $5.00 _____

   30+ pages: $2.00/10 pages _____

3. Flexicover (optional): $2.00/article _____

4. Postage & Handling: US: $1.00 for the first article/
   $.50 each additional article _____

   Federal Express: $25.00 _____

   Outside US: $2.00 for first article/
   $.50 each additional article _____

5. Same-day FAX service: $.35 per page _____

                           GRAND TOTAL: _____

METHOD OF PAYMENT: (please check one)

❑ Check enclosed ❑ Please ship and bill. PO # _____
    (sorry we can ship and bill to bookstores only! All others must pre-pay)

❑ Charge to my credit card: ❑ Visa; ❑ MasterCard; ❑ Discover;
                          ❑ American Express;

Account Number: _____ Expiration date: _____

Signature: ✗_____

Name: _____ Institution: _____

Address: _____

_____

City: _____ State: _____ Zip: _____

Phone Number: _____ FAX Number: _____

MAIL or *FAX* THIS ENTIRE ORDER FORM TO:

Haworth Document Delivery Service	**or FAX:** 1-800-895-0582
The Haworth Press, Inc.	**or CALL:** 1-800-342-9678
10 Alice Street	9am-5pm EST)
Binghamton, NY 13904-1580	